J. Herzberger (ed.)

Inclusion Methods for Nonlinear Problems

With Applications in Engineering, Economics and Physics

Computing
Supplement 16

Springer-Verlag Wien GmbH

Univ.-Prof. Dr. Jürgen Herzberger
Faculty of Mathematics, University of Oldenbourg, Oldenbourg, Germany

Originally published by Springer-Verlag Wien New York in 2003

Typesetting: Scientific Publishing Services (P) Ltd., Madras, India

Printed on acid-free and chlorine-free bleached paper
SPIN: 10879532

With 24 Figures

CIP-data applied for

ISSN 0344-8029

ISBN 978-3-211-83852-5 ISBN 978-3-7091-6033-6 (eBook)
DOI 10.1007/978-3-7091-6033-6

Foreword

This workshop was organized with the support of GAMM, the International Association of Applied Mathematics and Mechanics, on the occasion of J. Herzberger's 60th birthday. GAMM is thankful to him for all the time and work he spent in the preparation and holding of the meeting.

The talks presented during the workshop and the papers published in this volume are part of the field of Verification Numerics. The important subject is fostered by GAMM already since a number of years, especially also by the GAMM-Fachausschuß (special interest group) "Rechnerarithmetik und Wissenschaftliches Rechnen".

Karlsruhe, Dezember 2001

Götz Alefeld
(President of GAMM)

Preface

At the end of the year 2000, about 23 scientists from many countries gathered in the beautiful city of Munich on the occasion of the International GAMM-Workshop on "*Inclusion Methods for Nonlinear Problems with Applications in Engineering, Economics and Physics*" from December 15 to 18. The purpose of this meeting was to bring together representatives of research groups from Austria, Bulgaria, China, Croatia, Germany, Japan, Russia, Ukraine and Yugoslavia who in a wider sense work in the field of calculating numerical solutions with error-bounds. Most of those participants have already known each other from earlier occasions or closely cooperated in the past. Representatives from three Academies of Sciences were among the speakers of this conference: from the Bulgarian Academy, the Russian Academy and the Ukrainian Academy of Sciences.

The contributions covered a wide field of topics with emphasis on nonlinear problems and possible implications to the solution of practical problems. The pleasant atmosphere of the location of the conference, the Adolf-Kolping-House in the very center of the city, was stimulating many direct contacts among the participants and gave them the opportunity for intensive discussions and scientific exchange. This volume, extended by some contributions from authors who belong to these research groups but who were not able to attend the meeting due to the restricted number of invited guests, is intended to inform a broader audience about the result and scientific progress made in this area by a number of competent members of the international scientific community.

I would like to express my sincere thanks to all of those who have contributed to this volume and to all the numerous referees who helped to keep this volume on a high level. I am also grateful to my colleague Professor Dr. R. Albrecht, Innsbruck, for drawing the attention of Springer-Verlag in Vienna to this conference proceedings. My special thanks go to Ms. Silvia Schilgerius and Mag. Wolfgang Dollhäubl from the Springer-Verlag in Vienna who helped to make the appearance of this special volume possible and showed patience for this project.

Last but not least, I would like to thank Ms. Theresia Meyer from the University of Oldenbourg for promptly handling all the correspondence connected with my editorial work on this volume.

Oldenbourg, October 2002 *Jürgen Herzberger*

List of Participants of the International GAMM-Workshop on "Inclusion Methods for Nonlinear Problems with Applications in Engineering, Economics and Physics"

Munich and Oberschleißheim, December 15–18, 2000

Prof. Dr. R. Albrecht, Innsbruck (Austria)

Prof. Dr. G. Alefeld, Karlsruhe (Germany)

Prof. Dr. A. Andreev, Sofia (Bulgaria)

Dr. L. Atanassova, Munich (Germany)

Acad. Dr. F.L. Chernousko, Moscow (Russia)

Dr. H. Fischer, Munich (Germany)

Prof. Dr. V. Hari, Zagreb (Croatia)

Prof. Dr. G. Heindl, Wuppertal (Germany)

Prof. Dr. J. Herzberger, Oldenburg (Germany)

Prof. Dr. V. Hristov, Sofia (Bulgaria)

Prof. Dr. G. Iliev, Sofia (Bulgaria)

Prof. Dr. U. Kulisch, Karlsruhe (Germany)

Prof. Yong-xiang Ling, Xi'an (PR China)

Prof. Dr. G. Mayer, Rostock (Germany)

Prof. Shin'ichi Oishi, Tokyo (Japan)

Prof. K. Okomura, Kyoto (Japan)

Prof. Dr. M. Petković, Niš (Yugoslavia)

Prof. Dr. Lj. Petković, Niš (Yugoslavia)

Prof. Dr. N. Popivanov, Sofia (Bulgaria)

Acad. Dr. V.L. Rvachev, Kharkov (Ukraine)

Prof. Dr. S. Smelyakov, Kharkov (Ukraine)

Prof. Dr. Yu. Stoyan, Kharkov (Ukraine)

Prof. Dr. Shen Zuhe, Nanjing (PR China)

Contents

Computing [Suppl] 16, 1–22 (2002)

Computing

On Symmetric Solution Sets

G. Alefeld, V. Kreinovich and **G. Mayer**

Abstract

Given an $n \times n$ interval matrix $[A]$ and an interval vector $[b]$ with n components we present an overview on existing results on the solution set S_{sym} of linear systems of equations $Ax = b$ with symmetric matrices $A \in [A]$ and vectors $b \in [b]$. Similarly we consider the set E_{sym} of eigenpairs associated with the symmetric matrices $A \in [A]$. We report on characterizations of S_{sym} by means of inequalities, by means of intersection of sets, and by an approach which is generalizable to more general dependencies of the entries. We also recall two methods for enclosing S_{sym} by means of interval vectors, and we mention a characterization of E_{sym}.

AMS Subject Classifications: 65G10.

Keywords: Solution set of linear systems, symmetric solution set of linear systems, eigenpair set, symmetric eigenpair set, linear interval systems, Oettli–Prager theorem.

1. Introduction

With this paper we intend to give an overview on existing results for the symmetric solution set

$$S_{sym} := \{x \in \mathbb{R}^n \mid Ax = b,\ A = A^T \in [A],\ b \in [b]\}, \tag{1}$$

where $[A]$ is a given $n \times n$ interval matrix with $[A] = [A]^T$, and $[b]$ is a given interval vector with n components. This set obviously is a subset of the general solution set

$$S := \{x \in \mathbb{R}^n \mid Ax = b,\ A \in [A],\ b \in [b]\}, \tag{2}$$

where the restriction $A = A^T$ on $A \in \mathbb{R}^{n \times n}$ is not required. Knowing S and S_{sym} is particularly interesting in the following situations:

(a) Assume that one has to solve a linear system $\check{A}x = \check{b}$ on a computer using floating point arithmetic. Due to rounding errors, the computed result $\tilde{x}$ normally will not fulfill $\check{A}\tilde{x} = \check{b}$. If $\Delta A \in \mathbb{R}^{n \times n}$, $\Delta b \in \mathbb{R}^n$are given nonnegative tolerances one may view $\tilde{x}$ as an acceptable solution whenever $\tilde{x} \in S$ with S formed as in (2) with respect to $[A] := \check{A} + [-\Delta A, \Delta A]$, $[b] := \check{b} + [-\Delta b, \Delta b]$; in this case $\tilde{x}$ can be interpreted as *exact* solution of a linear system $\tilde{A}x = \tilde{b}$ with some $\tilde{A} \in [A]$, $\tilde{b} \in [b]$.

(b) In contrast to (a), where the linear system is known we assume now that one has to solve a linear system $Ax = b$ where A, b are not given exactly, but they are known to differ from some $\check{A} \in \mathbb{R}^{n\times n}, \check{b} \in \mathbb{R}^n$ by at most $\Delta A \in \mathbb{R}^{n\times n}$ and $\Delta b \in \mathbb{R}^n$, respectively ($\Delta A$, Δb nonnegative). Then $A \in [A] := \check{A} + [-\Delta A, \Delta A]$, $b \in [b] := \check{b} + [-\Delta b, \Delta b]$. Compute a solution x^* of $\check{A}x = \check{b}$. Since $x^* \in S$ one can accept x^* as a good approximation for the unknown solution of $Ax = b$. This situation can occur due to

- conversion errors (from decimal to binary or vice versa),
- errors in measurements,
- errors in adjusting the technical devices.

As we shall see S and S_{sym} are not so easy to handle. Therefore, enclosures of S and S_{sym} are important. For S such enclosures can be computed by means of interval arithmetic. Since such methods are contained in textbooks like [1], [20], e.g., we will omit them here. They trivially deliver also enclosures for $S_{\text{sym}} \subseteq S$. But there are also methods to enclose S_{sym}without bounding S at the same time. We will study such methods later on. Although we shall concentrate on S_{sym} in this paper we will give a short glance at S in order to work out particularities of S_{sym}. So we start in Sect. 3 with several equivalent statements for $x \in S$ and list some properties of S. In Sect. 4, we characterize the boundary ∂S_{sym} of S_{sym} by means of parts of hyperplanes and quadrics. In Sect. 5, we introduce two methods for *enclosing* S_{sym} and in Sect. 6, we report on the eigenpair set

$$E := \left\{ (x^T, \lambda)^T \in \mathbb{R}^{n+1} \mid Ax = \lambda x,\ x \neq 0,\ A \in [A] \right\} \tag{3}$$

and the symmetric eigenpair set

$$E_{\text{sym}} := \left\{ (x^T, \lambda)^T \in \mathbb{R}^{n+1} \mid Ax = \lambda x,\ x \neq 0,\ A = A^T \in [A] = [A]^T \right\}. \tag{4}$$

It turns out that quadrics are needed in order to describe E and algebraic inequalities of order at most three in order to describe E_{sym}.

2. Notations

In the sequel we denote intervals in square brackets, i.e., $[a] = [\underline{a}, \overline{a}]$, and identify point intervals $[a, a]$ by their unique element omitting the brackets. We assume that the reader is familiar with the elementary rules and basic facts of interval arithmetic as introduced in the first chapters of [1] or [20], e.g. We will write $\mathbb{IR}$, $\mathbb{IR}^n$, $\mathbb{IR}^{m\times n}$ for the set of real compact intervals, interval vectors with n components and $m \times n$ interval matrices, respectively. We apply the notation $[A] = [\underline{A}, \overline{A}] = ([a]_{ij}) = ([\underline{a}_{ij}, \overline{a}_{ij}])$ simultaneously for interval matrices and have a similar notation for interval vectors, real vectors and real matrices. An interval matrix $[A]$ is called regular if each $A \in [A]$ has this property. By $\check{A}$ we denote the midpoint of $[A]$, i.e., $\check{A} := \frac{1}{2}(\underline{A} + \overline{A})$, and by $\text{rad}[A] := \frac{1}{2}(\overline{A} - \underline{A})$ its radius which we wrote as ΔA in Sect. 1. In connection with interval vectors and intervals this notation is used analogously. If T is any bounded subset of $\mathbb{R}^n$ the symbol $[]T$ denotes the interval hull of T, i.e., the smallest enclosure of T by an interval vector. For real matrices $A \in \mathbb{R}^{n\times n}$ we write $\rho(A)$ for the spectral radius of A, and define the absolute value $|A| = (|a_{ij}|) \in \mathbb{R}^{n\times n}$. In addition, we introduce the entrywise defined partial ordering '$\leq$', and proceed similarly with vectors. For intervals [a], [c] we define $|[a]| := \max\{|a| | a \in [a]\}$, $\langle [a] \rangle := \min\{|a| | a \in [a]\}$ and $q([a],[c]) := \max\{|\underline{a} - \underline{c}|, |\overline{a} - \overline{c}|\}$ (= Hausdorff distance). For interval matrices $[A], [B] \in \mathbb{R}^{n\times n}$ we introduce the distance $q([A],[B]) := (q([a]_{ij},[b]_{ij})) \in \mathbb{R}^{n\times n}$ and the comparison matrix $\langle [A] \rangle = (c_{ij}) \in \mathbb{R}^{n\times n}$ which we define by

$$c_{ij} := \begin{cases} \langle [a]_{ij} \rangle, & \text{if } i = j, \\ |[a]_{ij}|, & \text{if } i \neq j. \end{cases}$$

We call a regular matrix $A \in \mathbb{R}^{n\times n}$ an M matrix if $a_{ij} \leq 0$ for $i \neq j$ and $A^{-1} \geq 0$. An $n \times n$ interval matrix $[A]$ is termed an interval M matrix if each element $A \in [A]$ is an M matrix. In particular, an interval M matrix is always regular, and it is an interval H matrix, i.e., it satisfies $\langle [A] \rangle^{-1} \geq 0$.

From interval analysis (cf. [1], e.g.) we use the following result:

Theorem 1. *Let $f(x_1, \ldots, x_n)$ be a rational expression in $x_1, \ldots, x_n$ in which each variable x_i occurs at most once. If $[x]_1, \ldots, [x]_n$ are given intervals then*

$$f([x]_1, \ldots, [x]_n) = \{f(x_1, \ldots, x_n) |\, x_1 \in [x]_1, \ldots, x_n \in [x]_n\},$$

i.e., replacing $x_1, \ldots, x_n$ in f by $[x]_1, \ldots, [x]_n$ and evaluating all appearing arithmetic operations according to their interval arithmetic definition yields to the range of f restricted to $[x]_1 \times [x]_2 \times \ldots \times [x]_n$. □

By I we denote the unit matrix, by O an orthant of $\mathbb{R}^n$ and by O_1 the first orthant, i.e., $O_1 = \{x \in \mathbb{R}^n |\, x_i \geq 0 \text{ for } i = 1, \ldots, n\}$.

3. The Solution Set *S*

In this section, we collect some well-known results on S. We begin with some equivalent formulations of the statement '$x \in S$' which we prove for completeness.

Theorem 2. *For $[A] \in \mathbb{R}^{n\times n}$ and $[b] \in \mathbb{R}^n$ the following assertions are equivalent:*

(a) $x \in S$;

(b) $|\check{b} - \check{A}x| \leq (\text{rad}[A]) \cdot |x| + \text{rad}[b]$; (Oettli and Prager, 1964 [21])

(c) $\exists D \in \mathbb{R}^{n\times n}: \quad |D| \leq I \wedge \check{A}x - \check{b} = D((\text{rad}[A])|x| + \text{rad}[b])$; (Rohn, 1984 [22])

(d) $[b] \cap [A]x \neq \emptyset$; (Beeck, 1972 [10])

(e) $0 \in [b] - [A]x$; (Beeck, 1972 [10])

(f) $\underline{b}_i - \sum_{j=1}^{n} a_{ij}^+ x_j \leq 0 \leq \bar{b}_i - \sum_{j=1}^{n} a_{ij}^- x_j, \quad i = 1, \ldots, n,$

where a_{ij}^-, a_{ij}^+ are defined by $[a]_{ij} = \begin{cases} [a_{ij}^-, a_{ij}^+] & \text{if } x_j \geq 0, \\ [a_{ij}^+, a_{ij}^-] & \text{if } x_j < 0. \end{cases}$ (Hartfiel, 1980 [12])

Proof:

(a) ⇒ (b)

Since $x \in S$ there are $A \in [A]$, $b \in [b]$ such that $Ax = b$. Therefore,

$$0 = Ax - b\{ \geq \check{A}x - (\text{rad}[A])|x| - \check{b} - \text{rad}[b], \ \leq \check{A}\,x + (\text{rad}[A])|x| - \check{b} + \text{rad}[b],$$

whence $-((\text{rad}[A])|x| + \text{rad}[b]) \leq \check{b} - \check{A}x \leq (\text{rad}[A])|x| + \text{rad}[b]$. This implies (b).

(b) ⇒ (c)

is seen immediately.

(c) ⇒ (d)

Let D_x be such that $|D_x| = I$ and $D_x x = |x|$. From (c) follows $(\check{A} - D(\text{rad}[A])D_x)x = \check{b} + D(\text{ rad}[b]) \in [b]$. Since $\check{A} - D(\text{rad}[A])D_x \in [A]$, the assertion (d) follows.

(d) ⇒ (e)

From the assumption (d) and Theorem 1 we obtain $A \in [A]$, $b \in [b]$ such that $Ax = b$ holds. Hence $0 = b - Ax \in [b] - [A]x$.

(e) ⇒ (f)

From (e) and the definition of $a_{ij}^{\pm}$ it follows:

$$\inf([A]x - [b])_i = -\bar{b}_i + \sum_{j=1}^{n} a_{ij}^- x_j \leq 0 \leq -\underline{b}_i + \sum_{j=1}^{n} a_{ij}^+ x_j = \sup([A]x - [b])_i, \quad i = 1, \ldots, n.$$

(f) ⇒ (a)

From (f) we obtain:

$$\underline{([A]x)}_i = \sum_{j=1}^{n} a_{ij}^- x_j \leq \bar{b}_i, \quad \overline{([A]x)}_i = \sum_{j=1}^{n} a_{ij}^+ x_j \geq \underline{b}_i.$$

Define $\alpha_{ij}(t) := t a_{ij}^+ + (1-t) a_{ij}^-$, $\beta_i(t) := \sum_{j=1}^{n} \alpha_{ij}(t) x_j$, $t \in [0, 1]$, $i, j = 1, \ldots, n$. Since $\beta_i(t)$ is continuous it assumes for $0 \leq t \leq 1$ all values between $\beta_i(0)$ and $\beta_i(1)$. From $\beta_i(0) \leq \bar{b}_i$, $\underline{b}_i \leq \beta_i(1)$ and $\underline{b}_i \leq \bar{b}_i$ there is some value $t = t_i \in [0, 1]$

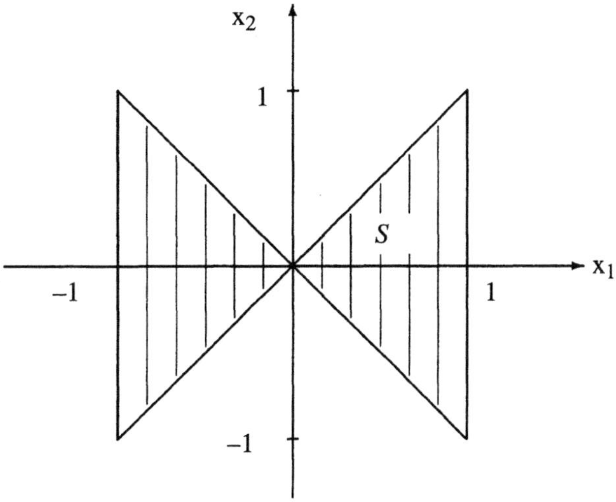

Fig. 1. The solution set S of Example 1

such that $\beta_i(t_i) \in [b]_i$. With $\hat{A} := A(t_1, \ldots, t_n) := (\alpha_{ij}(t_i))$, $\hat{b} := (\beta_i(t_i))$ we get $\hat{A}x = \hat{b}$, $\hat{A} \in [A]$, $\hat{b} \in [b]$, hence $x \in S$. □

Note that the equivalence $(a) \Leftrightarrow (b)$ is the famous Oettli–Prager criterion which was generalized by Fischer and Heindl in [13] (cf. also [14]). In the equivalence $(a) \Leftrightarrow (f)$ the coefficients of the inequalities remain fix as long as one stays in a fixed orthant. Therefore, this equivalence shows that the intersection of S with any fixed closed orthant O is the intersection of finitely many half–spaces. In particular, the boundary ∂S of S is composed of finitely many pieces of hyperplanes. As long as $[A]$ is regular, S is connected and compact which follows from the continuous mapping $f(A,b) = A^{-1}b$ on the connected and compact set $[A] \times [b]$. If $[A]$ contains a singular matrix, connectivity may be lost as Jansson's example $[A] = [-1,1]$, $[b] = 1$, $S = (-\infty,-1] \cup [1,\infty)$ in [17] shows. On the other hand it can be present as the example $[A] = [-1,1]$, $[b] = [0,1]$, $S = \mathbb{R}$ illustrates. Compactness is always absent if $[A]$ is not regular and $S \neq \emptyset$. In this case S is unbounded even if the linear systems with singular matrices $A \in [A]$ are not solvable for any $b \in [b]$. Use Cramer's rule, e.g., to prove this statement. Since half–spaces are convex and intersections of convex sets share this property one obtains it for $S \cap O$. The following example shows, however, that convexity need not hold for the whole solution set S.

Example 1. Let

$$[A] = \begin{pmatrix} 1 & 0 \\ [-1,1] & 1 \end{pmatrix}, \quad [b] = \begin{pmatrix} [-1,1] \\ 0 \end{pmatrix}.$$

From Theorem 2, we get the inequalities $|x_2| \leq |x_1| \leq 1$ which characterize the set S completely. Its position in $\mathbb{R}^2$ can be seen from the subsequent Fig. 1.

4. The Symmetric Solution Set S_{sym}

4.1. Historical Remarks and Properties

In contrast to S the symmetric solution set S_{sym} as defined in (1) is much more difficult to characterize. It occurs when the coefficients of a fixed linear system with a symmetric matrix A is perturbed by symmetric matrices only. Since $S_{\text{sym}} \subseteq S$ Neumaier advised in a letter to J. Rohn [23] dated on 23rd December, 1985, to consider this set for its own. In his book [20] which appeared in 1990 he mentioned S_{sym} in Example 3.4.2 . In the same year, Jansson gave a talk in Albena in which he presented inner and outer enclosures for S_{sym} (cf. Sect. 5 of the present paper). For more general dependencies such enclosures were given by Rump in his survey article [24] which appeared in 1994. In 1993 an interval version of the Cholesky method was introduced by the authors in [8] in order to enclose S_{sym}. Criteria of feasibility were proved there and a corresponding perturbation result was given in [9]. This method was recently extended to a block version by Schäfer in [26]. It was proved for the first time in 1995 in [9] that the boundary ∂S_{sym} can be curvilinear – in contrast to the boundary ∂S of S. The proof covered only 2×2 matrices and could unfortunately not be generalized to the n–dimensional case. By means of the Fourier–Motzkin elimination process of linear programming this was possible in 1996 in [2]; cf. also [3]. These results were even generalized two years later in [4] where now affine dependencies were allowed. In [7] the elimination process introduced in [2] and [4] was generalized in order to handle particular dependencies such as Toeplitz and Hankel matrices. While the approaches always took place in a fixed orthant it was possible in [19] to modify the elimination process by an approach similar to the equivalence (a) $\Leftrightarrow$ (e) in Theorem 2 due to Beeck; cf. Sect. 4.3 of the present paper.

We start by an instructive 2×2 example whose features remain true also in the $n \times n$ case.

Example 2. Let

$$[A] = \begin{pmatrix} 1 & [1,2] \\ [1,2] & [-1,0] \end{pmatrix}, \quad [b] = \begin{pmatrix} 4 \\ [1,2] \end{pmatrix}.$$

Considering $A^{-1}b$ one easily recognizes $S_{\text{sym}} \subseteq S \subseteq O_1$. Anticipating Sect. 4.2, we list the inequalities by which S_{sym} is described:

$$\begin{aligned} -4 + x_1 + x_2 &\le 0, \\ 4 - x_1 - 2x_2 &\le 0, \\ -2 + x_1 - x_2 &\le 0, \\ 1 - 2x_1 &\le 0, \\ x_1^2 - 4x_1 + x_2 &\le 0, \\ x_1^2 + x_2^2 - 4x_1 + 2x_2 &\ge 0. \end{aligned}$$

The first four inequalities characterize S while the last two restrict x to be contained in S_{sym}.

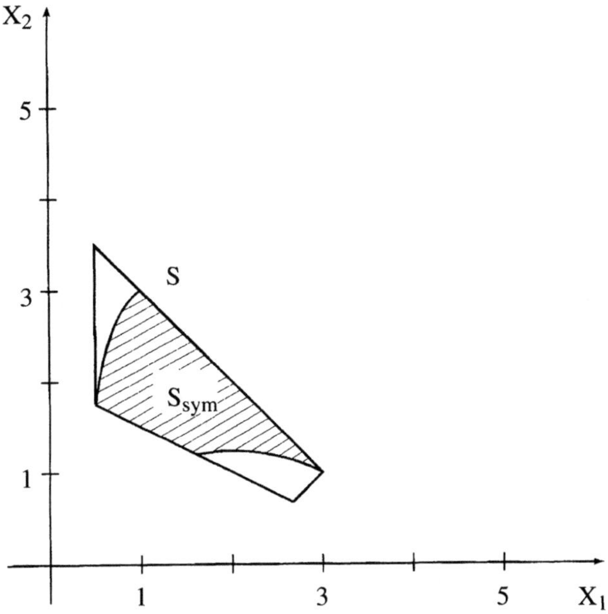

Fig. 2. The solution sets S and S_{sym} of Example 2

From Fig. 2, one sees that S_{sym} is connected and compact. This remains true also in the general case as long as $[A]$ is regular. The counterexamples for non-regular matrices are the same as in Sect. 3. As we shall see (Sects. 4.2 and 4.4) the boundary ∂S_{sym} of S_{sym} is composed by pieces of hyperplanes and quadrics; in particular, ∂S_{sym} can be curvilinear. In general, S_{sym} is not convex, and the same holds true for its intersection with an orthant. Up to now it is unknown whether $S_{\text{sym}} \cap O$ is always connected.

4.2. Characterization of S_{sym} by Means of a Fourier–Motzkin Elimination

In this section, we show that $S_{\text{sym}} \cap O$ can be characterized by means of quadrics and hyperplanes. To this end we aim at describing $S_{\text{sym}} \cap O$ by means of linear and quadratic inequalities whose coefficients consist of the bounds $\underline{b}_i$, $\overline{b}_i$ and $\underline{a}_{ij}$, $\overline{a}_{ij}$. Without loss of generality we choose the orthant $O = O_1$. We only indicate the algorithm by replacing b and a_{12} by $\underline{b}$, $\overline{b}$ and $\underline{a}_{12}$, $\overline{a}_{12}$, respectively. We get the following equivalences:

$$\begin{aligned} x \in S_{\text{sym}} \cap O_1 \Leftrightarrow & x \in O_1 \ \wedge \ \exists A = A^T \in [A],\ b \in [b] :\ Ax = b \\ & \Leftrightarrow x \in O_1 \ \wedge \ \exists a_{ij} \in \mathbb{R} \ \text{ for } i,j = 1,\ldots,n : \\ & \left(a_{ij} = a_{ji} \wedge \left\{ \begin{array}{c} \underline{b}_i \le \sum_{j=1}^{n} a_{ij}x_j \le \overline{b}_i \\ \underline{a}_{ij} \le a_{ij} \le \overline{a}_{ij} \end{array} \right\} \right) \\ & \Leftrightarrow x \in S \cap O_1 \ \wedge \ \exists a_{ij} \in \mathbb{R} \ \textit{for } i,j = 1,\ldots,n : \end{aligned}$$

$$\left(a_{ij} = a_{ji} \wedge \begin{Bmatrix} \underline{b}_i x_i \le \sum_{j=1}^{n} a_{ij} x_i x_j \le \overline{b}_i x_i \\ \underline{a}_{ij} x_i x_j \le a_{ij} x_i x_j \le \overline{a}_{ij} x_i x_j \end{Bmatrix}\right). \tag{5}$$

Only the implication '$\Leftarrow$' of the last assertion requires a proof since $x_i = 0$ or $x_j = 0$ is possible. Let $A = A^T$ be the matrix whose entries fulfill, by assumption, the inequalities in (5). If $x_k \neq 0$, $x_l \neq 0$ then obviously $a_{kl} \in [a]_{kl}$. Otherwise nothing can be said on a_{kl} in this respect. From $x \in S \cap O_1$ we get $\hat{a}_{ij} \in [a]_{ij}$ such that $\sum_{j=1}^{n} \hat{a}_{ij} x_j = b_i \in [b]_i$. Define $\tilde{A}$ by

$$\tilde{a}_{kl} := \begin{cases} a_{kl} & \text{if } x_k \neq 0,\ x_l \neq 0\ , \\ \hat{a}_{kl} & \text{if } x_k = 0 \text{ and } x_l \neq 0\ , \\ \hat{a}_{lk} & \text{if } x_k \neq 0 \text{ and } x_l = 0, \\ \check{a}_{kl} & \text{if } x_k = 0 \text{ and } x_l = 0. \end{cases}$$

Since we assumed $[A] = [A]^T$ when considering S_{sym}, the midpoint $\check{A}$ of $[A]$ is symmetric, and the definition of $\tilde{A}$ shows that $\tilde{A}$ has this property, too. Moreover we get $\tilde{A} \in [A]$ and $\underline{b} \le \tilde{A}x \le \overline{b}$ for the vector x under consideration. Note that

(a) whenever a_{ij} occurs in (5) it has the same factor $x_i x_j$;

(b) trivial inequalities like $0 \le 0$ can be omitted;

(c) there are only linear inequalities (cf. Theorem 2 (f)) and quadratic ones;

(d) no additional multiplication will be needed in the further stages of the algorithm.

We next isolate the a_{12}–term:

$x \in S_{\text{sym}} \cap O_1 \Leftrightarrow x \in S \cap O_1 \ \wedge\ \exists a_{ij} \in \mathbb{R}$ for $i, j = 1, \ldots, n$:

$$\left(a_{ij} = a_{ji} \ \wedge \begin{Bmatrix} \{\underline{b}_1 - \sum_{\substack{j=1 \\ j\neq 2}}^{n} a_{1j} x_j\} x_1 \le a_{12} x_1 x_2 \\ \le \{\overline{b}_1 - \sum_{\substack{j=1 \\ j\neq 2}}^{n} a_{1j} x_j\} x_1 \\ \{\underline{b}_2 - \sum_{j=2}^{n} a_{2j} x_j\} x_2 \le a_{12} x_1 x_2 \\ \le \{\overline{b}_2 - \sum_{j=2}^{n} a_{2j} x_j\} x_2 \\ \underline{a}_{12} x_1 x_2 \le a_{12} x_1 x_2 \le \overline{a}_{12} x_1 x_2 \\ \text{remaining} \quad a_{12} - \text{free inequalities} \end{Bmatrix}\right). \tag{6}$$

Now those inequalities written down explicitly in (6) hold if and only if the maximum of their left–hand sides is less or equal than the minimum of their right–hand sides and this again holds if and only if each of these left–hand sides is less or equal than each of the right–hand sides. Therefore we get:

$$x \in S_{\mathrm{sym}} \cap O_1 \Leftrightarrow x \in S \cap O_1 \wedge \exists a_{ij} \in \mathbb{R} \text{ for } i,j = 1,\dots,n,\ (i,j) \notin \{(1,2),(2,1)\} :$$

$$\left(a_{ij} = a_{ji} \ \wedge \left\{ \begin{array}{c} \{\underline{b}_1 - \sum\limits_{\substack{j=1 \\ j\neq 2}}^{n} a_{1j}x_j\}x_1 \le \overline{a}_{12}x_1x_2 \\ \{\underline{b}_2 - \sum\limits_{j=2}^{n} a_{2j}x_j\}x_2 \le \overline{a}_{12}x_1x_2 \\ \{\underline{b}_1 - \sum\limits_{\substack{j=1 \\ j\neq 2}}^{n} a_{1j}x_j\}x_1 \le \{\overline{b}_2 - \sum\limits_{j=2}^{n} a_{2j}x_j\}x_2 \\ \underline{a}_{12}x_1x_2 \le \{\overline{b}_1 - \sum\limits_{\substack{j=1 \\ j\neq 2}}^{n} a_{1j}x_j\}x_1 \\ \underline{a}_{12}x_1x_2 \le \{\overline{b}_2 - \sum\limits_{j=2}^{n} a_{2j}x_j\}x_2 \\ \{\underline{b}_2 - \sum\limits_{j=2}^{n} a_{2j}x_j\}x_2 \le \{\overline{b}_1 - \sum\limits_{\substack{j=1 \\ j\neq 2}}^{n} a_{1j}x_j\}x_1 \\ \text{remaining } a_{12} - \text{free inequalities} \end{array} \right\} \right).$$

No a_{12}-term appears now any longer in the set of inqualities, and the isolation and elimination process can be repeated for the other entries a_{ij}. At the end we are left with a set of inequalities as indicated above. This process which was given in a slightly modified form in [2] must be repeated for any of the 2^n orthants, so the amount of inequalities can increase tremendously with n although sometimes inequalities can be combined and turn out to be contained in other ones.

We end this section with another example whose inequalities were derived by the method above.

Example 3. Let

$$[A] := \begin{pmatrix} 1 & [0,1] \\ [0,1] & [-4,-1] \end{pmatrix}, \quad [b] := \begin{pmatrix} [0,2] \\ [0,2] \end{pmatrix}.$$

Then $S \cap O_1$ is characterized by

$$0 \le x_1 \le 2, \quad 0 \le x_2 \le x_1,$$

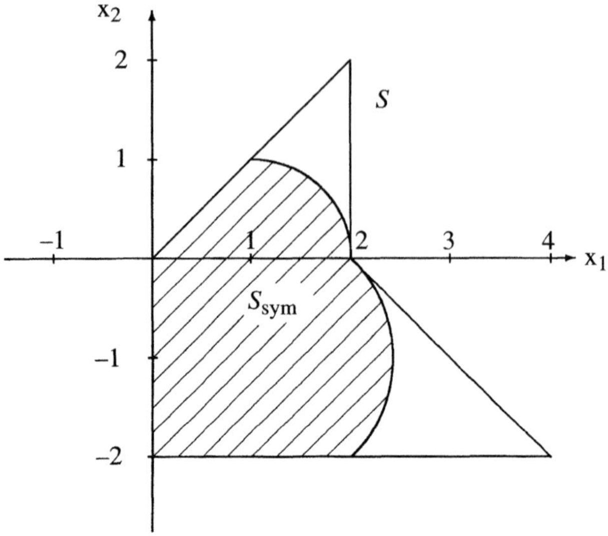

Fig. 3. The solution sets S and S_{sym} of Example 3

while $S \cap O_4$ with $O_4 := \{(x_1, x_2)^T \in \mathbb{R}^2 |\, x_1 \geq 0,\ x_2 \leq 0\}$ can be described by

$$-2 \leq x_2 \leq 0, \quad 0 \leq x_1 \leq 2 - x_2.$$

Note that we combined some of the inequalities omitting redundant ones and taking into account the respective orthant. For S_{sym} the following *additional* inequalities are needed

$$\begin{cases} (x_1 - 1)^2 + x_2^2 \leq 1 & \text{in the case } S_{sym} \cap O_1, \\ (x_1 - 1)^2 + (x_2 + 1)^2 \leq 2 & \text{in the case } S_{sym} \cap O_4, \end{cases}$$

where we omitted again redundant ones. Figure 3 shows S and S_{sym}.

4.3 Characterization of S_{sym} by Means of Sets

In this section, we do not stick to a fixed orthant but describe S_{sym} as a whole. This decreases the amount of relations drastically. For the algorithm we need a simple lemma which we formulate first.

Lemma 1.

(a) *Let* $[a]_i \in \mathbb{IR}$ *for* $i = 1, \ldots, n,\ n \geq 3$. *Then*

$$\bigcap_{i=1}^{n} [a]_i \neq \emptyset \;\Leftrightarrow\; [a]_i \cap [a]_j \neq \emptyset \quad \textit{for } i < j,\ \ i, j = 1, \ldots, n.$$

(b) *Let* $[a],\ [b],\ [c] \in \mathbb{IR}$. *Then*

$$([a] + [b]) \cap [c] \neq \emptyset \;\Leftrightarrow\; [a] \cap ([c] - [b]) \neq \emptyset. \qquad \square$$

The proof is easy. Like the whole approach in this section it can be found in [19]. As the referee pointed out part (a) of the lemma is the simplest case of Helly's theorem on intersections of convex sets; cf., e.g., Theorem 12 in [18], p. 166. He also mentioned that part (b) is often used in the equivalent form

$$([a]-[b])\cap[c]\neq\emptyset \;\Leftrightarrow\; ([a]-[c])\cap[b]\neq\emptyset.$$

Now we show how to replace the parameters $a_{ij},\ b_i$ by intersections involving $[a]_{ij}$, $[b]_j$.

$$x\in S_{\text{sym}}\Leftrightarrow \exists\ a_{ij}\in\mathbb{R} \text{ for } i,j=1,\dots,n:$$

$$\left(a_{ij}=a_{ji}\ \wedge\ \begin{Bmatrix} \left\{\sum_{j=1}^{n} a_{ij}x_j\right\}\cap[b]_i\neq\emptyset \\ \{a_{ij}\}\cap[a]_{ij}\neq\emptyset \end{Bmatrix}\right)$$

$$\Leftrightarrow x\in S\ \wedge\ \exists\ a_{ij}\in\mathbb{R} \text{ for } i,j=1,\dots,n:$$

$$\left(a_{ij}=a_{ji}\ \wedge\ \begin{Bmatrix} \left\{\sum_{j=1}^{n} a_{ij}x_{ij}\right\}\cap[b]_ix_i\neq\emptyset \\ \{a_{ij}x_ix_j\}\cap[a]_{ij}x_ix_j\neq\emptyset \end{Bmatrix}\right)$$

$$\Leftrightarrow x\in S\ \wedge\ \exists\ a_{ij}\in\mathbb{R} \text{ for } i,j=1,\dots,n:$$

$$\left(a_{ij}=a_{ji}\ \wedge\ \begin{Bmatrix} \{a_{12}x_1x_2\}\cap([b]_1x_1-\sum_{\substack{j=1\\ j\neq 2}}^{n} a_{1j}x_1x_j)\neq\emptyset \\ \{a_{12}x_1x_2\}\cap([b]_2x_2-\sum_{j=2}^{n} a_{2j}x_2x_j)\neq\emptyset \\ \{a_{12}x_1x_2\}\cap[a]_{12}x_1x_2\neq\emptyset \\ \text{remaining } a_{12}-\text{free intersections} \end{Bmatrix}\right)$$

$$\Leftrightarrow x\in S\ \wedge\ \exists\ a_{ij}\in\mathbb{R} \text{ for } i,j=1,\dots,n,\ (i,j)\notin\{(1,2),(2,1)\}:$$

$$\left(a_{ij}=a_{ji}\wedge\begin{Bmatrix} [a]_{12}x_1x_2\cap([b]_1x_1-\sum_{\substack{j=1\\ j\neq 2}}^{n} a_{1j}x_1x_j) \\ \cap([b]_2x_2-\sum_{j=2}^{n} a_{2j}x_2x_j)\neq\emptyset \\ \text{remaining}\quad a_{12}--\text{free intersections} \end{Bmatrix}\right).$$

The multiplication by x_i and x_j, respectively, is used as in Sect. 4.2. No additional multiplication is needed in the further course of the algorithm. In order to continue it is necessary to resolve the triple intersection by intersections with two operands. The key to this is given by Lemma 1 . We leave the details to the reader. The quadrics are seen here only at a second glance. They are hidden behind the intersections. However, since these intersections can be expressed by inequalities the approach in this section can be reformulated by formulas as in Sect. 4.2. Both approaches lack, however, a generalization to more than dependencies as they occur in symmetric ($a_{ij} = a_{ji}$), skew–symmetric ($a_{ij} = -a_{ji}$), and persymmetric matrices ($a_{ij} = a_{n+1-j,n+1-i}$), respectively. In order to generalize the elimination process in Sect. 4.2 we need a modification presented in Sect. 4.4.

We conclude the present section by the general 3×3 case.

Example 4. Let $[A] = [A]^T \in \mathbb{IR}^{3\times 3}$, $[b] \in \mathbb{IR}^3$ and consider again S_{sym}. Eliminating b_i, a_{ii} for $i = 1,2,3$ and a_{12} leads to

$$x \in S_{\text{sym}} \Leftrightarrow x \in S \ \wedge \ \exists \ a_{13},\ a_{23} \in \mathbb{R} : \Bigg(a_{13} = a_{31} \ \wedge \ a_{23} = a_{32}$$

$$\wedge \left\{ \begin{array}{ll} [a]_{12} x_1 x_2 & \cap \ ([b]_1 x_1 - [a]_{11} x_1^2 - a_{13} x_1 x_3) \\ & \cap \ ([b]_2 x_2 - [a]_{22} x_2^2 - a_{23} x_2 x_3) \neq \emptyset \\ \{a_{13} x_1 x_3\} & \cap \ ([b]_3 x_3 - a_{23} x_2 x_3 - [a]_{33} x_3^2) \neq \emptyset \\ \{a_{13} x_1 x_3\} & \cap \ [a]_{13} x_1 x_3 \neq \emptyset \\ \{a_{23} x_2 x_3\} & \cap \ [a]_{23} x_2 x_3 \neq \emptyset \end{array} \right\} \Bigg). \tag{7}$$

After having resolved the triple intersection in (7) into simple intersections the isolation of a_{13} and its elimination yield to

$$x \in S_{\text{sym}} \Leftrightarrow \ x \in S \ \wedge \ \exists \ a_{23} \in \mathbb{R} : \Bigg(a_{23} = a_{32}$$

$$\wedge \left\{ \begin{array}{ll} [a]_{13} x_1 x_3 & \cap \ ([b]_1 x_1 - [a]_{11} x_1^2 - [a]_{12} x_1 x_2) \\ & \cap \ ([b]_3 x_3 - a_{23} x_2 x_3 - [a]_{33} x_3^2) \\ & \cap \ ([b]_1 x_1 - [b]_2 x_2 - [a]_{11} x_1^2 + [a]_{22} x_2^2 + a_{23} x_2 x_3) \neq \emptyset \\ \{a_{23} x_2 x_3\} & \cap \ ([b]_2 x_2 - [a]_{12} x_1 x_2 - [a]_{22} x_2^2) \neq \emptyset \\ \{a_{23} x_2 x_3\} & \cap \ [a]_{23} x_2 x_3 \neq \emptyset \end{array} \right\} \Bigg).$$

Here again we omitted an intersection which can be deduced from $x \in S$. Eliminating the last parameter a_{23} the fourfold intersection produces $\binom{4}{2}$ simple intersections among which only one is a_{23}–free. This one turns out to be a consequence of $x \in S$. Therefore, it can be omitted, and we are finally left with the equivalence

$$
\begin{aligned}
x \in S_{\text{sym}} \;\Leftrightarrow\; x \in S & \\
\wedge\; [a]_{23}x_2x_3 \;\cap\; & ([b]_2x_2 - [a]_{12}x_1x_2 - [a]_{22}x_2^2) \\
\cap\; & ([b]_3x_3 - [a]_{13}x_1x_3 - [a]_{33}x_3^2) \\
\cap\; & (-[b]_1x_1 + [b]_2x_2 + [a]_{11}x_1^2 - [a]_{22}x_2^2 + [a]_{13}x_1x_3) \\
\cap\; & (-[b]_1x_1 + [b]_2x_2 + [b]_3x_3 + [a]_{11}x_1^2 - [a]_{22}x_2^2 - [a]_{33}x_3^2)/2 \\
\cap\; & (-[b]_1x_1 + [b]_3x_3 + [a]_{11}x_1^2 + [a]_{12}x_1x_2 - [a]_{33}x_3^2) \\
\cap\; & \{([b]_1 - [b]_1)x_1 + [b]_2x_2 + ([a]_{11} - [a]_{11})x_1^2 \\
& \quad - [a]_{12}x_1x_2 - [a]_{22}x_2^2\} \\
\neq\; & \emptyset.
\end{aligned}
$$

Using Lemma 1 (a) this multiple intersection can equivalently be written as $\binom{7}{2} = 21$ simple ones. Two of them can be deduced from the requiry $x \in S$. The remaining 19 ones together with the three occuring from Beeck's criterion in Theorem 2 (d) result in a total of 22 which all can be rewritten in the form "$0 \in \ldots$". Introducing the interval bounds, fixing an orthant O and taking into account

$$0 \in [a] \Leftrightarrow \underline{a} \leq 0 \leq \overline{a}$$

finally yields to 44 inequalities for the characterization of $S_{\text{sym}} \cap O$ as was already mentioned without proof in [2].

4.4. *A Generalizable Approach for Characterizing S_{sym}*

The subsequent modification of Sect. 4.2 is based on the following theorem which is proved in [7].

Theorem 3. *Let $f_{\lambda\mu}$, g_λ, $\lambda = 1, \ldots, k$ (≥ 2), $\mu = 1, \ldots, m$, be real valued functions of $x = (x_1, \ldots, x_n)^T$ on some subset $D \subseteq \mathbb{R}^n$. Assume that there is a positive integer $k_1 < k$ such that*

$$f_{\lambda 1}(x) \not\equiv 0 \textit{ for all } \lambda \in \{1, \ldots, k\}, \tag{8}$$

$$f_{\lambda 1}(x) \geq 0 \textit{ for all } x \in D \textit{ and all } \lambda \in \{1, \ldots, k\}, \tag{9}$$

$$\textit{for each } x \in D \textit{ there is an index } \beta^* = \beta^*(x) \in \{1, \ldots, k_1\} \textit{ with } f_{\beta^* 1}(x) > 0$$
$$\textit{and an index } \gamma^* = \gamma^*(x) \in \{k_1 + 1, \ldots, k\} \textit{ with } f_{\gamma^* 1}(x) > 0. \tag{10}$$

For m parameters $u_1, \ldots, u_m$ varying in $\mathbb{R}$ and for x varying in D define the sets S_1, S_2 by

$$S_1 := \{x \in D | \exists u_\mu \in \mathbb{R},\ \mu = 1, \dots, m:\ (11),\ (12)\ hold\},$$
$$S_2 := \{x \in D | \exists u_\mu \in \mathbb{R},\ \mu = 2, \dots, m:\ (13)\ holds\},$$

where the inequalities (11), (12) and (13), respectively, are given by

$$g_\beta(x) + \sum_{\mu=2}^{m} f_{\beta\mu}(x)u_\mu \le f_{\beta 1}(x)u_1, \quad \beta = 1, \dots, k_1, \tag{11}$$

$$f_{\gamma 1}(x)u_1 \le g_\gamma(x) + \sum_{\mu=2}^{m} f_{\gamma\mu}(x)u_\mu, \quad \gamma = k_1 + 1, \dots, k, \tag{12}$$

and

$$g_\beta(x)f_{\gamma 1}(x) + \sum_{\mu=2}^{m} f_{\beta\mu}(x)f_{\gamma 1}(x)u_\mu \le g_\gamma(x)f_{\beta 1}(x) + \sum_{\mu=2}^{m} f_{\gamma\mu}(x)f_{\beta 1}(x)u_\mu, \tag{13}$$
$$\beta = 1, \dots, k_1,\ \gamma = k_1 + 1, \dots, k.$$

(Trivial inequalities such as $0 \le 0$ can be omitted.) Then

$$S_1 = S_2. \qquad \square$$

The assertion of Theorem 3 even holds if $f_{\beta 1}(x)$, $f_{\gamma 1}(x)$ are replaced in (13) (but not in (11), (12)) by $\tilde{f}_{\beta 1}(x)$ and $\tilde{f}_{\gamma 1}(x)$, respectively, where

$$f_{\beta 1}(x) = h_{\beta\gamma}(x)\tilde{f}_{\beta 1}(x), \quad f_{\gamma 1}(x) = h_{\beta\gamma}(x)\tilde{f}_{\gamma 1}(x)$$

with nonnegative functions $\tilde{f}_{\beta 1}$, $\tilde{f}_{\gamma 1}$, $h_{\beta\gamma}$ defined on D.

In contrast to the procedure in Sect. 4.2 there is no need in Theorem 3 to find factors such that the expressions to be eliminated have the same shape in *all* inequalities after multiplication. This allows the application of this theorem also for more general dependencies of the entries of A and b, respectively. We apply now the theorem to symmetric matrices with $D := O_1$ starting as in Sect. 4.2. Just for illustration, we choose a_{12} as the entry to be eliminated.

$$x \in S_{\text{sym}} \cap O_1 \Leftrightarrow x \in O_1 \ \wedge\ \exists\ a_{ij} \in \mathbb{R} \text{ for } i, j = 1, \dots, n:$$
$$\left(a_{ij} = a_{ji} \wedge \left\{ \begin{array}{c} \underline{b}_i \le \sum_{j=1}^{n} a_{ij}x_j \le \overline{b}_i \\ \underline{a}_{ij} \le a_{ij} \le \overline{a}_{ij} \end{array} \right\}\right)$$

$$\Leftrightarrow x \in O_1 \ \wedge \ \exists\, a_{ij} \in \mathbb{R} \text{ for } i,j = 1,\ldots,n :$$

$$\left(a_{ij} = a_{ji} \ \wedge \ \left\{\begin{array}{rl} \underline{b}_1 - \sum\limits_{\substack{j=1\\ j\neq 2}}^{n} a_{1j}x_j \leq & a_{12}x_2 \\ & a_{12}x_2 \leq \overline{b}_1 - \sum\limits_{\substack{j=1\\ j\neq 2}}^{n} a_{1j}x_j \\ \underline{b}_2 - \sum\limits_{j=2}^{n} a_{2j}x_j \leq & a_{12}x_1 \\ & a_{12}x_1 \leq \overline{b}_2 - \sum\limits_{j=2}^{n} a_{2j}x_j \\ \underline{a}_{12} \leq & a_{12} \\ & a_{12} \leq \overline{a}_{12} \\ \multicolumn{2}{c}{\text{remaining } a_{12} - \text{free inequalities}} \end{array}\right\}\right)$$

$$\Leftrightarrow x \in O_1 \ \wedge \ \exists\, a_{ij} \in \mathbb{R} \text{ for } i,j = 1,\ldots,n,\ (i,j) \notin \{(1,2),(2,1)\} :$$

$$\left(a_{ij} = a_{ji} \ \wedge \ \left\{\begin{array}{c} \underline{b}_1 - \sum\limits_{\substack{j=1\\ j\neq 2}}^{n} a_{1j}x_j \leq \overline{a}_{12}x_2, \\ \underline{a}_{12}x_2 \leq \overline{b}_1 - \sum\limits_{\substack{j=1\\ j\neq 2}}^{n} a_{1j}x_j, \\ \underline{b}_2 - \sum\limits_{j=2}^{n} a_{2j}x_j \leq \overline{a}_{12}x_1, \\ \underline{a}_{12}x_1 \leq \overline{b}_2 - \sum\limits_{j=2}^{n} a_{2j}x_j, \\ \underline{b}_1x_1 - \sum\limits_{\substack{j=1\\ j\neq 2}}^{n} a_{1j}x_1x_j \leq \overline{b}_2x_2 - \sum\limits_{j=2}^{n} a_{2j}x_2x_j, \\ \underline{b}_2x_2 - \sum\limits_{j=2}^{n} a_{2j}x_2x_j \leq \overline{b}_1x_1 - \sum\limits_{\substack{j=1\\ j\neq 2}}^{n} a_{1j}x_1x_j, \\ \text{remaining } a_{12} - \text{free inequalities} \end{array}\right\}\right)$$

The first four inequalities coincide with those in the elimination process for $S \cap O_1$. The next two are new; they are apparently caused by the symmetry and contain quadratic polynomials. When eliminating a_{ij} for $(i,j) \notin \{(1,2),(2,1)\}$

according to the remark below Theorem 3, no additional multiplication is needed in inequalities which contain quadratic polynomials. This is true because the function $f_{\lambda\mu}$ in front of a_{ij} reads $f_{\lambda\mu}(x) = x_i x_j$ in these inequalities, and in the remaining (non–quadratic) inequalities they are given by $f_{\lambda\mu}(x) = x_i$, $f_{\lambda\mu}(x) = x_j$ and $f_{\lambda\mu}(x) = 1$, respectively. Note that the sign of these functions remains constant over the orthant O_1. This is the reason, why no splitting is needed for $D = O_1$ during the elimination process. Pursuing this process shows that the final inequalities for $S_{\text{sym}} \cap O_1$ consist of the inequalities which characterize $S \cap O_1$, and quadratic inequalities.

5. Enclosures for S_{sym}

As the preceding sections show it is generally not easy to decide whether a given vector x belongs to S_{sym} for given $[A] = [A]^T \in I\mathbb{R}^{n \times n}$, $[b] \in \mathbb{R}$. Therefore, it is interesting to look for bounds of S_{sym}. We restrict ourselves to bounds which are not at the same time bounds of S, i.e., we look for bounds which should be better than the latter ones. Up to now there are essentially two methods with this property: a direct one and an iterative one. The first generalizes the well–known Cholesky method and uses the definitions

$$[a]^2 := \{a^2 | a \in [a]\}, \quad \sqrt{[a]} := [a]^{\frac{1}{2}} := \{\sqrt{a} | a \in [a]\},$$

where $\underline{a} \geq 0$ is assumed for the square root. The algorithm which was first considered in [8] reads

Step 1: “LL^T–decomposition”

for $j := 1$ to n do

$$[l]_{jj} := \left([a]_{jj} - \sum_{k=1}^{j-1} [l]_{jk}^2 \right)^{1/2};$$

for $i := j + 1$ to n do

$$[l]_{ij} := \left([a]_{ij} - \sum_{k=1}^{j-1} [l]_{ik}[l]_{jk} \right) / [l]_{jj};$$

Step 2: Forward substitution‘

for $i := 1$ to n do

$$[y]_i := \left([b]_i - \sum_{j=1}^{i-1} [l]_{ij}[y]_j \right) / [l]_{ii};$$

Step 3: Backward substitution

for $i := n$ downto 1 do

$$[x]_i^C := \left([y]_i - \sum_{j=i+1}^{n} [l]_{ji}[x]_j^C \right) / [l]_{ii}.$$

It is easy to see by the inclusion monotonicity of the interval arithmetic operations that $S_{\text{sym}} \subseteq [x]^C$ holds. One also sees directly that the feasibility of the preceding interval Cholesky method does not depend on $[b]$ and requires that each matrix $A \in [A]$ is symmetric and positive definite. Unfortunately, this property is not sufficient. As was shown in [8] the feasibility is guaranteed if $[A]$ is a symmetric H matrix satisfying $\underline{a}_{ii} > 0$ for $i = 1, \ldots, n$. Interval matrices $[A]$ with a strictly diagonally dominant comparison matrix $\langle [A] \rangle$ and $\underline{a}_{ii} > 0$, $i = 1, \ldots, n$, belong to this class. Moreover, if the algorithm is feasible for $[A]$ and if $\rho(|[A]^C| q([A], [B])) < 1$ then it is feasible for $[B]$, too, as was proved in [9]. Here, $|[A]^C| := \langle [L]^T \rangle^{-1} \langle [L] \rangle^{-1}$ where $[L] = ([l]_{ij})$ is the lower triangular interval matrix whose non zero entries $[l]_{ij}$, $i \geq j$, are defined by the algorithm above. Replacing $[A]$, $[B]$ by $\check{A}$, $[A]$, respectively, proves that the interval Cholesky method is feasible whenever $\check{A}$ is positive definite and satisfies $\rho(|\check{A}^C| \text{rad}([A])) < 1$. The following example shows that the interval Cholesky method is tailored to enclose S_{sym}, but not S. However, it does not always yield to enclosures for S_{sym} which are better than general enclosures for S. In particular, for larger n it may show a bad behavior due to rounding errors and data dependencies.

Example 5. Let

$$[A] := \begin{pmatrix} 4 & [-1, 1] \\ [-1, 1] & 4 \end{pmatrix}, \quad [b] := \begin{pmatrix} 6 \\ 6 \end{pmatrix}.$$

Setting $A := \begin{pmatrix} 4 & \alpha \\ \beta & 4 \end{pmatrix}$ for $A \in [A]$, we get

$$A^{-1}b = \frac{6}{16 - \alpha\beta} \begin{pmatrix} 4 - \alpha \\ 4 - \beta \end{pmatrix} \text{ with } \alpha, \beta \in [-1, 1].$$

Since $\beta = \alpha$ in the case $A = A^T \in [A]$ one obtains

$$\begin{aligned} []S_{sym} &= ([\tfrac{18}{15}, 2], [\tfrac{18}{15}, 2])^T, & []S &= ([\tfrac{18}{17}, 2], [\tfrac{18}{17}, 2])^T, \\ [x]^C &= ([1, 2], [\tfrac{18}{16}, 2])^T, & [x]^G &= ([1, 2], [\tfrac{18}{17}, 2])^T, \end{aligned}$$

where $[x]^G$ denotes the vector resulting from the interval Gaussian algorithm (cf. [1]). The sets

$$S_{sym} = \left\{ \frac{6}{4 + \alpha} \begin{pmatrix} 1 \\ 1 \end{pmatrix} \middle| -1 \leq \alpha \leq 1 \right\} = \left\{ \gamma \cdot \begin{pmatrix} 1 \\ 1 \end{pmatrix} \middle| \frac{6}{5} \leq \gamma \leq 2 \right\}$$

and (see [12])

$$S = \text{convex hull} \left(\{ (\frac{6}{5}, \frac{6}{5})^T, (2, 2)^T, (\frac{18}{17}, \frac{30}{17})^T, (\frac{30}{17}, \frac{18}{17})^T \} \right)$$

can be seen in Fig. 4.

Example 5 illustrates that the following properties can occur.

(i) $\square S_{sym} \neq \square S$ (cf. also [20], p. 95),

(ii) $\square S_{sym} \neq [x]^C$,

(iii) $\square S \neq [x]^G$,

(iv) $S \not\subseteq [x]^C$ (but $S_{sym} \subseteq [x]^C$),

(v) $[x]^C \subseteq [x]^G$ *with* $[x]^C \neq [x]^G$.

Unfortunately, $[x]^G \subseteq [x]^C$ with $[x]^G \neq [x]^C$ can also occur as was shown by an example in [8].

The second method which we want to recall was introduced by Jansson [15], [16] and is a modification of the following well–known iterative method for general linear interval systems

$$[x]^{k+1} = \tilde{x} + C([b] - [A]\tilde{x}) + (I - C[A])([x]^k - \tilde{x}), \quad k = 0, 1, \ldots,$$

where $C \in \mathbb{R}^{n \times n}$ is any preconditioning matrix and $\tilde{x}$ is any fixed vector from $\mathbb{R}^n$. Usually, C is an approximate inverse of the midpoint $\check{A}$ and $\tilde{x}$ is an approximate solution of the midpoint equation $\check{A}x = \check{b}$. Exploiting the symmetry of $A \in [A]$ and of $[A] = [A]^T$ and substituting $[x]^k_\Delta := [x]^k - \tilde{x}$ finally yields to

$$[x]^{k+1}_\Delta = [z]^{\text{sym}} + (I - C[A])[x]^k_\Delta, \quad k = 0, 1, \ldots \tag{14}$$

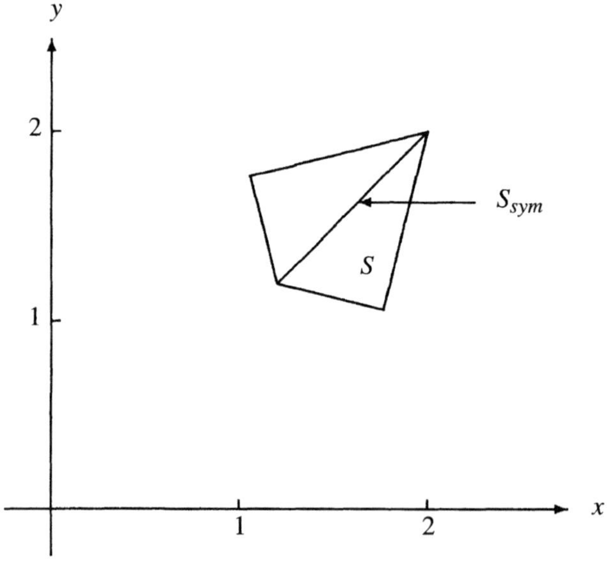

Fig. 4. The solution sets S and S_{sym} of Example 5

with

$$[z]_i^{\text{sym}} := \sum_{j=1}^{n} c_{ij}([b]_j - [a]_{jj}\tilde{x}_j) - \sum_{j=1}^{n}\sum_{l=1}^{j-1}(c_{ij}\tilde{x}_l + c_{il}\tilde{x}_j)[a]_{jl}$$
$$= \{(C(b - A\tilde{x}))_i | A = A^T \in [A], b \in [b]\} \quad \text{for } i = 1, \ldots, n.$$

Note that the last equality holds by virtue of Theorem 1 . Thus $[z]^{\text{sym}}$ is the interval hull of the set $\{C(b - A\tilde{x}) | A = A^T \in [A],\ b \in [b]\}$ and is therefore optimal. The subsequent results were proved in [15], [16].

Theorem 4. *Let $(S_{\text{sym}})_i$ denote the projection of S_{sym} onto the x_i coordinate axis and let $[\Delta]^k := (I - C[A])[x]_\Delta^k$. If*

$$([x]_\Delta^{k_0+1})_i \subsetneq ([x]_\Delta^{k_0})_i \quad \textit{for } i = 1, \ldots, n$$

then the following assertions hold.

(a) $[A]$ *and* C *are regular.*

(b) $S_{\text{sym}} \subseteq \tilde{x} + [x]_\Delta^{k+1} \subseteq \tilde{x} + [x]_\Delta^k$ *for* $k \geq k_0$.

(c) $\tilde{x}_i + \underline{z}_i^{\text{sym}} + \underline{\Delta}_i^{k_0} \leq \min(S_{\text{sym}})_i \leq \tilde{x}_i + \underline{z}_i^{\text{sym}} + \overline{\Delta}_i^{k_0}$,

$\tilde{x}_i + \overline{z}_i^{\text{sym}} + \underline{\Delta}_i^{k_0} \leq \max(S_{\text{sym}})_i \leq \tilde{x}_i + \overline{z}_i^{\text{sym}} + \overline{\Delta}_i^{k_0}$. □

By Theorem 4 one sees at once that the relations

$$\inf S_{\text{sym}} \in \tilde{x} + \underline{z}^{\text{sym}} + [\Delta]^{k_0},$$
$$\sup S_{\text{sym}} \in \tilde{x} + \overline{z}^{\text{sym}} + [\Delta]^{k_0}$$

hold. Therefore, if $\text{rad}[\Delta]^{k_0}$ is small then the enclosure for S_{sym} is good. This can be expected, in particular, if $\text{rad}[A], \text{rad}[b]$ are small, $C \approx \check{A}^{-1}$ and $\tilde{x} \approx \check{A}^{-1}\check{b}$. Then $\|\text{rad}[x]_\Delta^{k_0}\|_\infty \ll 1$ is possible whence $[\Delta]^{k_0}$ is quadratically small.

6. The Symmetric Eigenpair Set

Our final section is devoted to the eigenpair sets E and E_{sym} from (3) and (4), respectively. In order to characterize E we apply the Oettli–Prager theorem to the equation $(A - \lambda I)x = 0$ which leads to

$$(x^T, \lambda)^T \in E \Leftrightarrow |(\check{A} - \lambda I)x| \leq (\text{rad}[A])|x| \wedge x \neq 0. \tag{15}$$

There is a slight difficulty in comparison with Sect. 3 in so far as the matrix $A - \lambda I$ is not fixed but depends on the last component of the vector $(x^T, \lambda)^T$ tested to belong to E. This difficulty can be overcome thinking λ to be arbitrary but fixed.

Then $A - \lambda I$ is a fixed matrix for which Theorem 2 can be applied without any restrictions. Rewriting (15) as in Theorem 2(f) by means of inequalities shows that $E \cap O$ can be characterized by a variety of inequalities which – by virtue of the λx term are at most quadratic. That means that E is the union of finitely many intersections of sets whose boundaries are pieces of hyperplanes and quadrics. We mention that the inequalities for E can also be obtained by applying Theorem 3 . Using any approach of Sect. 4 shows that the symmetric eigenpair set E_{sym} can be described by means of inequalities with polynomials of order three at most. The starting point is now formed by the double inequality

$$(A - \lambda I)x \leq 0 \leq (A - \lambda I)x$$

which is equivalent to $(A - \lambda I)x = 0$. For details we refer to [6]. We conclude our paper with an illustrative example for E and E_{sym}.

Example 6. Let

$$[A] = \begin{pmatrix} 1 & [-1, 1] \\ [-1, 1] & 1 \end{pmatrix}.$$

Then E is characterized by

$$x_1 \cdot x_2 \neq 0 \ \wedge \ \begin{cases} x_1 - |x_2| \leq \lambda x_1 \leq x_1 + |x_2| \\ -|x_1| + x_2 \leq \lambda x_2 \leq |x_1| + x_2 \end{cases}.$$

If $\lambda = 1$ then any vector $(x_1, x_2)^T \neq 0$ occurs as an eigenvector hence $E \subseteq \mathbb{R}^3$ contains the plane $\lambda = 1$ punctured at $(0, 0, 1)^T$. If $\lambda \neq 1$ then $x_1 \cdot x_2 \neq 0$ and $(x^T, \lambda)^T \in E$ if and only if the subsequent statement holds.

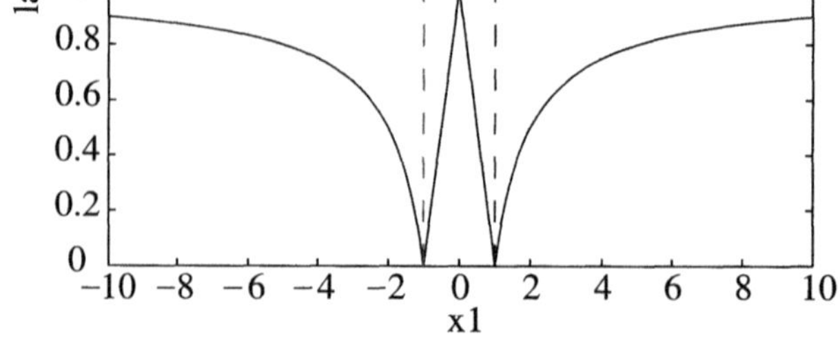

Fig. 5. $E_1 := E \cap P_1$ and $E_{\mathrm{sym}} \cap P_1$ (dashed) of Example 6

$$\lambda \neq 1 \wedge x_1 \cdot x_2 \neq 0 \wedge \begin{cases} 1 - \left|\frac{x_2}{x_1}\right| \leq \lambda \leq 1 + \left|\frac{x_2}{x_1}\right| \\ 1 - \left|\frac{x_1}{x_2}\right| \leq \lambda \leq 1 + \left|\frac{x_1}{x_2}\right| \end{cases}.$$

Restricting $A \in [A]$ to be symmetric yields to any pair $(x_1, x_2)^T \neq 0$ as eigenvector in the case $\lambda = 1$. This is true by virtue of $I = I^T \in [A]$. If $\lambda \neq 1$ then E_{sym} is described by $|x_1| = |x_2| > 0$, $0 \leq \lambda \leq 2$, $\lambda \neq 1$, which together with E leads to Fig. 5 in which we intersected these sets with the plane $P_1 : x_2 = 1$.

References

[1] Alefeld, G., Herzberger, J.: Introduction to interval computations. New York: Academic Press 1983.

[2] Alefeld, G., Kreinovich, V., Mayer, G.: The shape of the symmetric solution set. In: Applications of interval computations (Kearfott, R. B., and Kreinovich, V., eds.). Boston: Kluwer 1996, pp. 61–79.

[3] Alefeld, G., Kreinovich, V., Mayer, G.: Symmetric linear systems with perturbed input data. In: Numerical methods and error bounds (Alefeld, G., and Herzberger, J., eds.). Berlin: Akademie Verlag 1996, pp. 16–22.

[4] Alefeld, G., Kreinovich, V., Mayer, G.: On the shape of the symmetric, persymmetric, and skew–symmetric solution set. SIAM J. Matrix Anal. Appl. *18*, 693–705 (1997).

[5] Alefeld, G., Kreinovich, V., Mayer, G.: The shape of the solution set of linear interval equations with dependent coefficients. Math. Nachr. *192*, 23–26 (1998).

[6] Alefeld, G., Kreinovich, V., Mayer, G.: Modifications of the Oettli–Prager theorem with application to the algebraic eigenvalue problem. In: Symbolic algebraic methods and verification methods – Theory and applications (Alefeld, G., Rohn, J., Rump, S. M., and Yamamoto, T., eds.). Wien: Springer 2001.

[7] Alefeld, G., Kreinovich, V., Mayer, G.: On the solution set of particular classes of linear interval systems. J. Comp. Appl. Math. (forthcoming)

[8] Alefeld G., Mayer, G.: The Cholesky method for interval data. Linear Algebra Appl. *194*, 161–182 (1993).

[9] Alefeld, G., Mayer, G.: On the symmetric and unsymmetric solution sets of interval systems. SIAM J. Matrix Anal. Appl. *16*, 1223–1240 (1995).

[10] Beeck, H.: Über Struktur und Abschätzungen der Lösungsmenge von linearen Gleichungssystemen mit Intervallkoeffizienten. Computing *10*, 231–244 (1972)

[11] Golub, G. H., van Loan, C. F.: Matrix computations. Baltimore, Maryl.: The Johns Hopkins University Press 1983.

[12] Hartfiel, D. J.: Concerning the solution set of $Ax = b$ where $P \leq A \leq Q$ and $p \leq b \leq q$. Numer. Math. *35*, 355–359 (1980).

[13] Heindl, G.: Computable hypernorm bounds for the error of an approximate solution to a system of nonlinear equations. In: Computer arithmetic and enclosure methods (Atanassova, L., and Herzberger, J., eds.). Amsterdam: North–Holland 1992, pp. 451–461.

[14] Heindl, G.: Some inclusion results based on a generalized version of the Oettli–Prager theorem. ZAMM *76*, 263–266 (1996).

[15] Jansson, C.: Rigorous sensitivity analysis for real symmetric matrices with uncertain data. In Computer arithmetic, scientific computation and mathematical modelling (Kaucher, E., Markov, S. M., and Mayer, G., eds.). Basel: Baltzer 1991, pp. 293–316.

[16] Jansson, C.: Interval linear systems with symmetric matrices, skew-symmetric matrices and dependencies in the right-hand side. Computing *46*, 265–274 (1991).

[17] Jansson, C.: Calculation of exact bounds for the solution set of linear interval systems. Linear Algebra Appl. *251*, 321–340 (1997).

[18] Lax, P. D.: Linear algebra. New York: Wiley 1997.

[19] Mayer, G.: A new way to describe the symmetric solution set S_{sym} of linear interval systems. In: Topics in numerical analysis with special emphasis on nonlinear problems (Alefeld, G., and Chen, X., eds.). Computing Suppl. *15*, 151–163 (2001).

[20] Neumaier, A.: Interval methods for systems of equations. Cambridge: Cambridge University Press 1990.

[21] Oettli, W., Prager, W.: Compatibility of approximate solution of linear equations with given error bounds for coefficients and right-hand sides. Numer. Math. *6*, 405–409 (1964).
[22] Rohn, J.: Interval linear systems. Freiburger Intervall-Berichte *84/7*, 33–58 (1984).
[23] Rohn, J.: Personal communication. Berlin 2001.
[24] Rump, S. M.: Verification methods for dense and sparse systems of equations. In: Topics in validated computations (Herzberger, J., ed.). Amsterdam: Elsevier (1994), pp. 63–135.
[25] Schrijver, A.: Theory of linear and integer programming. New York: Wiley 1986.
[26] Schäfer, U.: Two ways to extend the Cholesky decomposition to block matrices with interval entries. Reliable Computing *8*, 1–20 (2002).

Götz Alefeld
Institut für Angewandte Mathematik
Universität Karlsruhe
D-76128 Karlsruhe
Germany

Vladik Kreinovich
Department of Computer Science
University of Texas, El Paso
El Paso, TX 79968
USA

Günter Mayer
Fachbereich Mathematik
Universität Rostock
D-18051 Rostock
Germany

Computing [Suppl] 16, 23–35 (2002)

Computing

Methods for Computing All Roots of a Polynomial Simultaneously – Known Results and Open Problems*

L. Atanassova, N. Kyurkchiev and **T. Yamamoto**

Mathematics Subject Classification: 65H05, 30C15.

Keywords: Zeros of polynomials, simultaneous methods.

Let

$$f(x) = x^n + a_{n-1}x^{n-1} + \cdots + a_1 x + a_0,\ a_0, \ldots, a_{n-1} \in C,$$

be a polynomial with the simple real or complex zeros $x_i, i = 1, 2, \ldots, n$, and $x_i^0, i = 1, 2, \ldots, n$ be distinct reasonably close approximations of these zeros.

Then the iterates

$$x_i^{k+1} = x_i^k - \frac{f(x_i^k)}{\prod_{j \neq i}^{n}(x_i^k - x_j^k)},$$

$$i = 1, \ldots, n;\ k = 0, 1, 2, \ldots \tag{1}$$

are well-defined and x_i^k converges to $x_i, i = 1, \ldots, n$. This method for approximating the x_i was introduced by Weierstrass [34] in connection with a proof of the fundamental theorem of algebra. It should be remarked here that Dochev [3] first proved a local quadratic convergence of the method. Later, Kerner [11] proved that the method is Newton's method applied to the system of nonlinear equations

$$f_i(\xi_1, \ldots, \xi_n) = (-1)^i \phi_i(\xi_1, \ldots, \xi_n) - a_{n-i} = 0,$$

$$i = 1, 2, \ldots, n,$$

---

* Dedicated to Prof. Dr. J. Herzberger on the occasion of his 60th birthday.

where ϕ_i denotes the i-th elementary symmetric function

$$\phi_i(\xi_1, \ldots, \xi_n) = \sum_{1 \leq j_1 < \cdots < j_i \leq n} \xi_{j_1} \xi_{j_2} \ldots \xi_{j_i}.$$

For the first time, conditions for the choice of initial approximations that guarantee the convergence of methods of type (1) and other Newton type methods appeared in the papers of Iliev [8], and Iliev and Dochev [9].

Let P be a nonlinear mapping from an open subset D of the real or complex Banach space E to a space F of the same type. Kantorovich's classical work is ever an important source for proving the existence of solutions of the equation

$$P(X) = 0$$

as well as the convergence of Newton's method

$$X^{k+1} = X^k - P'(X^k)^{-1} P(X^k),$$

$k = 0, 1, 2, \ldots : X^k \in D$ and $P'(X^k)^{-1}$ exists.

In 1986, Smale first proposed to use the analytic property of the mapping P corresponding to the method of Weierstrass to replace the domain condition in Kantorovich's theorem method by thoroughly making use of the information about P at the initial point $X^0 \in D$.

The Weierstrass method inspired a number of other contributions. Some basic papers are dedicated to these important investigations: Durand [5], Werner [35], Alefeld and Herzberger [1], Petkovic [26], Sendov, Andreev and Kyurkchiev [28].

Many authors have observed that the Weierstrass method converges in practice with almost every starting point $(x_1^0, \ldots, x_n^0)^T \in C^n$.

Therefore one can expect that the assertion to be proved in the following problem is true:

Problem. (In: Problems and Solutions – SIAM Rev. 18, 501–502 (1976)), "Simultaneous iteration towards all roots of a complex polynomial", by M.W. Green, A.J. Korsak and M.C. Pease:

"Show that the Weierstrass procedure is globally convergent outside of a set of measure zero in the starting space and describe this set, for $n = 2$. Show the same result, if possible for arbitrary n."

For quadratic polynomials, the Weierstrass method is globally convergent outside of a set of measure zero in C^2. This was proved by R. Small (University of New Brunswick, New Brunswick, Canada, see [7]).

For the cubic polynomial $f(x) = x^3$, the set of initial vectors in C^3 for which the Weierstrass method does not converge has measure zero too. This was proved by Yamagishi [36], [37].

Pasquini and Trigiante [25] proposed a method with global convergence when it is applied to polynomials with real and simple roots. Their results follow from results by Barna [2] and Smale [29].

However a proof for the general case is missing. On the other hand no counter example is known of polynomials for which the method of Weierstrass is not convergent for a set of starting points with nonzero measure.

Related Problems

Problem K1. (Kyurkchiev [12])

(1) Find sufficient conditions for the divergence of the Weierstrass method and describe the set of starting approximations $x_i^0, i = 1, 2, \ldots, n$, satisfying these conditions.

(2) Are there other initial approximations, such that the sequence of approximations produced by the Weierstrass algorithm does not converge?

Let $a_{n-1} = 0$. The iteration (1) will fail at the 2 nd step if [12]:

$$\begin{aligned} &\sum_{l<s}^{n} x_l^0 x_s^0 + a_{n-2} = 0, \\ &2\sum_{l<s<t}^{n} x_l^0 x_s^0 x_t^0 - a_{n-3} = 0, \\ &\ldots \\ &(n-1)\prod_{j=1}^{n} x_j^0 + (-1)^n a_0 = 0. \end{aligned} \tag{2}$$

This implies that for any monic polynomial $f(x)$ of degree n, there exists a non-empty set $M_f \subset C^n$ such that the method (1) starting from $x^0 = x \in M_f$ does not converge to the zeros of $f(x)$. Let us note that the relations (2) can be regarded as a system of nonlinear equations of the form $H(x) = 0$. According to this it follows that M_f is the set of solutions of this system, that is, $M_f = \{x^0 | H(x^0) = 0\}$.

Numerical Example

The exact zeros of the polynomial

$$f(x) = x^4 - 5x^2 + 4$$

are $x_{1,2} = \pm 1, x_{3,4} = \pm 2$. The critical initial approximations $x_i^0, i = 1, 2, 3, 4$, obtained by Eq. (2), are given by

x_1^0	x_2^0	x_3^0	x_4^0
$-\alpha$	β	α	$-\beta$
β	$-\alpha$	α	$-\beta$
α	β	$-\alpha$	$-\beta$
β	α	$-\alpha$	$-\beta$
α	$-\beta$	β	$-\alpha$
$-\beta$	α	β	$-\alpha$
α	β	$-\beta$	$-\alpha$
β	α	$-\beta$	$-\alpha$
$-\alpha$	$-\beta$	β	α
$-\beta$	$-\alpha$	β	α
$-\alpha$	β	$-\beta$	α
β	$-\alpha$	$-\beta$	α
$-\alpha$	$-\beta$	α	β
$-\beta$	$-\alpha$	α	β
α	$-\beta$	$-\alpha$	β
$-\beta$	α	$-\alpha$	β,

where

$$\alpha = \sqrt{\sqrt{\frac{91}{12}} + \frac{5}{2}}i, \quad \beta = \sqrt{\sqrt{\frac{91}{12}} - \frac{5}{2}}.$$

We can not answer the second question.

Problem Y1. (Yamamoto [40]) In this problem it is not assumed that the zeros of f are simple.

Let $x'_1, \ldots, x'_m$ be the m distinct zeros of the polynomial f with multiplicities $s_1, \ldots, s_m$ such that $s_1 + \cdots + s_m = n$. Let

$$I_h = \{i \in \{1, 2, \ldots, n\} | x_i = x'_h\}, \quad h = 1, 2, \ldots, m,$$

and i_h be the smallest index of I_h. For simplicity, assume that $s_i > 1$ for $i \le p$ and $s_i = 1$ for $i > p$. Furthermore, assume that for each $h < p$, s_h's x_i^k with $i \in I_h$ are (approximately) evenly spaced on some circle

$$C_h : |x - x'_h| = r_h, \quad h = 1, 2, \ldots, p$$

and other x_i^k have converged to simple zeros $x'_{p+1}, \ldots, x'_m$.

If r_h is sufficiently small and x'_h is well separated from other zeros, then it is shown [38], [40] that

$$x_i^{k+1} - x_h' \approx \left(1 - \frac{1}{s_h}\right)(x_i^k - x_h'), \quad i \in I_h \tag{3}$$

under the above assumptions which are empirically observed.

Other results have been found by Fraigniaud [6] and Miyakoda [22], [23], [24].

Remark. Dochev and Burnev [4], make use of the following recurrent formula relating three systems of consecutive approximations $x_i^{k+1}, x_i^k, x_i^{k-1}, i = 1, 2, \ldots, n$, obtained from (1)

$$x_i^{k+1} = x_i^k - D_i^{k-1} \prod_{j \neq i}^{n} \frac{x_i^k - x_j^{k-1}}{x_i^k - x_j^k},$$

$$i = 1, 2, \ldots, n; \;\; k = 1, 2, \ldots,$$

where

$$D_i^{k-1} = x_i^k - x_i^{k-1} + \sum_{j \neq i}^{n} \frac{x_i^{k-1} - x_j^{k-1}}{x_i^k - x_j^{k-1}} (x_j^k - x_j^{k-1}).$$

Maybe, these results can be applied also to prove the property (3).

There is an SOR – like modification of the Weierstrass method [10]:

$$x_i^{k+1} = x_i^k - \omega . \frac{f(x_i^k)}{\prod_{j<i}^{n} (x_i^k - x_j^{k+1}) \prod_{j>i}^{n} (x_i^k - x_j^k)},$$

$$i = 1, \ldots, n; \;\; k = 0, 1, 2, \ldots, \tag{4}$$

where ω is an acceleration parameter. Kanno, Kyurkchiev and Yamamoto [10] showed that if $|\omega - 1| < 1$ and the x_i are simple, then the method (4) locally converges to the zeros. In particular, if ω is restricted to a real parameter, then (4) converges for $0 < \omega < 2$.

Furthermore, it is shown by Yamamoto [39] that (4) diverges if $|\omega - 1| > 1$. More general and detailed discussions on the convergence of some modifications on the Weierstrass methods including

$$x_i^{k+1} = x_i^k - \omega . \frac{f(x_i^k)}{\prod_{j<i}^{n} (x_i^k - x_j^{k+1,W}) \prod_{j>i}^{n} (x_i^k - x_j^k)},$$

are given there.

Problem KKY1. (Kanno, Kyurkchiev and Yamamoto [10]).

(1) Find other sufficient conditions for the divergence of the method (4).

(2) What is an optimal value of ω, which gives fastest convergence ?

The iteration (4) can not be defined at the second step, if the sequence of initial approximations $x_i^0, i = 1, 2, \dots, n$ satisfies [10]:

$$\sum_{i=1}^{n} x_i^0 \left(\frac{1}{\omega} \prod_{j}^{i-1} \frac{x_i^0 - x_j^{1,W}}{x_i^0 - x_j^0} - 1 \right) - a_{n-1} = 0,$$

$$\sum_{i=1}^{n} x_i^0 \left(\frac{1}{\omega} \prod_{j}^{i-1} \frac{x_i^0 - x_j^{1,W}}{x_i^0 - x_j^0} - 1 \right) \sum_{j \neq i}^{n} x_j^0 + \sum_{l<s}^{n} x_l^0 x_s^0 + a_{n-2} = 0, \tag{5}$$

$$\dots$$

$$\sum_{i=1}^{n} x_i^0 \left(\frac{1}{\omega} \prod_{j}^{i-1} \frac{x_i^0 - x_j^{1,W}}{x_i^0 - x_j^0} - 1 \right) \prod_{j \neq i}^{n} x_j^0 + (n-1) \prod_{j=1}^{n} x_j^0 + (-1)^n a_0 = 0.$$

The optimal value of ω is not known. However, it is known Yamamoto [39, p. 88 and the proof of Theorem 3.1] that $\omega = 1$ is locally best.

For a numerical solution of the equation $f(x) = 0$, we consider the following method

$$x_i^{k+1} = x_i^k - h_k . \frac{f(x_i^k)}{\prod_{j \neq i}^{n} (x_i^k - x_j^k)},$$

$$i = 1, \dots, n; \quad k = 0, 1, 2, \dots, \tag{6}$$

where $h_k \in (0, 1]$ is an acceleration parameter. This method is an obvious generalization of Weierstrass' method which is obtained from (1) for $h_k = 1$.

Problem KP1. (Kyurkchiev and Petkovic [19]).

What is an optimal value of h_k which gives guaranteed convergence ?

Kyurkchiev and Petkovic [19] showed that the SOR method (6) will fail if the approximations $x_i^0, i = 1, 2, \dots, n$ satisfy the system

$$\left(\frac{1}{h_0} - 1 \right) \sum_{i=1}^{n} x_i^0 - a_{n-1} = 0,$$

$$\left(\frac{2}{h_0} - 1 \right) \sum_{l<s}^{n} x_l^0 x_s^0 + a_{n-2} = 0, \tag{7}$$

$$\dots$$

$$\left(\frac{n}{h_0} - 1 \right) \prod_{j=1}^{n} x_j^0 + (-1)^n a_0 = 0.$$

Investigations of divergent starting points for the method (6) show that for any monic polynomial $f(x)$ of degree n there exists a nonempty set $G_f \subset C^n$ such that the SOR method, starting from $x^0 = x \in G_g$ does not converge to the roots of f. This set yielding divergent starting points and obtained as the set of solutions of a nonlinear systems of n equations will be called NS-divergent set. The study of the distribution and measure of NS-divergent sets is very complicated and is not carried out in literature. It is only certain that for the considered methods the NS-divergent sets consist of an infinite number of divergent starting vectors but have the measure zero in C^n. Moreover, in general, divergent sets of NS' type are not the only ones. For instance, solving polynomial equations with real coefficients having conjugate complex roots, any method will not converge if all starting approximations are real numbers. However, this problem restricted to NS-divergent sets is easy to solve for the SOR method (6) taking $0 < h_k < 1$, thus, excluding Weierstrass' method (1) ($h_k = 1$).

Hence we have [19]:

Let f be a monic polynomial of degree n with simple roots. There is only one NS-divergent starting vector $G_f = x = (x_1, \dots, x_n) \in C^n$ for the method (6) of Weierstrass' type, not counting the permutations of the components of x, and it is given by the set of roots of the algebraic polynomial

$$F(x) = x^n - \frac{h_0 a_{n-1}}{1-h_0} x^{n-1} - \frac{h_0 a_{n-2}}{2-h_0} x^{n-2} - \cdots - \frac{h_0 a_k}{n-k-h_0} x^k - \cdots - \frac{h_0 a_0}{n-h_0}.$$

We have the following

Theorem. *Let*

$$R = \left(\max_{0 \le k \le n-1} \frac{|a_k|}{n-k-h_0} \right)^{\frac{1}{n}}.$$

All the critical points $z_i^0, i = 1, \dots, n$ lie in the ring-shaped region

$$\frac{h_0 |a_0| (T-1)}{(n-h_0)(T^{n-1}(T^2 - 1 + h_0 R^n T + h_0 R^n) - 2h_0 R^n)} \le |z| \le 1 + h_0 R^n = T.$$

Proof: Let z be a zero of $F(z)$ such that $|z| > 1$. We have

$$|z|^n \le h_0 \left(\frac{|a_{n-1}|}{1-h_0} |z|^{n-1} + \cdots + \frac{|a_1|}{n-1-h_0} |z| + \frac{|a_0|}{n-h_0} \right) \le h_0 R^n \frac{|z|^n - 1}{|z| - 1},$$

i.e.

$$\frac{h_0 R^n}{|z|-1} \ge \frac{|z|^n}{|z|^n - 1} > 1$$

and

$$|z| \leq 1 + h_0 R^n = T.$$

Next, we prove that $F(z)$ has no zero in

$$|z| < \frac{h_0|a_0|(T-1)}{(n-h_0)(T^{n-1}(T^2-1+h_0R^nT+h_0R^n)-2h_0R^n)}.$$

Let $r(z) = (1-z)F(z)$. Then

$$\begin{aligned} r(z) &= -\frac{h_0 a_0}{n-h_0} - h_0 \sum_{k=1}^{n-1}\left(\frac{a_k}{n-k-h_0} - \frac{a_{k-1}}{n-k+1-h_0}\right) z^k + z^n + \frac{h_0}{1-h_0} a_{n-1} z^n - z^{n+1} \\ &= -\frac{h_0 a_0}{n-h_0} + \mu(z), \end{aligned}$$

$$\begin{aligned} \max_{|z|=T}|\mu(z)| &\leq T^{n+1} + T^n + h_0R^nT^n + h_0 2R^nT(1+T+T^2+\cdots+T^{n-2}) \\ &= T^{n+1} + T^n + h_0R^nT^n + h_0 2R^nT\frac{T^{n-1}-1}{T-1} \\ &= \frac{T}{T-1}(T^n(T-1) + T^{n-1}(T-1) + h_0R^nT^{n-1}(T-1) + h_0 2R^n(T^{n-1}-1)) \\ &= \frac{T}{T-1}(T^{n-1}(T^2-1+h_0R^nT+h_0R^n) - h_0 2R^n). \end{aligned}$$

Hence on $|z| \leq T$

$$|r(z)| \geq \frac{h_0}{n-h_0}|a_0| - |\mu(z)| \geq \frac{h_0}{n-h_0}|a_0| - \frac{|z|}{T}\max_{|z|=T}|\mu(z)|$$

by Schwarz's lemma,

$$|r(z)| \geq \frac{h_0}{n-h_0}|a_0| - \frac{|z|}{T-1}(T^{n-1}(T^2-1+h_0R^nT+h_0R^n) - 2h_0R^n),$$

and $|r(z)| > 0$ if

$$|z| < \frac{h_0|a_0|(T-1)}{(n-h_0)(T^{n-1}(T^2-1+h_0R^nT+h_0R^n)-2h_0R^n)}.$$

This completes the proof of the theorem. □

Euler – Chebyshev's well-known method for finding individually roots of non linear equations motives the following method for computing all zeros of f simultaneously:

$$x_i^{k+1} = x_i^k - \frac{f(x_i^k)}{\prod_{j\neq i}^n (x_i^k - x_j^k)} \left(1 + \sum_{j\neq i}^n \frac{f(x_j^k)}{x_j^k - x_i^k} \prod_{s\neq j}^n \frac{1}{x_j^k - x_s^k}\right), \tag{8}$$

$$i = 1, 2, \ldots, n; \quad k = 0, 1, 2, \ldots.$$

The convergence order of this method is three.

Problem K2. (Kyurkchiev [14]) Show that the Euler – Chebyshev method is globally convergent outside of a set of measure zero in the starting space and describe this set, for arbitrary n.

The following result is more readily applied (see Tanabe [30], [31]):

If $x_i^0, i = 1, 2, \ldots, n$ are evenly spaced on the circle

$$\left|x + \frac{a_{n-1}}{n}\right| = \rho,$$

for sufficiently large ρ, then

$$x_i^1 + \frac{a_{n-1}}{n} = (1 - 3/(2n) + 1/(2n^2))(x_i^0 + \frac{a_{n-1}}{n}).$$

Introducing the abbreviation $x_i^{k+1,W} = x_i^k - f(x_i^k)/\ \prod_{j\neq i}^n (x_i^k - x_j^k), i = 1, \ldots, n;$ $k = 0, 1, \ldots$ corresponding to the Euler – Chebyshev iteration (8), we have [14]:

Let $a_{n-1} = 0$. The iteration (8) is not defined, if

$$\begin{aligned}
2\sum_{l<s}^n x_l^0 x_s^0 - \sum_{l<s}^n (x_l^{1,W} x_s^{1,W} + x_l^0 x_s^0) + 2a_{n-2} = 0, \\
6\sum_{l<s<t}^n x_l^0 x_s^0 x_t^0 - \sum_{l<s}^n (x_l^{1,W} x_s^{1,W} + x_l^0 x_s^0) \sum_{t\neq l,s}^n x_t^0 - 3a_{n-3} = 0, \\
\ldots \\
n(n-1)\prod_{j=1}^n x_j^0 - \sum_{l<s}^n (x_l^{1,W} x_s^{1,W} + x_l^0 x_s^0) \prod_{t\neq l,s}^n x_t^0 + (-1)^n n a_0 = 0.
\end{aligned} \tag{9}$$

Numerical Example

The exact zeros of the polynomial

$$f(x) = x^2 + 1 = 0$$

are $x_{1,2} = \pm i$. The critical initial approximations $x_i^0, i = 1, 2$, obtained by (9), for which the method (8) fails are given in the next table

x_1^0	x_2^0
$\sqrt{1+2/3\sqrt{3}}$	$-\sqrt{1+2/3\sqrt{3}}$
$-\sqrt{1+2/3\sqrt{3}}$	$\sqrt{1+2/3\sqrt{3}}$
$i\sqrt{2/3\sqrt{3}-1}$	$-i\sqrt{2/3\sqrt{3}-1}$
$-i\sqrt{2/3\sqrt{3}-1}$	$i\sqrt{2/3\sqrt{3}-1}$.

The conjecture for global convergence of the method is still unsolved.

The following method with rate of convergence $R+2$ is known (see [28] for the related literature):

$$x_i^{k+1} = x_i^k + \Delta_i^{R+1,k},$$
$$i = 1, 2, \ldots, n; \quad k = 0, 1, \ldots, \tag{10}$$

where

$$\Delta_j^{p,k} = -\frac{f(x_j^k)}{\prod_{s\neq j}^{n}(x_j^k - x_s^k - \Delta_s^{p-1,k})},$$

$$\Delta_j^{0,k} = 0,$$

$$0 < p \leq R+1$$

and R is fixed integer.

Problem KA1. (Kyurkchiev and Atanassova [18]) We conjecture that the Weierstrass type method (10) is globally convergent outside of a set of measure zero in C^n

However, we are not able to prove this rigorously.

The iteration (10) is not defined, if the sequence of approximations $x_i^0, i = 1, 2, \ldots, n$ satisfies [18]:

$$\sum_{i=1}^{n} x_i^0 \left(\prod_{j\neq i}^{n} \frac{x_i^0 - x_j^0 - \Delta_j^{R,0}}{x_i^0 - x_j^0} - 1 \right) - a_{n-1} = 0,$$

$$\sum_{i=1}^{n} x_i^0 \sum_{j\neq i}^{n} x_j^0 \left(\prod_{j\neq i}^{n} \frac{x_i^0 - x_j^0 - \Delta_j^{R,0}}{x_i^0 - x_j^0} - 1 \right) + \sum_{l<s}^{n} x_l^0 x_s^0 + a_{n-2} = 0, \tag{11}$$

$$\ldots$$

$$\sum_{i=1}^{n} x_i^0 \prod_{j\neq i}^{n} x_j^0 \left(\prod_{j\neq i}^{n} \frac{x_i^0 - x_j^0 - \Delta_j^{R,0}}{x_i^0 - x_j^0} - 1 \right) + (n-1) \prod_{j=1}^{n} x_j^0 + (-1)^n a_0 = 0.$$

Remark. The conditions for the choice of the not atractive (divergent) initial approximations for some iterative methods can be found in Yamamoto, Kanno and Atanassova [38], Kanno, Kyurkchiev and Yamamoto [10], Kyurkchiev [13], [14], [15], Kyurkchiev and Mahdi [16], [17], Kyurkchiev and Atanassova [18], Zheng and Kyurkchiev [32], Wang and Zhao [33], Petkovic and Kyurkchiev [27], Kyurkchiev and Moskona [20]. Different aspects of this field can be found in [21]. The distribution of the critical initial approximations of Weierstrass' method is considered in [42].

Note Added in Proof. After writing this paper, Hristov and Kyurkchiev [43] showed that if $1 \leq i < j \leq n$ and the sequence of approximations $x_i^k, i = 1, \ldots n$ satisfies

$$\omega_{ij}(x_j^k)(F_n(x_i^k) + g_n(x_i^k)) + \omega_{ij}(x_i^k)\Big(F_n(x_j^k) + g_n(x_j^k)\Big) = 0, \tag{12}$$

where

$$\begin{aligned} g_n(x) &= a_{n-1}x^{n-1} + \cdots + a_0, \\ F_n(x) &= \sum_{m=1}^{n}(-1)^m(m-1)S_m x^{n-m}, \\ S_m &= \sum_{1\leq i_1<\cdots<i_m\leq n} x_{i_1}^k x_{i_2}^k \ldots x_{i_m}^k, \quad m = 1, \ldots, n, \\ \omega_{ij}(x) &= (x - x_1^k)\ldots(x - x_{i-1}^k)(x - x_{i+1}^k)\ldots(x - x_{j-1}^k)(x - x_{j+1}^k)\ldots(x - x_n^k) \end{aligned}$$

then $x_i^{k+1} = x_j^{k+1}$, and the Weierstrass method (1) will fail at the $(k+2)$-th step.

References

[1] Alefeld, G., Herzberger, J.: Introduction to interval computations. New York: Academic Press 1983.

[2] Barna, B.: Über die Divergenzpunkte des Newton'schen Verfahrens zur Bestimmung von Wurzeln algebraischer Gleichungen II. Publ. Math. Debrecen *4*, 384–397 (1956).

[3] Dochev, K.: A modified Newton method for simultaneous approximate solution of algebraic equations. Phys. Math. J. Bulg. Acad. Sci. *5*, 136–139 (1962).

[4] Dochev, K., Burnev, P.: Certain modifications of Newtons method for the approximate solution of algebraic equations. Z. Vycisl. Mat. i Fiz. *4*, 915–920 (1964).

[5] Durand, E.: Solution numerique des equations algebraiques, Tome I. Paris: Masson 1960.

[6] Fraigniaud, P.: The Durand – Kerner polynomial roots finding method in case of multiple roots. BIT *31*, 112–123 (1991).

[7] Green, M., Korsak, A., Pease, M.: Simultaneous iteration towards all roots of a complex polynomial. SIAM Rev. *18*, 501–502 (1976).

[8] Iliev, L.: Über Newtonsche Iterationen. Jb. Univ. Sofia *1*, 167–171 (1949).

[9] Iliev, L., Dochev, K.: Über Newtonsche Iterationen. Wiss. Z. Tech. Univ. Dresden *12*, 117–118 (1963).

[10] Kanno, S., Kyurkchiev, N., Yamamoto, T.: On some methods for the simultaneous determination of polynomial zeros. Japan J. Industrial Appl. Math. *13(2)*, 267–288 (1996).

[11] Kerner, I.: Ein Gesamtschrittverfahren zur Berechnung der Nullstellen von Polynomen. Numer. Math. *8*, 290–294 (1966).

[12] Kyurkchiev, N.: Some remarks on Weierstrass root-finding method. C. R. Acad. Bulg. Sci. *46*, 17–20 (1993).
[13] Kyurkchiev, N.: Some remarks on the parallel Alefeld-Herzberger iteration method. In: Math. and Educ. in Math., Proc 23th Conf. of the Union of Bulgarian Mathematicians, pp. 284–292 (1994).
[14] Kyurkchiev, N.: Initial approximations in Euler-Chebyshev's method. J. Comput. Appl. Math. *58*, 233–236 (1995).
[15] Kyurkchiev, N.: On the methods for finding roots of algebraic equations with Weierstrass' corrections. Num. Math, A J. of Chinese Univ. (English Series) *4*, 49–53 (1995).
[16] Kyurkchiev, N., Mahdi, K.: Some remarks on Dvorcuk root-finding method. BIT *34*, 318–322 (1994).
[17] Kyurkchiev, N., Mahdi, K.: A note on the divergent starting points for Euler – Chebyshev's type methods. Facta Universitatis (NIS), Ser. Math. Inform. *9*, 95–98 (1994).
[18] Kyurkchiev, N., Atanassova, L.: A note on the divergent starting points for Weierstrass type algorithms. ZAMM *76(5)*, 301–302 (1995).
[19] Kyurkchiev, N., Petkovic, M.: On the behavior of approximations of the SOR Weierstrass method. Comput. Math. Appl. *32*, 117–121 (1996).
[20] Kyurkchiev, N., Moskona, E.: On the critical points of Aberth's method. In: Math. and Educ. in Math., Proc 26th Conf. of the Union of Bulgarian Mathematicians, pp. 131–138 (1997).
[21] Kyurkchiev, N.: Initial approximations and root finding methods Berlin: Wiley 1998.
[22] Miyakoda, T.: Iterative methods for multiple zeros of a polynomial by clustering. Comp. Appl. Math. *28*, 315–326 (1989).
[23] Miyakoda, T.: Balanced convergence of iterative methods to a multiple zero of a complex polynomial. J. Comp. Appl. Math. *39*, 201–212 (1992).
[24] Miyakoda, T.: Cluster analysis for the approximate solution of a complex polynomial and its application. Ph.D. Thesis, Nagoya Univ. 1992.
[25] Pasquini, L., Trigiante, D.: A globally convergent method for simultaneously finding polynomial roots. Math. Comp. *44*, 135–149 (1985).
[26] Petkovic, M.: Iterative methods for simultaneous inclusion of polynomial zeros. Berlin: Springer 1989.
[27] Petkovic, M., Kyurkchiev, N.: A note on the convergence of the Weierstrass SOR method for polynomial roots. JCAM *80*: 163–168 (1997).
[28] Sendov, Bl., Andreev, A., Kyurkchiev, N.: Numerical solution of polynomial equations. Handbook of numerical analysis (Ciarlet, P., Lions, J., eds.), Vol. 3. Amsterdam: Elsevier 1993.
[29] Smale, S.: The fundamental theorem of algebra and complexity theory. Bull. Amer. Math. Soc. (n.S.) *4*, 1–36 (1981).
[30] Tanabe, K.: A new algorithm of simultaneous iteration towards all roots of a complex polynomial. Symp. on Numerical Algebra Computation and Analysis, Japan, Kyoto (1982).
[31] Tanabe, K.: Technical report on information processing. Num. Anal. *4(2)*, 1–6 (1983).
[32] Zheng, S., Kyurkchiev, N.: Initial approximations in Wang-Zheng's root finding method. C. R. Acad. Bulg. Sci. *49(3)*, 17–20 (1996).
[33] Wang, D., Zhao, F.: On the determination of the safe initial approximation for the Durand-Kerner algorithm. J. Comput. Appl. Math. *38*, 447–456 (1991).
[34] Weierstrass, K.: Neuer Beweis des Satzes, dass jede ganze rationale Funktion einer Veränderlichen dargestellt werden kann als ein Product aus linearen Funktionen derselben Veränderlichen. Ges. Werke *3*, 251–269 (1903).
[35] Werner, W.: On the simultaneous determination of polynomial roots. Lecture Notes in Mathematics 953. Berlin: Springer 1982.
[36] Yamagishi, Y.: Sc. Disertation, Kyoto Univ. (1991).
[37] Yamagishi, Y.: Global convergence of the Durand-Kerner method applied to the equation $z^3 = 0$. J. Comput. Appl. Math. *70*, 67–73 (1996).
[38] Yamamoto, T.: A method for solving algebraic equations and a posteriori error estimates Japanese). Suuri Kagaku (Math. Sci.) *157*, 52–57 (1976).
[39] Yamamoto, T.: On nonlinear SOR – like methods, I – Applications to simultaneous methods for polynomial zeros. Japan J. Indust. Appl. Math. *14*, 87–97 (1977).
[40] Yamamoto, T., Furukante, V., Nogura, K.: The Durand-Kerner and the Aberth methods for solving algebraic equations (Japanese). Information Processing *18*, 566–571 (1977).
[41] Yamamoto, T., Kanno, S., Atanassova, L.: Validated computation of polynomial zeros by the Durand-Kerner method. In: Topics in validated computations. Amsterdam: North-Holland, 1994.

[42] Kyurkchiev, N., Hristov, V.: Distribution of the critical points of Weierstrass procedure. Mathematics and education in mathematics. Proc. of 29th Spring Conf. of the Union of Bulgarian Mathematicians, Lovetsch, April 3–6, 2000, pp. 192–197.

[43] Hristov, V., Kyurkchiev, N.: A note on the globally convergent properties of the Weierstrass – Dochev method. In: Approximation theory: A volume dedicated to Blagovest Sendov (B. Bojanov, ed.), DARBA, Sofia, 2002, pp. 231–240.

Dr. Lydia Atanassova
T-Systems International GmbH
Hansastr. 24
D-80674 München
Germany

Prof. Dr. N. Kyurkchiev
Institute of Mathematics
Bulgarian Academy of Sciences
"Acad. G. Bonchev Str.", bl. 8
BG - 1113 Sofia
Bulgaria

Prof. Tetsuro Yamamoto
Department of Information and Computer Science
School of Science and Engineering
Waseda University
Tokyo 169 - 8555
Japan

Computing [Suppl] 16, 37–43 (2002)

Computing

Narrow Bounds for the Effective Rate of Return Concerning a Special Problem for Annuities

F. Detmers and **J. Herzberger**

Abstract

We consider the calculation of the effective rate of an annuity which was derived by changing the conditions of a given annuity in a certain manner. In this note we will derive narrow bounds for the effective rate of the changed annuity in terms of the (given) rate of the original annuity and the coefficients of the polynomial involved. It is demonstrated by some numerical examples that the bounds achieved for this purpose are of sufficient accuracy for practical applications and thus an application of an iterative method for solving the equation can be avoided.

Mathematics Subject Classification: 26C05, 65H05, 90A09.

Keywords: Mathematics in finance, polynomial roots.

1. Introduction

We consider a simple annuity with annual payments and a fixed term (see for example [10]). In this case one pays back a debt of an amount of S during the term of say n years by regular annual payments of equal amount A including the interest due. At the end of the term the whole debt will be zero and is payed back. Let the contracted interest rate be i (as decimal fraction), then we denote by $q = 1 + i$ the accumulation factor. Using the well-known principle "*The sum of the compounded payments at the end of the term has to be equal to the compounded debt (applying in both cases the same interest rate i)*" we get immediately the simple polynomial equation

$$Sq^n - A\sum_{\nu=0}^{n-1} q^{n-1-\nu} = 0. \tag{1}$$

Given the amount of debt S, the amount of annual payments A and the number of payments n (equal to the term), then, according to Descartes' rule of sign, the existing unique positive root q^* of Eq. (1) determines the effective rate of return of the annuity i^* by $i^* = q^* - 1$. Assuming annual payments like here, the calculations according to the US-rule and the German rule (Preisangabeverordnung) are the same.

Let us now formulate Eq. (1) in a different form as

$$q^n - a\sum_{\nu=0}^{n-1} q^{n-1-\nu} = 0, \qquad a = A/S. \tag{1a}$$

Such a class of polynomial equations is looking quite similar to some in numerical analysis which were already investigated in connection with the determination of the order of convergence of iterative processes. In particular, for so-called interpolation iteration methods for the calculation a zero of a nonlinear equation (for example the secant method or Newton method) one can find results in [11]. These classes were later generalized (see [3] or [6]) and we want to refer to the paper of M. Petković and Lj. Petković [9] which gives useful results even for our applications. The mentioned generalizations are given in such manner that the treatment of the polynomial equation for the effective rate of annuities with geometrically growing payments is also possible. In this paper we are not dealing with these cases.

In the above quoted paper of M. Petković and Lj. Petković [9] we can find a lower bound for the positive root q^* of Eq. (1) reducing in our more simple case to the formula

$$q^* > \frac{2a + a(n-1)(n+2)}{2 + a(n-1)n}. \tag{2}$$

2. Problem

The mathematical problem treated in this paper should be formulated now. First, let us explain the problem in the setting of finance. Considering simple annual annuities like in Sect. 1, we want to change the conditions for the annuity in the following way:

Keeping the term of the original annuity – here n years – the amount of regular payments should be lowered to A/k ($k > 1$) with a simultaneous increase of the number of payments to the k-fold number, which means a total number of $k \cdot n$ payments during the same term. How about the effective rate of return of this new and changed annuity?

For $k = 2$ this problem was communicated by H. Lichtenberg of the Federal Ministry of Finance in Bonn/Germany [7].

Since we take the German rule (Preisangabeverordnung) as basis for solving this problem, the resulting polynomial equation for calculating the effective rate of interest cannot immediately be written down. As the changed annuity has several payments each year and as the compounding intervals according to the German rule are whole years, no matter how many payments are made during the year, the defining equation for the effective rate of return has to be derived by some calculations in contrast to the US-rule, where the compounding intervals are determined by the payments made. In both cases the interest rate applied to subannual intervals is the linear proportional one. In case of the German rule all payments during a year are used to lower the debt and their interest is calculated with this linear proportional rate. At the end of each year these payments including the interest due are balanced. This means that we can first calculate an annual modified payment $\bar{A}$ as the sum of all payments plus their interest due in order to apply our old formulas. Thus we get

$$\bar{A} = \sum_{\nu=0}^{k-1} \left(\frac{A}{k}\right)(1 + i_k \nu) = \left(\frac{A}{2k}\right)\left((k+1) + (k-1) \cdot q\right),$$

where i_k is the linear proportional interest rate $i_k = i/k$. If we plug the formula for $\bar{A}$ into the polynomial Eq. (1), then we get after some elementary rearrangements the new polynomial equation in normal form

$$q^n - \frac{A}{\left(S - \left(\frac{A}{2k}\right)(k-1)\right)} \sum_{\nu=1}^{n-1} q^{n-\nu} - \frac{\left(\frac{A}{2k}\right)(k+1)}{\left(S - \left(\frac{A}{2k}\right)(k-1)\right)} = 0. \tag{3}$$

This equation can be simplified by using some abbreviations to

$$q^n - \bar{a} \sum_{\nu=1}^{n-1} q^{n-\nu} - \bar{b} = 0, \qquad \bar{a} = \frac{2Ak}{2kS - A(k-1)}$$

$$\text{and} \quad \bar{b} = \frac{A(k+1)}{2kS - A(k-1)}. \tag{3a}$$

Obviously, for an annuity we always have the inequalities $\bar{a}, \bar{b} > 0$ and $\bar{a} > \bar{b}$.

Equation (3) has the same degree like Eq. (1) but the representation of its coefficients are different from those of Eq. (1). By the way, the application of the US-rule to this problem would result in a polynomial equation of much more higher degree compared with Eq. (1).

After these preparations we are now able to formulate exactly the mathematical task raised by this problem in finance:

For $k > 1$ we want to establish a formula linking the unique positive root of both polynomial equations

$$q^n - a \sum_{\nu=0}^{n-1} q^{n-1-\nu} = 0 \quad \text{and} \quad q^n - \bar{a} \sum_{\nu=1}^{n-1} q^{n-\nu} - \bar{b} = 0,$$

with $a, \bar{a}$ and $\bar{b}$ like in (1a) *and* (3a), *respectively.*

As solution it would be sufficient to derive explicitly narrow lower and upper bounds for the positive root q^* of Eq. (3) in terms of the known root $i + 1$ of Eq. (1), the coefficients of the polynomial and the number k.

3. Bounds for the Effective Rate of Interest

Let us first denote the interest rate of the original annuity by i^* and the interest rate of the changed and new annuity, defined in Eq. (3), by j^*.

As a first step we can derive rather crude bounds for j^* motivated by the following more heuristic considerations.

First, we rewrite Eq. (3a) in the form of

$$q^n - \bar{a} \cdot \frac{q^n - q}{q - 1} - \bar{b} = 0$$

and after doing this we divide both sides of this equation by $q^n > 0$ which gives the equation

$$1 - \frac{\bar{a}\left(1 - \frac{1}{q^{n-1}}\right)}{(q-1)} - \frac{\bar{b}}{q^n} = 0.$$

Since for annuities we can always assume $j^* > 0$ and thus for $n \to \infty$ follows $q^n \to \infty$ and the last equation will be transformed into

$$1 - \frac{\bar{a}}{(q-1)} = 0.$$

But this simple equation has the unique solution $q^* = \bar{a} + 1$. Thus we get $\lim_{n\to\infty} j^* = \lim_{n\to\infty} \bar{a}$. Remember that the quantity A in the expression for $\bar{a}$ itself depends on n and Eq. (1) shows, that

$$\lim_{n\to\infty} A = \frac{Sq^n(q-1)}{q^n - 1} = Si^*,$$

provided $i^* > 0$. Finally, we derive by this considerations the limit

$$\lim_{n\to\infty} j^* = \frac{2ki^*}{2k - i^*(k-1)}.$$

Numerical examples show that in all these cases

$$j^* > \frac{2ki^*}{2k - i^*(k-1)}$$

was valid. This conjecture will rigorously be proved in the sequel.

The necessary proof can be done by verifying that

$$\left(q^n - \bar{a}\sum_{\nu=1}^{n-1} q^{n-\nu} - \bar{b}\right)\Bigg|_{q=\frac{2ki^*}{2k-i^*(k-1)}+1} < 0$$

holds, ensuring that the lower bound in question is really located on the left-hand side of q^*. Substituting the argument

$$q = 1 + \frac{2ki^*}{2k - i^*(k-1)}$$

into the polynomial of Eq. (3a) and using the definition of

$$A = S \cdot \frac{(1+i^*)^n i^*}{(1+i^*)^n - 1}$$

leads after simplification of some lengthy expressions to

$$\left(q^n - \bar{a}\sum_{\nu=1}^{n-1} q^{n-\nu} - \bar{b}\right)\Bigg|_{q=\frac{2ki^*}{2k-i^*(k-1)}+1} = \frac{2i^*kS(1+i^*)^n - 2i^*kS\left(1+\frac{2i^*k}{2k-i^*(k-1)}\right)^n}{(1+i^*)^n - 1}.$$

Since $k > 1$ and $0 < i^* \leq 1$ holds true, the right-hand side is surely proved to be less than 0. Therefore the argument is located on the left-hand side of the root q^*. Hence we have verified the (crude) initial bounds

$$\frac{2ki^*}{2k - i^*(k-1)} < j^* < \frac{2Ak}{2kS - A(k-1)}. \tag{4}$$

Numerical examples are showing that these first bounds in (4) are not narrow enough for practical applications. This makes it necessary to improve these bounds once more.

Observing, that the polynomial of Eq. (3) is convex in the interval under consideration, we apply one explicit step of the secant method using the bounds given in (4) as initial approximations and get an improved lower bound for q^*. A lengthy but elementary calculation results in the following formula

$$j^* > \frac{2ki^*}{2k - i^*(k-1)} - \frac{\frac{2ki^*}{2k-i^*(k-1)} - \frac{2Ak}{2kS-A(k-1)}}{1 - \frac{i^*S}{\left[\left(1+\frac{2ki^*}{2k-i^*(k-1)}\right)^n (i^*S-A)\right]+A}}. \tag{5}$$

But if we do one explicit step of Newton's method with the upper bound in (4) as initial approximation, then we get – again due to the convexity property of the polynomial – a new and improved upper bound for q^*. Again after some lengthy but elementary calculations we get the result

$$j^* < t - \frac{\frac{(1+t)^n 2(i^* kS - kA) + 2Ak}{i^*(2kS - A(k-1))}}{(1+t)^{n-1}\left[n - \frac{2Ak}{2kS - A(k-1)} \cdot \frac{(n-1)(1+t) - n}{t^2}\right] - \frac{\frac{2Ak}{2kS - A(k-1)}}{t^2}}, \qquad (6)$$

where

$$t = \frac{2ki^*}{2k - i^*(k-1)}.$$

4. Examples and Conclusions

In order to demonstrate the quality of bounds derived here for j^* we want to present now two numerical examples.

(i) $n = 12$, $k = 2$, $i^* = 0.005$. For these datas we get the interval

$$0.0051128166954 < j^* < 0.0052031802759.$$

The exact value of j^* calculated by Newton's method in rounded arithmetic was $j^* = 0.0052028941$ showing that in this example the upper bound is nearly equal to the properly rounded value of j^*.

(ii) $n = 5$, $k = 2$, $i^* = 0.0375$. Here we obtain the interval

$$0.039853937512 < j^* < 0.041167113248.$$

The exact value of j^* again by Newton's method in rounded arithmetic was $j^* = 0.0411382538$.

For polynomials like in Eq. (3) it is also possible to derive an approximation in a manner similar to the lower bound given by M. Petković and Lj. Petković in [9]. This is here of the form

$$q^* > \frac{\frac{(n-1)n}{2}\bar{a} + n\bar{b}}{1 + \frac{(n-1)(n-2)}{2}\bar{a} + (n-1)\bar{b}}, \qquad \text{if } (n-1)\bar{a} + \bar{b} \neq 1.$$

A numerical comparison in [1] showed that the lower bound modified in this way and the corresponding bound given in (5) give for some concrete datas approximately similar values.

The application of the US-rule to our financial problem in Sect. 2 was already treated in [4] and had to deal with polynomials of a more simpler form.

References

[1] Detmers, F.: Schranken für den Effektivzinssatz nach der PangV bei einem speziellen Annuitätenproblem. Diplomthesis, Universität Oldenburg 1998.
[2] Detmers, F., Herzberger, J.: Enge Schranken für den Effektivzinssatz nach der PAngV bei einem Annuitätenproblem. ZAMM *81*, 711–712 (2001).
[3] Herzberger, J.: On the *R*-order of some recurrences with applications to inclusion methods. Computing *36*, 175–180 (1986).
[4] Herzberger, J.: Bounds for the positive root of a certain class of polynomials with applications. BIT *39(2)*, 366–372 (1999).
[5] Herzberger, J., Backhaus, K.: Explicit bounds and approximations for polynomial roots in mathematics of finance. In: Recent advances in numerical methods and applications II (O. P. Iliev et al., eds.). Singapore: World Scientific 1999, pp. 742–747.
[6] Kjurkchiev, N.: Note on the estimation of the order of convergence of classes of iterative processes. BIT *32*, 525–528 (1992).
[7] Lichtenberg, H.: private communication 1998.
[8] Locarek, H.: Finanzmathematik. München: R Oldenburg 1991.
[9] Petković, M. S., Petković, L. D.: On the bounds of the *R*-order of some iteration methods. ZAMM *69*, T197–T198 (1989).
[10] Shao, S. P., Shao, L. P.: Mathematics for management and finance, 6th ed. Cincinatti: South-Western Publishing 1990.
[11] Traub, J. F.: Iterative methods for the solution of equations. Englewood-Cliffs, N.J.: Prentice-Hall 1964.

F. Detmers
Goerdelerstr. 1
D-26129 Oldenburg
Germany

J. Herzberger
Fachbereich Mathematik
Universität Oldenburg
D-26111 Oldenburg
Germany

Computing [Suppl] 16, 45–65 (2002)

Computing

Algorithmic Differentiation with Intervals

H. Fischer

Abstract

The computational solution of many mathematical problems involves derivatives. Furthermore, in search of verified computer results, it is necessary to construct algorithms for interval inclusions of derivatives. For "small" functions hand-coded formulas for derivatives and appropriate algorithms for interval inclusions are possible, but for "large" functions algorithmic methods have to be employed. In this lecture we screen the complexity of algorithmic differentiation methods, and we investigate the quality of corresponding interval inclusions of derivatives.

Mathematics Subject Classification: 65D25, 65G40, 68U01.

Keywords: Algorithmic differentiation, automatic differentiation, interval inclusion.

1. Introduction

In the context of classical numerical methods, truncation errors or existence/convergence assertions are often phrased in terms of bounds for certain derivatives. Hence, interval inclusions of derivatives can be used to give guaranteed bounds for quantities of concern to classical methods. This slightly modified quotation from Corliss and Rall [5] shall serve as incentive for investigating algorithmic differentiation with intervals.

There has been a long-standing rumor that the derivative is cheap. In fact, for an explicitly given rational function a function value and the corresponding derivative value together cost at most *four* times as much as the function value alone, where costs are measured in terms of rational operations of addition, subtraction, multiplication, division of two real numbers, or changing the sign of a real number. This point of view agrees with Baur and Strassen [2], Theorem 2. The cheap derivative can be obtained by *algorithmic differentiation*, in particular by the *reverse mode*. Approximating the derivative by divided differences costs at least $n + 1$ times as much as the function value alone, where n is the number of variables. Hence, for an explicitly given rational function with $n > 3$ we have the startling situation: The approximate derivative, computed with divided differences, costs more than the exact derivative, computed with the *reverse mode* of algorithmic differentiation.

It is difficult to locate the beginning of algorithmic differentiation, which sometimes is called *automatic differentiation*, *symbolic differentiation*, or *computational differentiation*. Investigations lead back to a report written 1959 by Beda, Korolev, Sukkikh and Frolova [3]. In 1981 Rall's book [21] appeared, which now is a standard reference. Further references are the article Fischer [6], the conference proceedings [4], [9], the books of Kearfott [12], and Griewank [8]. It seems that Linnainmaa [15] was the first who described the *reverse mode* (1970). Since then the reverse mode has been rediscovered many times, 1971 by Ostrowski, Wolin and Borisow [20], 1972 by Tienari [28], 1974 by Werbos [29], 1980 by Miller and Wrathall [16], and by Speelpenning [27], 1983 by Baur and Strassen [2], 1984 by Iri [11], by Kim, Nesterov and Cherkasskĭ [13], and by Sawyer [25].

Let f be a rational function of n variables and let $\#(f, \mathrm{FUN})$ denote the number of rational operations for evaluating $f(x)$ with some algorithm FUN. Assume that FUN′ is an algorithm for computing both $f(x)$ and $f'(x)$, requiring $\#(f, f', \mathrm{FUN}')$ rational operations. We relate $\#(f, \mathrm{FUN})$ and $\#(f, f', \mathrm{FUN}')$ by

$$\#(f, f', \mathrm{FUN}') \leq K(n) \cdot \#(f, \mathrm{FUN}).$$

Using the *forward mode* we can achieve $K(n) = 3n + 1$. The earlier papers on the *reverse mode* claimed or showed that the algorithm FUN′ can be set up such that $K(n)$ is a constant, independent of n. Baur and Strassen [2] proved that $K(n) = 4$ for a suitably designed algorithm FUN′. Surprisingly, later papers state $K(n) = 5$. Even more surprising, there exist papers on the reverse mode which claim $K(n) = 4$ or $K(n) = 5$, but the described software actually requires $K(n) = 7$ or higher. We resolve this somewhat dubious situation as we state several versions of FUN′ with different complexities.

The transition from algorithmic differentiation with reals to algorithmic differentiation with intervals is easy: Replace reals and operations by intervals and interval operations at proper places. Of course, as in almost every algorithm with intervals, resulting intervals are likely to overestimate. We point out ways to tighten inclusions.

Normally one starts with a function f, proceeds to a formula for $f(x)$, steps forward to discuss different formulas for $f(x)$, and ends up with an algorithm for computing $f(x)$, which hopefully produces good inclusions when applied to intervals. This paper is different; we start with an algorithm.

2. Theoretical Background

Let us consider an algorithm of the form:

	FUN
(0)	input: $x \in \mathbb{R}^n$
(1)	for $k = 1, \ldots, n$
	$y_k \leftarrow x_k = k$-th component of x
(2)	for $k = n+1, \ldots, n+t$
	$y_k \leftarrow \Phi_k(y_1, \ldots, y_{k-1})$
(3)	output: y_{n+t}

where $\Phi_{n+1}, \ldots, \Phi_{n+t}$ denote given functions, $\Phi_k : D_{k-1} \subseteq \mathbb{R}^{k-1} \longrightarrow \mathbb{R}$. FUN breaks down as soon as some Φ_k is applied to $(y_1, \ldots, y_{k-1}) \notin D_{k-1}$. Therefore we define

$$D := \{x | x \in \mathbb{R}^n, (y_1, \dots, y_{k-1}) \in D_{k-1} \quad \text{for} \quad k = n+1, \dots, n+t\},$$

and we restrict the input of FUN to $x \in D$. Of course, in practice one does not know in advance whether $x \in D$ or $x \notin D$, and one has to provide emergency exits. FUN defines a function

$$f : D \subseteq \mathbb{R}^n \longrightarrow \mathbb{R} \quad \text{with} \quad f(x) := < \text{output of FUN} >.$$

We depict the situation by the diagram

$$f(x) \longleftarrow \boxed{\text{FUN}} \longleftarrow x.$$

We assume that the functions $\Phi_{n+1}, \dots, \Phi_{n+t}$ are differentiable. Then the function f is differentiable too. Our first goal is an algorithm FUN′ that accepts $x \in D$ and produces the function value $f(x)$ and the derivative value $f'(x)$,

$$f(x), f'(x) \longleftarrow \boxed{\text{FUN}'} \longleftarrow x.$$

Furthermore, the transition from FUN to FUN′ should be accomplished algorithmically, that means FUN′ should be the output of some algorithm DIFF,

$$\text{FUN}' \longleftarrow \boxed{\text{DIFF}} \longleftarrow \text{FUN}.$$

For this purpose we construct a one-line formula for f'. And that formula will yield several versions of FUN′ and consequently several versions of DIFF. Define the functions

$$B : D \subseteq \mathbb{R}^n \to \mathbb{R}^n \quad \text{with} \quad B(x) := x,$$

$$E_k : D_{k-1} \subseteq \mathbb{R}^{k-1} \longrightarrow \mathbb{R}^k \quad \text{with} \quad E_k(v) := \begin{bmatrix} v \\ \Phi_k(v) \end{bmatrix},$$

$$L : \mathbb{R}^{n+t} \to \mathbb{R} \quad \text{with} \quad L(v) := \text{last component of } v.$$

Then we get a representation of f in the form

$$f(x) = L(E_{n+t}(E_{n+t-1}(\dots E_{n+2}(E_{n+1}(B(x)))\dots))). \tag{1}$$

Let us compare formula (1) with our algorithm FUN. The function B corresponds to assigning values to $y_1, \dots, y_n$, the function E_k copies all previous intermediate values $y_1, \dots, y_{k-1}$ and performs the computation of y_k, and the function L picks the last intermediate result y_{n+t}, which is $f(x)$. Hence, formula (1) expresses the stepwise computation of $f(x)$ with FUN.

For shorter notation we set $z_k := \begin{bmatrix} y_1 \\ \vdots \\ y_k \end{bmatrix}$ for $k = 1, \ldots, n + t$.

Then we find $z_n = B(x)$ and $z_k = E_k(z_{k-1})$ for $k = n + 1, \ldots, n + t$.

Using the chain-rule of differentiation we obtain from (1)

$$f'(x) = L'(z_{n+t}) \cdot E'_{n+t}(z_{n+t-1}) \cdot \ldots \cdot E'_{n+2}(z_{n+1}) \cdot E'_{n+1}(z_n) \cdot B'(x). \tag{2}$$

So, formula (2) expresses $f'(x)$ as product of Jacobian matrices (see also Rall [22]). Since matrix-multiplication is associative, there are various ways to do the multiplications in formula (2). Working from right to left yields a *forward mode* algorithm, and working from left to right yields a *reverse mode* algorithm. Other ways of exploiting formula (2) yield *mixed mode* algorithms.

The functions $B, E_{n+1}, \ldots, E_{n+t}, L$ serve to express $f(x)$ and $f'(x)$ in a concise form. Since formula (2) involves the derivatives $B', E'_{n+1}, \ldots, E'_{n+t}, L'$, we have to know these derivatives. $B'(x)$ is an identity-matrix, $E'_k(z_{k-1})$ is a matrix with k rows and $k - 1$ columns, where the upper square part is an identity matrix and its last row is $\Phi'_k(z_{k-1})$, and $L'(z_{n+t}) = (0, \ldots, 0, 1)$, a matrix with 1 row and $n + t$ columns.

3. The Forward Mode

The algorithm FUN defines the function f. In addition, FUN defines intermediate functions $f_1, f_2, \ldots, f_{n+t} : D \subseteq \mathbb{R}^n \longrightarrow \mathbb{R}$, namely

for $k = 1, \ldots, n$

$f_k(x) := y_k = k$-th component of x

for $k = n + 1, \ldots, n + t$

$f_k(x) := y_k = \Phi_k(y_1, \ldots, y_{k-1}) = \Phi_k(f_1(x), \ldots, f_{k-1}(x))$.

Of course $f_{n+t} = f$. For shorter notation we set

$$z'_k := \begin{bmatrix} y'_1 \\ \vdots \\ y'_k \end{bmatrix} := \begin{bmatrix} f'_1(x) \\ \vdots \\ f'_k(x) \end{bmatrix}.$$

Then we find $z'_n = B'(x)$ and $z'_k = E'_k(z_{k-1}) \cdot z'_{k-1}$ for $k = n + 1, \ldots, n + t$.

We use formula (2) and perform the multiplications from right to left. The successive products have names as shown in the diagram.

$$f'(x) = L'(z_{n+t}) \cdot E'_{n+t}(z_{n+t-1}) \cdot \ldots \cdot E'_{n+2}(z_{n+1}) \cdot E'_{n+1}(z_n) \cdot B'(x)$$

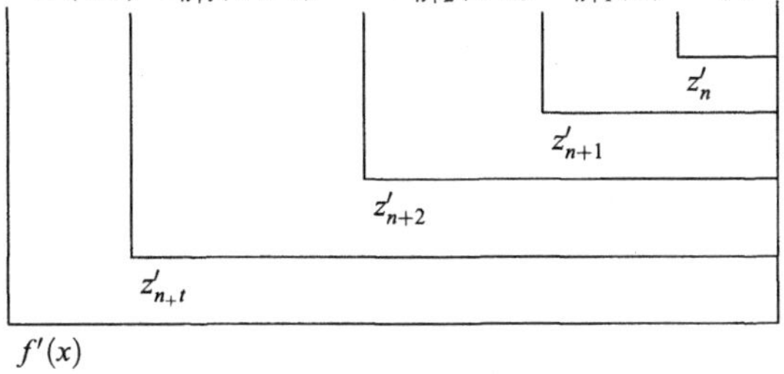

Now we exploit the special form of the functions $B, E_{n+1}, \ldots, E_{n+t}, L$, and the sparsity of the corresponding Jacobian matrices.

(1) The vector z_n has the components $y_1, \ldots, y_n$, and the matrix z'_n has the rows $y'_1, \ldots, y'_n$.

(2) Computation of z_k from z_{k-1} requires computation of $y_k = \Phi_k(z_{k-1})$, and computation of z'_k from z_{k-1} and z'_{k-1} requires computation of $y'_k = \Phi'_k(z_{k-1}) \cdot z'_{k-1}$.

(3) $f(x) = L(z_{n+t}) = y_{n+t}$ is obvious, and $f'(x) = (0, \ldots, 0, 1) \cdot z'_{n+t} = y'_{n+t}$.

This yields the forward mode algorithm FM.

	forward mode algorithm FM
(1)	for $k = 1, \ldots, n$ $y_k \leftarrow k$-th component of x $y'_k \leftarrow k$-th row of the $n \times n$-identity-matrix
(2)	for $k = n+1, \ldots, n+t$ $y_k \leftarrow \Phi_k(y_1, \ldots, y_{k-1})$ $y'_k \leftarrow \Phi'_k(y_1, \ldots, y_{k-1}) \cdot \begin{bmatrix} y'_1 \\ \vdots \\ y'_{k-1} \end{bmatrix}$
(3)	$f(x) \leftarrow y_{n+t}$ $f'(x) \leftarrow y'_{n+t}$

Obviously, the algorithm FM could have been derived directly from the algorithm FUN using well-known differentiation rules, but then it would have been necessary to start anew for deriving the reverse mode. Our framework seems to be a uniform basis for both the forward mode and the reverse mode, and for mixed modes as well.

Compare FUN and FM, and you see how DIFF has to transform FUN into FM.

Univ.-Prof. Dr. Jürgen Herzberger
Faculty of Mathematics, University of Oldenbourg, Oldenbourg, Germany

Originally published by Springer-Verlag Wien New York in 2003

Typesetting: Scientific Publishing Services (P) Ltd., Madras, India

Printed on acid-free and chlorine-free bleached paper
SPIN: 10879532

With 24 Figures

CIP-data applied for

ISSN 0344-8029

ISBN 978-3-211-83852-5 ISBN 978-3-7091-6033-6 (eBook)
DOI 10.1007/978-3-7091-6033-6

$$(p_{k,1},\ldots,p_{k,k-1}) = (p_{k+1,1},\ldots,p_{k+1,k-1}) + p_{k+1,k}\cdot\Phi'_k(z_{k-1}). \tag{3}$$

This formula suggests overwriting. We introduce a working row U of length $n+t$,

$$U = \boxed{U_1\,.\,.\,.\,.\,.\,.\,.\,.\,U_n\,.\,.\,.\,.\,.\,.\,.\,.\,.\,.\,.\,.\,.\,.\,U_{n+t}}$$

in which we store current values of p_k. (3) then reads for $\mu = 1,\ldots,k-1$

$$U_\mu \longleftarrow U_\mu + U_k\cdot\Phi'_k(z_{k-1})_\mu. \tag{4}$$

If for instance $y_k = \Phi_k(z_{k-1}) = y_i\cdot y_j$ with $i\neq j$, then the row $\Phi'_k(z_{k-1})$ has only 2 significant entries, and the transition from p_{k+1} to p_k causes an *update* of only 2 entries of U. After completion of block (2.2) the row U contains the values

$$U = \boxed{p_{n+1,1}\cdots p_{n+1,n}}\boxed{p_{n+2,n+1}\cdots p_{n+t+1,n+t}}.$$

Finally, since $B'(x)$ is an identity-matrix, we get $f'(x) = (U_1,\ldots,U_n)$, yielding the reverse mode algorithm RM.

	reverse mode algorithm RM
(1)	compute $y_1,\ldots,y_{n+t}$ and $f(x)$ with FUN
(2.1)	$U \leftarrow (0,\ldots,0,1)$ with length $n+t$
(2.2)	for $k = n+t,\ldots,n+1$ in this order
	for $\mu = 1,\ldots,k-1$
	$U_\mu \leftarrow U_\mu + U_k\cdot\Phi'_k(y_1,\ldots y_{k-1})_\mu$
(3)	$f'(x) \leftarrow (U_1,\ldots,U_n)$

Compare FUN and RM, and you see how DIFF has to transform FUN into RM.

5. Special Form of FUN

From now on we assume that the functions $\Phi_{n+1},\ldots,\Phi_{n+t}$ have a special form, and consequently FUN has a special form.

	FUN
(1)	for $k = 1,\ldots,n$
	$y_k \leftarrow x_k = k$-th component of x
(2)	for $k = n+1,\ldots,n+t$
	$y_k \leftarrow c_i =$ constant, or
	$y_k \leftarrow y_i * y_j$ with $*\in\{+,-,\cdot,/\}$
(3)	$f(x) \leftarrow y_{n+t}$

Then f is a rational function, and any rational function can be written in the above form. Our algorithm FM specializes to FM.A.

	forward mode algorithm FM.A
(1)	for $k = 1, \ldots, n$
	$y_k \leftarrow k$-th component of x
	$y'_k \leftarrow k$-th row of the $n \times n$-identity-matrix
(2)	for $k = n+1, \ldots, n+t$
case C:	constant
	$y_k \leftarrow c_i$
	$y'_k \leftarrow (0, \ldots, 0)$
case A:	addition
	$y_k \leftarrow y_i + y_j$
	$y'_k \leftarrow y'_i + y'_j$
case S:	subtraction
	$y_k \leftarrow y_i - y_j$
	$y'_k \leftarrow y'_i - y'_j$
case M:	multiplication
	$y_k \leftarrow y_i \cdot y_j$
	$y'_k \leftarrow y_j \cdot y'_i + y_i \cdot y'_j$
case D:	division
	$y_k \leftarrow y_i / y_j$
	$y'_k \leftarrow (y'_i - y_k \cdot y'_j)/y_j$
(3)	$f(x) \leftarrow y_{n+t}$
	$f'(x) \leftarrow y'_{n+t}$

Example 5.1. Here we illustrate the forward mode FM.A.

	FUN
(1)	$y_1 \leftarrow x_1$
	$y_2 \leftarrow x_2$
(2)	$y_3 \leftarrow 25$
	$y_4 \leftarrow y_1 \cdot y_2$
	$y_5 \leftarrow y_4 - y_3$
	$y_6 \leftarrow y_1 + y_2$
	$y_7 \leftarrow y_5 / y_6$
(3)	$f(x) \leftarrow y_7$

	FUN′ via FM.A	numbers
(1)	$y_1 \leftarrow x_1$	-1
	$y'_1 \leftarrow (1, 0)$	$(1, 0)$
	$y_2 \leftarrow x_2$	2
	$y'_2 \leftarrow (0, 1)$	$(0, 1)$
(2)	$y_3 \leftarrow 25$	25
	$y'_3 \leftarrow (0, 0)$	$(0, 0)$
	$y_4 \leftarrow y_1 \cdot y_2$	-2
	$y'_4 \leftarrow y_2 \cdot y'_1 + y_1 \cdot y'_2$	$(2, -1)$
	$y_5 \leftarrow y_4 - y_3$	-27
	$y'_5 \leftarrow y'_4 - y'_3$	$(2, -1)$
	$y_6 \leftarrow y_1 + y_2$	1
	$y'_6 \leftarrow y'_1 + y'_2$	$(1, 1)$
	$y_7 \leftarrow y_5 / y_6$	-27
	$y'_7 \leftarrow (y'_5 - y_7 \cdot y'_6)/y_6$	$(29, 26)$
(3)	$f(x) \leftarrow y_7$	-27
	$f'(x) \leftarrow y'_7$	$(29, 26)$

We mention two modifications of the algorithm FM.A. The first modification FM.B concerns storing the pairs (y_k, y'_k). In FM.A we need space for $n + t$ pairs. We can reduce this storage-requirement considerably by *overwriting* storage locations containing pairs which are no longer relevant. For doing this we need to know at which stage k we can *forget* which pairs. The necessary information is inherent in the algorithm FUN. As soon as a pair is "forgotten", we can use its storage location to store another pair. That, of course, requires some book-keeping.

The second modification FM.C concerns avoiding rational operations with system-zeros. With a row y'_k we associate an index-set I_k that contains the indices of the significant entries of y'_k. For instance for y'_k with $k \in \{1, 2, \ldots, n\}$ we have $I_k = \{k\}$. The index-sets are built up iteratively. Assume that y'_i, I_i and y'_j, I_j are already available, and that $y_k = y_i + y_j$. Then the *sparse* addition $y'_k = y'_i + y'_j$ is done as follows:

$$y'_{k\mu} = \left\{ \begin{array}{lll} y'_{i\mu} + y'_{j\mu} & \text{for} & \mu \in I_i \cap I_j \\ y'_{i\mu} & \text{for} & \mu \in I_i \backslash I_j \\ y'_{j\mu} & \text{for} & \mu \in I_j \backslash I_i \\ 0 & \text{for} & \mu \notin I_i \cup I_j \end{array} \right\} \quad \text{for } \mu = 1, \ldots, n.$$

For the new index set we take $I_k = I_i \cup I_j$. The other *sparse* operations with derivatives are defined in an analogous way. Replacing in FM.A the entry-wise operations by *sparse* operations yields the algorithm FM.C.

Now we turn to the reverse mode. Our algorithm RM specializes to RM.A.

	reverse mode algorithm RM.A
(1)	compute $y_1, \ldots, y_{n+t}$ and $f(x)$ with FUN
(2.1)	$U \leftarrow (0, \ldots, 0, 1)$ with length $n + t$
(2.2)	for $k = n + t, \ldots, n + 1$ in this order
case C:	constant $(y_k \leftarrow c_i)$
	no action necessary
case A:	addition $(y_k \leftarrow y_i + y_j)$
(A,i)	$U_i \leftarrow U_i + U_k \cdot 1$
(A,j)	$U_j \leftarrow U_j + U_k \cdot 1$
case S:	subtraction $(y_k \leftarrow y_i - y_j)$
(S,i)	$U_i \leftarrow U_i + U_k \cdot 1$
(S,j)	$U_j \leftarrow U_j + U_k \cdot (-1)$
case M:	multiplication $(y_k \leftarrow y_i \cdot y_j)$
(M,i)	$U_i \leftarrow U_i + U_k \cdot y_j$
(M,j)	$U_j \leftarrow U_j + U_k \cdot y_i$
case D:	division $(y_k \leftarrow y_i / y_j)$
(D,i)	$U_i \leftarrow U_i + U_k \cdot (1/y_j)$
(D,j)	$U_j \leftarrow U_j + U_k \cdot (-y_i/(y_j \cdot y_j))$
(3)	$f'(x) \leftarrow (U_1, \ldots, U_n)$

Example 5.2. We choose FUN as in Example 5.1 and illustrate the reverse mode RM.A.

	FUN
(1)	$y_1 \leftarrow x_1$
	$y_2 \leftarrow x_2$
(2)	$y_3 \leftarrow 25$
	$y_4 \leftarrow y_1 \cdot y_2$
	$y_5 \leftarrow y_4 - y_3$
	$y_6 \leftarrow y_1 + y_2$
	$y_7 \leftarrow y_5/y_6$
(3)	$f(x) \leftarrow y_7$

	FUN′ via RM.A	
k=1	$y_1 \leftarrow x_1$	−1
k=2	$y_2 \leftarrow x_2$	2
k=3	$y_3 \leftarrow 25$	25
k=4	$y_4 \leftarrow y_1 \cdot y_2$	−2
k=5	$y_5 \leftarrow y_4 - y_3$	−27
k=6	$y_6 \leftarrow y_1 + y_2$	1
k=7	$y_7 \leftarrow y_5/y_6$	−27
	$f(x) \leftarrow y_7$	−27
	$U \leftarrow (0,0,0,0,0,0,1)$	
k=7	$U_5 \leftarrow U_5 + U_7 \cdot (1/y_6)$	1
	$U_6 \leftarrow U_6 + U_7 \cdot (-y_5/(y_6 \cdot y_6))$	27
k=6	$U_1 \leftarrow U_1 + U_6 \cdot 1$	27
	$U_2 \leftarrow U_2 + U_6 \cdot 1$	27
k=5	$U_4 \leftarrow U_4 + U_5 \cdot 1$	1
	$U_3 \leftarrow U_3 + U_5 \cdot (-1)$	−1
k=4	$U_1 \leftarrow U_1 + U_4 \cdot y_2$	29
	$U_2 \leftarrow U_2 + U_4 \cdot y_1$	26
	$f'(x) \leftarrow (U_1, U_2)$	(29, 26)

Obviously, the update formulas in RM.A can be modified such that the number of operations decreases. This leads to the algorithms RM.B, RM.C, RM.D, RM.E.

RM.B	everything as in RM.A, except
	(D,j) $U_j \leftarrow U_j + U_k \cdot (-y_k/y_j)$

RM.C	everything as in RM.B, except
	(A,i) $U_i \leftarrow U_i + U_k$
	(A,j) $U_j \leftarrow U_j + U_k$
	(S,i) $U_i \leftarrow U_i + U_k$
	(S,j) $U_j \leftarrow U_j + (-U_k)$
	(D,i) $U_i \leftarrow U_i + U_k/y_j$
	(D,j) $U_j \leftarrow U_j + (-U_k \cdot y_k/y_j)$

RM.D	everything as in RM.C, except
	Case D: division ($y_k \leftarrow y_i/y_j$)
	$Q \leftarrow U_k/y_j$
	(D,i) $U_i \leftarrow U_i + Q$
	(D,j) $U_j \leftarrow U_j + (-Q \cdot y_k)$

In the algorithm RM.D we recognize that a sign-change in combination with addition can be replaced by subtraction. This yields algorithm RM.E.

RM.E	everything as in RM.D, except
	(S,j) $U_j \leftarrow U_j - U_k$
	(D,j) $U_j \leftarrow U_j - Q \cdot y_k$

In RM.E it may happen that prior to the update, one or the other of U_i, U_j is a system zero. For $k = n + t$ all entries of U except U_{n+t} are system zeros. As k runs down from $n + t$ to $n + 1$, these entries remain system zeros unless they get updated. We want to avoid rational operations with such system zeros. Therefore we introduce a control row σ of length $n + t$ with entries 1,0 that keeps track of system zeros. This yields the algorithm RM.F.

Let us look at RM.F, in particular at the updates

(S,j)	if $\sigma_j = 1$ then $U_j \leftarrow U_j - U_k$ else $U_j \leftarrow -U_k$, $\sigma_j \leftarrow 1$
(D,j)	if $\sigma_j = 1$ then $U_j \leftarrow U_j - Q \cdot y_k$ else $U_j \leftarrow -Q \cdot y_k$, $\sigma_j \leftarrow 1$

Here some savings seem to be possible. We suspend the sign change, because in a later update of U_j it may be subsumed by replacing an addition by a subtraction or vice versa, or it may stay pending. Our aim now is to do explicit sign changes as late as possible. Instead of the row U we use a row W which agrees with U up to signs. The control row σ with entries $1, 0, -1$ keeps track of system zeros and of the signs. We update W and σ such that $U_\mu = \sigma_\mu \cdot W_\mu$. Our directive *sign change as late as possible* leads from RM.F to the algorithm RM.G.

Example 5.3. We choose FUN as in the Examples 5.1 and 5.2 and illustrate RM.G.

FUN	
(1)	$y_1 \leftarrow x_1$
	$y_2 \leftarrow x_2$
(2)	$y_3 \leftarrow 25$
	$y_4 \leftarrow y_1 \cdot y_2$
	$y_5 \leftarrow y_4 - y_3$
	$y_6 \leftarrow y_1 + y_2$
	$y_7 \leftarrow y_5 / y_6$
(3)	$f(x) \leftarrow y_7$

	FUN′ via RM.G	
k = 1	$y_1 \leftarrow x_1$	−1
k = 2	$y_2 \leftarrow x_2$	2
k = 3	$y_3 \leftarrow 25$	25
k = 4	$y_4 \leftarrow y_1 \cdot y_2$	−2
k = 5	$y_5 \leftarrow y_4 - y_3$	−27
k = 6	$y_6 \leftarrow y_1 + y_2$	1
k = 7	$y_7 \leftarrow y_5 / y_6$	−27
	$f(x) \leftarrow y_7$	−27
	$W \leftarrow (0,0,0,0,0,0,1)$	
k = 7	$Q \leftarrow W_7 / y_6$	1
	$W_5 \leftarrow Q$	1
	$W_6 \leftarrow Q \cdot y_7$	−27
k = 6	$W_1 \leftarrow W_6$	−27
	$W_2 \leftarrow W_6$	−27
k = 5	$W_4 \leftarrow W_5$	1
	$W_3 \leftarrow W_5$	1
k = 4	$W_1 \leftarrow W_4 \cdot y_2 - W_1$	29
	$W_2 \leftarrow W_4 \cdot y_1 - W_2$	26
	$f'(x) \leftarrow (W_1, W_2)$	(29, 26)

In the introduction we posed the relation $\#(f, f', \text{FUN}') \leq K(n) \cdot \#(f, \text{FUN})$, where the factor $K(n)$ is of particular interest. For our 10 versions of FUN′ we obtain the following results.

Theorem 1. For the forward mode algorithms FM.A, FM.B, and FM.C we have

$$\begin{aligned}
&(A)\quad \#(f, f', \text{FM.A}) \leq (3n+1) \cdot \#(f, \text{FUN})\\
&(B)\quad \#(f, f', \text{FM.B}) \leq (3n+1) \cdot \#(f, \text{FUN})\\
&(C)\quad \#(f, f', \text{FM.C}) \leq (3n+1) \cdot \#(f, \text{FUN})
\end{aligned}$$

Theorem 2. For the reverse mode algorithms RM.A to RM.G we have

$$\begin{aligned}
&(A)\quad \#(f, f', \text{RM.A}) \leq 9 \cdot \#(f, \text{FUN})\\
&(B)\quad \#(f, f', \text{RM.B}) \leq 8 \cdot \#(f, \text{FUN})\\
&(C)\quad \#(f, f', \text{RM.C}) \leq 7 \cdot \#(f, \text{FUN})\\
&(D)\quad \#(f, f', \text{RM.D}) \leq 6 \cdot \#(f, \text{FUN})\\
&(E)\quad \#(f, f', \text{RM.E}) \leq 5 \cdot \#(f, \text{FUN})\\
&(F)\quad \#(f, f', \text{RM.F}) < 5 \cdot \#(f, \text{FUN})\\
&(G)\quad \#(f, f', \text{RM.G}) \leq 4 \cdot \#(f, \text{FUN})
\end{aligned}$$

Proofs can be found in Fischer and Warsitz [7]. The factors in the two theorems cannot be reduced.

6. Intervalize

We started with an algorithm FUN that defines a function $f : D \subseteq \mathbb{R}^n \to \mathbb{R}$. In the previous section we established 10 versions of FUN′, 3 forward mode algorithms FM.A, FM.B, FM.C, and 7 reverse mode algorithms RM.A to RM.G. Each of these can be used for computing the function-value $f(x)$ and the derivative-value $f'(x)$. Our next goal is an algorithm that produces interval extensions for f and f'. This is an easy task: Replace in FUN′ reals by intervals and real operations by interval operations.

$$\text{FUN'I} \longleftarrow \boxed{\text{intervalize}} \longleftarrow \text{FUN'}$$

The algorithm FUN′ has input $x \in \mathbb{R}^n$ and output $f(x) \in \mathbb{R}$, $f'(x) \in \mathbb{R}^{1\times n}$ (and emergency exits). The new algorithm FUN′I has input $X \in I(\mathbb{R})^n$ and output $F(X) \in I(\mathbb{R})$, $F'(X) \in I(\mathbb{R})^{1\times n}$ (and emergency exits).

$$F(X), F'(X) \longleftarrow \boxed{\text{FUN'I}} \longleftarrow X$$

The function F has the properties

$$\begin{aligned}
&f(x_1,\ldots,x_n) = F([x_1,x_1],\ldots,[x_n,x_n])\\
&x_\nu \in X_\nu \in I(\mathbb{R}) \text{ for } \nu = 1,\ldots,n \Longrightarrow f(x_1,\ldots,x_n) \in F(X_1,\ldots,X_n)\\
&X_\nu \subseteq Y_\nu \in I(\mathbb{R}) \text{ for } \nu = 1,\ldots,n \Longrightarrow F(X_1,\ldots,X_n) \subseteq F(Y_1,\ldots,Y_n),
\end{aligned}$$

and the function F' has the properties

$$\begin{aligned}
&f'(x_1,\ldots,x_n) = F'([x_1,x_1],\ldots,[x_n,x_n])\\
&x_\nu \in X_\nu \in I(\mathbb{R}) \text{ for } \nu = 1,\ldots,n\\
&\qquad \Longrightarrow f'(x_1,\ldots,x_n)_\nu \in F'(X_1,\ldots,X_n)_\nu \text{ for } \nu = 1,\ldots,n\\
&X_\nu \subseteq Y_\nu \in I(\mathbb{R}) \text{ for } \nu = 1,\ldots,n\\
&\qquad \Longrightarrow F'(X_1,\ldots,X_n)_\nu \subseteq F'(Y_1,\ldots,Y_n)_\nu \text{ for } \nu = 1,\ldots,n.
\end{aligned}$$

In other words, F and F' are inclusion isotone interval extension of f and f'.

The tightness of F and F' may depend on our choice of FUN′, but the situation is not as complex as one might expect. From the 10 versions of FUN′ we obtain 10 versions of FUN′I. All define the same function F. The 3 forward mode versions of FUN′I agree in F' which we call $F'_{forward}$. A closer investigation reveals that the 7 reverse mode versions of FUN′I agree in F' which we call $F'_{reverse}$. Now the question arises: Which of $F'_{forward}$ and $F'_{reverse}$ is better in terms of interval width? Let us consider some demonstrative examples. Since here we focus our interest not on complexity, but rather on tightness, we restrict ourselves to short examples with $n = 1$ or $n = 2$.

Example 6.1. Consider the algorithm

FUN		
(1) $y_1 \leftarrow x$	-4	-3
(2) $y_2 \leftarrow 2$	2	2
$y_3 \leftarrow y_1 \cdot y_2$	-8	-6
$y_4 \leftarrow y_1 + y_2$	-2	-1
$y_5 \leftarrow y_3/y_4$	4	6
(3) $f(x) \leftarrow y_5$	4	6

and the corresponding algorithms for $F'_{forward}$ and $F'_{reverse}$.

	FUN'I via FM.A	
(1)	$Y_1 \leftarrow X$	[−4, −3]
	$Y_1' \leftarrow 1$	1
(2)	$Y_2 \leftarrow 2$	2
	$Y_2' \leftarrow 0$	0
	$Y_3 \leftarrow Y_1 \cdot Y_2$	[−8, −6]
	$Y_3' \leftarrow Y_2 \cdot Y_1' + Y_1 \cdot Y_2'$	2
	$Y_4 \leftarrow Y_1 + Y_2$	[−2, −1]
	$Y_4' \leftarrow Y_1' + Y_2'$	1
	$Y_5 \leftarrow Y_3/Y_4$	[3, 8]
	$Y_5' \leftarrow (Y_3' - Y_5 \cdot Y_4')/Y_4$	[0.5, 6]
(3)	$F(X) \leftarrow Y_5$	[3, 8]
	$F'_{forward}(X) \leftarrow Y_4'$	[0.5, 6]

	FUN'I via RM.E	
k = 1	$Y_1 \leftarrow X$	$[-4, -3]$
k = 2	$Y_2 \leftarrow 2$	2
k = 3	$Y_3 \leftarrow Y_1 \cdot Y_2$	$[-8, -6]$
k = 4	$Y_4 \leftarrow Y_1 + Y_2$	$[-2, -1]$
k = 5	$Y_5 \leftarrow Y_3/Y_4$	$[3, 8]$
	$F(X) \leftarrow Y_5$	$[3, 8]$
	$U \leftarrow (0, 0, 0, 0, 1)$	
k = 5	$Q \leftarrow U_5/Y_4$	$[-1, -0.5]$
	$U_3 \leftarrow U_3 + Q$	$[-1, -0.5]$
	$U_4 \leftarrow U_4 - Q \cdot Y_5$	$[1.5, 8]$
k = 4	$U_1 \leftarrow U_1 + U_4$	$[1.5, 8]$
	$U_2 \leftarrow U_2 + U_4$	$[1.5, 8]$
k = 3	$U_1 \leftarrow U_1 + U_3 \cdot Y_2$	$[-0.5, 7]$
	$U_2 \leftarrow U_2 + U_3 \cdot Y_1$	[3, 12]
	$F'_{reverse}(X) \leftarrow U_1$	$[-0.5, 7]$

Here it can be shown that $F'_{forward}(X) \subseteq F'_{reverse}(X)$ for any admissible X.

In particular we have $F'_{forward}([-4,-3]) = [0.5, 6] \neq [-0.5, 7] = F'_{reverse}([-4,-3])$.

So, in this example $F'_{forward}$ is better.

It is tempting to conclude from Example 6.1, that $F'_{forward}$ is better for any FUN, but that would be the wrong conclusion!

Example 6.2. Consider the algorithm

FUN			
(1)	$y_1 \leftarrow x$	0	1
(2)	$y_2 \leftarrow y_1 \cdot y_1$	0	1
	$y_3 \leftarrow y_1 \cdot y_2$	0	1
	$y_4 \leftarrow y_3 - y_2$	0	0
(3)	$f(x) \leftarrow y_4$	0	0

and the corresponding algorithms for $F'_{forward}$ and $F'_{reverse}$.

FUN'I via FM.A		
(1)	$Y_1 \leftarrow X$	[0, 1]
	$Y'_1 \leftarrow 1$	1
(2)	$Y_2 \leftarrow Y_1 \cdot Y_1$	[0, 1]
	$Y'_2 \leftarrow Y_1 \cdot Y'_1 + Y_1 \cdot Y'_1$	[0, 2]
	$Y_3 \leftarrow Y_1 \cdot Y_2$	[0, 1]
	$Y'_3 \leftarrow Y_2 \cdot Y'_1 + Y_1 \cdot Y'_2$	[0, 3]
	$Y_4 \leftarrow Y_3 - Y_2$	$[-1, 1]$
	$Y'_4 \leftarrow Y'_3 - Y'_2$	$[-2, 3]$
(3)	$F(X) \leftarrow Y_4$	$[-1, 1]$
	$F'_{forward}(X) \leftarrow Y'_4$	$[-2, 3]$

	FUN'I via RM.E	
k = 1	$Y_1 \leftarrow X$	[0, 1]
k = 2	$Y_2 \leftarrow Y_1 \cdot Y_1$	[0, 1]
k = 3	$Y_3 \leftarrow Y_1 \cdot Y_2$	[0, 1]
k = 4	$Y_4 \leftarrow Y_3 - Y_2$	$[-1, 1]$
	$F(X) \leftarrow Y_4$	$[-1, 1]$
	$U \leftarrow (0, 0, 0, 1)$	
k = 4	$U_3 \leftarrow U_3 + U_4$	1
	$U_2 \leftarrow U_2 - U_4$	-1
k = 3	$U_1 \leftarrow U_1 + U_3 \cdot Y_2$	[0, 1]
	$U_2 \leftarrow U_2 + U_3 \cdot Y_1$	$[-1, 0]$
k = 2	$U_1 \leftarrow U_1 + U_2 \cdot Y_1$	$[-1, 1]$
	$U_1 \leftarrow U_1 + U_2 \cdot Y_1$	$[-2, 1]$
	$F'_{reverse}(X) \leftarrow U_1$	$[-2, 1]$

Here it can be shown that $F'_{reverse}(X) \subseteq F'_{forward}(X)$ for any admissible X.

In particular we have $F'_{reverse}([0,1]) = [-2,1] \neq [-2,3] = F'_{forward}([0,1])$.

So, in this example $F'_{reverse}$ is better.

The answer to our question so far is: For one example $F'_{forward}$ is better, and for the other example $F'_{reverse}$ is better.

We classify the algorithms of the form FUN as follows:

class 1 := $\{\text{FUN} \mid F'_{forward}(X) \subseteq F'_{reverse}(X)$ for any admissible $X\}$

class 2 := $\{\text{FUN} \mid F'_{reverse}(X) \subseteq F'_{forward}(X)$ for any admissible $X\}$

class 3 := $\{\text{FUN} \mid F'_{forward}(X) = F'_{reverse}(X)$ for any admissible $X\}$

class 4 := the rest

Example 6.3. Consider the algorithm

	FUN		
(1)	$y_1 \leftarrow x$	-4	-3
(2)	$y_2 \leftarrow 2$	2	2
	$y_3 \leftarrow y_2/y_1$	$-\frac{1}{2}$	$-\frac{2}{3}$
	$y_4 \leftarrow 1$	1	1
	$y_5 \leftarrow y_4 + y_3$	$\frac{1}{2}$	$\frac{1}{3}$
	$y_6 \leftarrow y_2/y_5$	4	6
(3)	$f(x) \leftarrow y_6$	4	6

and the corresponding algorithms for $F'_{forward}$ and $F'_{reverse}$.

	FUN'I via FM.A	
(1)	$Y_1 \leftarrow X$	$[-4, -3]$
	$Y'_1 \leftarrow 1$	1
(2)	$Y_2 \leftarrow 2$	2
	$Y'_2 \leftarrow 0$	0
	$Y_3 \leftarrow Y_2/Y_1$	$[-\frac{2}{3}, -\frac{1}{2}]$
	$Y'_3 \leftarrow (Y'_2 - Y_3 \cdot Y'_1)/Y_1$	$[-\frac{2}{9}, -\frac{1}{8}]$
	$Y_4 \leftarrow 1$	1
	$Y'_4 \leftarrow 0$	0
	$Y_5 \leftarrow Y_4 + Y_3$	$[\frac{1}{3}, \frac{1}{2}]$
	$Y'_5 \leftarrow Y'_4 + Y'_3$	$[-\frac{2}{9}, -\frac{1}{8}]$
	$Y_6 \leftarrow Y_2/Y_5$	$[4, 6]$
	$Y'_6 \leftarrow (Y'_2 - Y_6 \cdot Y'_5)/Y_5$	$[1, 4]$
(3)	$F(X) \leftarrow Y_6$	$[4, 6]$
	$F'_{forward}(X) \leftarrow Y'_6$	$[1, 4]$

	FUN'I via RM.E	
k = 1	$Y_1 \leftarrow X$	$[-4, -3]$
k = 2	$Y_2 \leftarrow 2$	2
k = 3	$Y_3 \leftarrow Y_2/Y_1$	$[-\frac{2}{3}, -\frac{1}{2}]$
k = 4	$Y_4 \leftarrow 1$	1
k = 5	$Y_5 \leftarrow Y_4 + Y_3$	$[\frac{1}{3}, \frac{1}{2}]$
k = 6	$Y_6 \leftarrow Y_2/Y_5$	$[4, 6]$
	$F(X) \leftarrow Y_6$	$[4, 6]$
	$U \leftarrow (0,0,0,0,0,1)$	
k = 6	$Q \leftarrow U_6/Y_5$	$[2, 3]$
	$U_2 \leftarrow U_2 + Q$	$[2, 3]$
	$U_5 \leftarrow U_5 - Q \cdot Y_6$	$[-18, -8]$
k = 5	$U_4 \leftarrow U_4 + U_5$	$[-18, -8]$
	$U_3 \leftarrow U_3 + U_5$	$[-18, -8]$
k = 3	$Q \leftarrow U_3/Y_1$	$[2, 6]$
	$U_2 \leftarrow U_2 + Q$	$[4, 9]$
	$U_1 \leftarrow U_1 - Q \cdot Y_3$	$[1, 4]$
	$F'_{reverse}(X) \leftarrow U_1$	$[1, 4]$

Here it can be shown that $F'_{forward}(X) = F'_{reverse}(X)$ for any admissible X, and hence, FUN belongs to class 3.

Example 6.4. We construct an algorithm FUN for a function $f : D \subseteq \mathbb{R}^2 \to \mathbb{R}$ with $f(x_1, x_2) = f_1(x_1) + f_2(x_2)$, where $f_1 = f$ from Example 6.1 and $f_2 = f$ from Example 6.2. Then $f'(x_1, x_2) = (f_1'(x_1), f_2'(x_2))$.

Using the results of Example 6.1 and Example 6.2 we find for the present example: In the first component forward is better, and in the second component reverse is better, and hence, FUN belongs to class 4.

7. Tighten Inclusions

Algorithms for computing interval inclusions tend to overestimate. In Example 6.1 we found $f'(X) \subseteq F'_{forward}(X) \subseteq F'_{reverse}(X)$. Though in this example $F'_{forward}$ provides the better inclusion, it is not optimal. In Example 6.2 we found $f'(X) \subseteq F'_{reverse}(X) \subseteq F'_{forward}(X)$, where $F'_{reverse}$ provides the better inclusion, but $F'_{reverse}$ is not optimal. Reasons for overestimation are

- the subdistributive law $A \cdot (B + C) \subseteq A \cdot B + A \cdot C$ for $A, B, C \in I(\mathbb{R})$,
- $A - A \neq 0$ and $A/A \neq 1$ for $A = [\underline{A}, \overline{A}] \in I(\mathbb{R})$ with $\underline{A} \neq \overline{A}$,
- multiple occurrence of interval variables,
- the wrapping effect.

Means to tighten inclusions are: subinterval strategies, exploiting monotonicity, use of the mean value theorem, and others, see [1], [5], [10], [12], [14], [17], [18], [19], [23], [24], [26], [30].

Computer Algebra seems to have had little influence on Interval Analysis so far, in particular when functions are not represented by closed formulas but rather by algorithms. We sketch an idea to tighten inclusions in a computer-algebra-like manner.

Let FUN1 be an algorithm defining a function $f : D \subseteq \mathbb{R}^n \to \mathbb{R}$. From FUN1 we obtain FUN1'I. This algorithm defines inclusion isotone interval extensions $F1$ and $F1'$ of f and f'.

We modify FUN1 by certain rules, such that the modified algorithm FUN2 also defines f.

FUN2 ⟵ `MODIFY` ⟵ FUN1

From FUN2 we obtain FUN2'I. This algorithm defines inclusion isotone interval extensions $F2$ and $F2'$ of f and f'. Then we can state an intermediate result, further research is in progress.

Theorem 3. If `MODIFY` uses proper rules, then

$$F2(X) \subseteq F1(X),$$
$$F2'(X) \subseteq F1'(X).$$

Of course, this theorem is of use only if we have proper rules. We present two examples.

The first rule scans FUN for the distributive law $a \cdot b + a \cdot c = a \cdot (b + c)$.

Rule 1 : If FUN comprises three lines of the form

line i: $y_i \leftarrow y_a \cdot y_b$
line j: $y_j \leftarrow y_a \cdot y_c$
line k: $y_k \leftarrow y_i + y_j$

then

replace line k by two lines
new line: $y_{k1} \leftarrow y_b + y_c$
new line: $y_k \leftarrow y_a \cdot y_{k1}$
if line i is obsolete, delete it,
if line j is obsolete, delete it,
renumber the variables $y_1, y_2, \ldots$

Example 7.1. Let us consider the algorithm FUN1. Rule 1 leads from FUN1 to FUN2.

FUN1	
(1)	$y_1 \leftarrow x$
(2)	$y_2 \leftarrow 1$
	$y_3 \leftarrow y_1 \cdot y_1$
	$y_4 \leftarrow y_1 + y_2$
	$y_5 \leftarrow y_2 / y_4$
	$y_6 \leftarrow y_1 \cdot y_5$
	$y_7 \leftarrow y_3 + y_6$
(3)	$f(x) \leftarrow y_7$

FUN2	
(1)	$y_1 \leftarrow x$
(2)	$y_2 \leftarrow 1$
	$y_3 \leftarrow y_1 + y_2$
	$y_4 \leftarrow y_2 / y_3$
	$y_5 \leftarrow y_1 + y_4$
	$y_6 \leftarrow y_1 \cdot y_5$
(3)	$f(x) \leftarrow y_6$

`MODIFY` using Rule 1 provides

$$F2(X) \subseteq F1(X) \quad \text{for any admissible } X$$
$$F2'_{forward}(X) \subseteq F1'_{forward}(X) \quad \text{for any admissible } X$$
$$F2'_{reverse}(X) \subseteq F1'_{reverse}(X) \quad \text{for any admissible } X$$

In particular for $X = [-0.5, 0]$, we obtain

with FUN1	with FUN2
$f(X) \subseteq F1(X) = [-1, 0.25]$	$f(X) \subseteq F2(X) = [-1, 0]$
$f'(X) \subseteq F1'_{forward}(X) = [0, 4]$	$f'(X) \subseteq F2'_{forward}(X) = [0.5, 3.5]$
$f'(X) \subseteq F1'_{reverse}(X) = [0, 4]$	$f'(X) \subseteq F2'_{reverse}(X) = [0, 4]$

In this example, `MODIFY` tightens the inclusion for f, and it tightens the inclusion for f' in the forward mode.

The second rule scans FUN for the cancellation law $\dfrac{a \cdot b}{a + c} = \dfrac{b}{1 + \frac{c}{a}}$.

Rule 2 : If FUN comprises three lines of the form

line i: $y_i \leftarrow y_a \cdot y_b$
line j: $y_j \leftarrow y_a + y_c$
line k: $y_k \leftarrow y_i / y_j$

then

replace line k by four lines
new line: $y_{k1} \leftarrow y_c / y_a$
new line: $y_{k2} \leftarrow 1$
new line: $y_{k3} \leftarrow y_{k2} + y_{k1}$
new line: $y_k \leftarrow y_b / y_{k3}$
if line i is obsolete, delete it,
if line j is obsolete, delete it,
renumber the variables $y_1, y_2, \ldots$

Example 7.2. Let us consider the algorithm FUN1. Rule 2 leads from FUN1 to FUN2.

FUN1	
(1)	$y_1 \leftarrow x$
(2)	$y_2 \leftarrow 2$
	$y_3 \leftarrow y_1 \cdot y_2$
	$y_4 \leftarrow y_1 + y_2$
	$y_5 \leftarrow y_3/y_4$
(3)	$f(x) \leftarrow y_5$

FUN2	
(1)	$y_1 \leftarrow x$
(2)	$y_2 \leftarrow 2$
	$y_3 \leftarrow y_2/y_1$
	$y_4 \leftarrow 1$
	$y_5 \leftarrow y_4 + y_3$
	$y_6 \leftarrow y_2/y_5$
(3)	$f(x) \leftarrow y_6$

`MODIFY` using Rule 2 provides

$$F2(X) \subseteq F1(X) \quad \text{for any admissible } X$$
$$F2'_{forward}(X) \subseteq F1'_{forward}(X) \quad \text{for any admissible } X$$
$$F2'_{reverse}(X) \subseteq F1'_{reverse}(X) \quad \text{for any admissible } X$$

In particular for $X = [-4, -3]$ we obtain

with FUN1	with FUN2
$f(X) \subseteq F1(X) = [3, 8]$	$f(X) \subseteq F2(X) = [4, 6]$
$f'(X) \subseteq F1'_{forward}(X) = [0.5, 6]$	$f'(X) \subseteq F2'_{forward}(X) = [1, 4]$
$f'(X) \subseteq F1'_{reverse}(X) = [-0.5, 7]$	$f'(X) \subseteq F2'_{reverse}(X) = [1, 4]$

As can be seen, `MODIFY` drastically tightens inclusions for f and its derivative f'.

8. Conclusion

We described algorithmic differentiation methods for computing the derivative value $f'(x)$ for a function $f : D \subseteq \mathbb{R}^n \to \mathbb{R}$. In particular we investigated the complexity of the forward mode and of several versions of the reverse mode. Applied to an interval $X \in I(D)$, these methods produce interval inclusions for $f'(X)$. It turned out that for one class of functions the forward mode is better than the reverse mode in terms of interval width, and for another class the reverse mode yields better results. So we may pose the question: How can we predict the class a function belongs to? Independent of the classification of a function, we presented ideas to tighten results in a computer-algebra-like manner.

References

[1] Alefeld, G., Herzberger, J.: Introduction to interval computations. New York: Academic Press 1983.

[2] Baur, W., Strassen, V.: The complexity of partial derivatives. Theor Comp Sci *22*, 317–330 (1983).

[3] Beda, L. M., Korolev, L. N., Sukkikh, N. V., Frolova, T. S.: Programs for automatic differentiation for the machine BESM. (in Russian) Technical Report, Institute for Precise Mechanics and Computation Techniques, Academy of Science, USSR, Moscow 1959.

[4] Berz, M., Bischof, Ch., Corliss, G. F., Griewank, A. (eds.): Computational differentiation: Techniques, applications, and tools. Philadelphia: SIAM 1996.
[5] Corliss, G. F., Rall, L. B.: Computing the range of derivatives. In: Computer arithmetic, scientific computation and mathematical modelling (E. Kaucher, S. M. Markov and G. Mayer, eds.) (IMACS Annals on Computing and Applied Mathematics, vol. 12). Basel: Baltzer J. C., 1991, Switzerland, pp. 195–212.
[6] Fischer, H.: Automatisches Differenzieren. In: Wissenschaftliches Rechnen – Eine Einführung in das Scientific Computing (J. Herzberger, ed.). Berlin: Akademie Verlag 1995, 53–104.
[7] Fischer, H., Warsitz, H.: Complexity investigations concerning derivatives generated by symbolic differentiation. Technical Report 168, Fakultät für Mathematik, Technische Universität München, 2000.
[8] Griewank, A.: Evaluating derivatives: Principles and techniques of algorithmic differentiation. Philadelphia: SIAM 2000.
[9] Griewank, A., Corliss, G. F. (eds.): Automatic differentiation of algorithms: Theory, implementation, and application. Philadelphia: SIAM 1991.
[10] Hansen, E.: Sharpness in interval computations. Reliable Comp *3*, 17–29 (1997).
[11] Iri, M.: Simultaneous computation of functions, partial derivatives and estimates of rounding errors, complexity and practicality. Japan J. Appl. Math. *1*, 223–252 (1984).
[12] Kearfott, R. B.: Rigorous global search: Continuous problems. Dordrecht: Kluwer 1996.
[13] Kim, K. V., Nesterov, Yu. E., Cherkasski, B. V.: An estimate of the effort in computing the gradient. Soviet Math. Dokl. *29*, 384–387 (1984).
[14] Krückeberg, F.: Arbitrary accuracy with variable precision arithmetic. In: Interval mathematics 1985 (K. Nickel, ed.) LNCS 212, Springer 1986, pp. 95–102.
[15] Linnainmaa, S.: The representation of the cumulative rounding error of an algorithm as a Taylor expansion of the local rounding errors. (in Finnish) Master's Thesis, Department of Computer Science, University of Helsinki, 1970.
[16] Miller, W., Wrathall, C.: Software for roundoff analysis of matrix algorithms. New York: Academic Press 1980.
[17] Moore, R. E.: Interval analysis. Englewood Cliffs, NJ: Prentice-Hall 1966.
[18] Neumaier, A.: Interval methods for systems of equations. Cambridge: Cambridge University Press 1990.
[19] Oliveira, J. B.: New slope methods for sharper interval functions and a note on Fischer's acceleration method. Reliable Comput *2*, 299–320 (1996).
[20] Ostrowski, G. M., Wolin, Yu. M., Borisow, W. W.: Über die Berechnung von Ableitungen. Wissenschaftliche Zeitschrift der Technischen Hochschule für Chemie, Leuna-Merseburg, Vol. 13, 1971, pp. 382–384.
[21] Rall, L. B.: Automatic differentiation: Techniques and applications. Lecture Notes in Computer Science 120, Berlin: Springer 1981.
[22] Rall, L. B.: Differentiation in PASCAL-SC: Type gradient. ACM Trans. Math. Software *10*, 161–184 (1984).
[23] Rall, L. B.: Improved interval bounds for ranges of functions. In: Interval mathematics 1985 (K. Nickel, ed.) LNCS 212, Springer 1986, pp. 143–155.
[24] Ratschek, H., Rokne, J.: Computer methods for the range of functions. Chichester: Ellis Horwood 1984.
[25] Sawyer, J. W.: First partial differentiation by computer with an application to categorical data analysis. The American Statistician *38*, 300–308 (1984).
[26] Skelboe, S.: Computation of rational interval functions. BIT *14*, 87–95 (1974).
[27] Speelpenning, B.: Compiling fast partial derivatives of functions given by algorithms. Ph.D. Thesis, Department of Computer Science, University of Illinois, Urbana-Champaign 1980.
[28] Tienari, M.: On some topological properties of numerical algorithms. BIT *12*, 409–433 (1972).
[29] Werbos, P.: Beyond regression: New tools for prediction and analysis in the behavioral sciences. Ph.D. Thesis, Harvard University, Cambridge, Mass., 1974.
[30] Wolfe, M. A.: An interval algorithm for bounding the ranges of real-valued functions of one real variable. Reliable Comp *3*, 31–50 (1997).

Dr. Herbert Fischer
Fakultät für Mathematik
Technische Universität München
D-80290 München
Germany
e-mail: fischer@mathematik.tu-muenchen.de

Computing [Suppl] 16, 67–82 (2002)

Computing

Computation of a Family of Non-cosymmetrical Equilibria in a System of Nonlinear Parabolic Equations

K. Frischmuth, Rostock and **V. G. Tsybulin,** Rostov

Abstract

We study a system of two nonlinear parabolic equations with cosymmetry and find a continuous family of equilibria with nonconstant spectrum. Basing on the finite-difference approach, we develop a numerical method for solving the partial differential equations and calculating the continuous family of non-cosymmetrical equilibria. An application of the proposed system to population kinetics is demonstrated. Numerically computed families are presented. The dependence on model parameters is discussed.

AMS Subject Classifications: 35B20, 65M06, 35K60, 73Q05, 90A16.

Keywords: Cosymmetry, parabolic systems, stationary states.

1. Introduction

Recently V. Yudovich [1], [2] introduced the notion of a *cosymmetry* in order to explain the onset of a continuous family of equilibria with variable spectra in certain problems of mathematical physics. At present, the theory of cosymmetry and its applications are intensively studied, cf. [1]–[10].

For a differential equation $\dot{u} = F(u)$ in a Hilbert space H, a cosymmetry is an nontrivial operator L which is orthogonal to F at each point of the phase space, i.e. $(F(u), L(u)) = 0$, where $(\cdot,\cdot)$ denotes the inner product. Assume a state of equilibrium $u_0 \in H$, $F(u_0) = 0$ to be non-cosymmetric, i.e., $L(u_0) \neq 0$. Then, as a consequence of the Implicit Function Theorem, cf. [1], [4], u_0 belongs to a continuous one-parameter family $\{u(\tau) : F(u(\tau)) = 0, \tau \in \Gamma\}$ with $u(0) = u_0$, provided there are no additional degeneracies. In degenerate cases there exists a k-dimensional sub-manifold of equilibria that contains the equilibrium u_0.

We construct below such a system of nonlinear parabolic equations which admits a nontrivial cosymmetry and a family of non-cosymmetrical equilibria. Problems of that type were first observed and studied in mathematical physics, in particular in diffusion-convection problems in porous media. Numerical solutions for such systems with arbitrary schemes usually do not find a continuous family of equilibria but rather a number of isolated equilibria.

We intend to model a situation in population kinetics where infinitely many stationary states are possible. We assume two populations sharing a common space, e.g., a river bank or a traffic route, and suppose a dominant role of the unique spatial coordinate x. Looking for best opportunities for their respective trades, each population may move in dependence on the current concentrations of their own kind and the other one. However, the first population (concentration function $u(x,t)$) thrives on relocation of the second ($v(x,t)$), while the second suffers from motion of the first.

Thus, we consider a system for a time dependent vector field $w(x,t) = (u(x,t), v(x,t))^\top$ defined on a domain $\Omega = [0,a] \subset \mathbb{R}$ with Dirichlet conditions applied on $\partial\Omega$

$$\dot{w} = \epsilon w'' + \kappa w' + f(w, w'), \tag{1}$$

$$w(x,0) = w_0(x) \quad \text{on } \Omega, \tag{2}$$

$$w(x,t) = 0 \quad \text{on } \partial\Omega \times [0,T]. \tag{3}$$

Here the dot and prime mean differentiation with respect to time t and space coordinate x. The solution is sought on the time span T, in particular T may be $+\infty$. Square matrixes κ and ϵ comprise the known constants ϵ_u, ϵ_v, κ_u, κ_v,

$$\epsilon = \begin{pmatrix} \epsilon_u & 0 \\ 0 & \epsilon_v \end{pmatrix}, \quad \kappa = \begin{pmatrix} 0 & \kappa_u \\ \kappa_v & 0 \end{pmatrix}. \tag{4}$$

It is essential that the flux of the v-component results in a source of the u-component, and vice versa. Both sources may be interpreted as some preferred direction of a structured medium. It is of great interest to consider the case of opposite signs, i.e., when the product $\kappa_u \kappa_v$ is negative. Diffusion terms, given by second order derivatives (one-dimensional Laplace operator), often play merely the role of a regularization. In particular, the parameters ϵ_u, ϵ_v may be important beyond their physical meaning. The nonlinear term $f = (f_u(w,w'), f_v(w,w'))^\top$ describes some sort of activity depending simultaneously on transportation and availability of the components.

The system (1)–(3) with zero initial conditions has the trivial solution $w = 0$. More interesting is that there also exist nontrivial stationary solutions for suitable values of the parameters κ_u and κ_v and certain types of nonlinearities. Here, we concentrate on the case when in fact a whole one-parameter family of stationary solutions exists, but they cannot be obtained by action of an invariance as translation or rotation. This is exactly the situation for which the theory of cosymmetry and the notion of families of non-cosymmerical equilibria were introduced, [1], [2].

In the next section, we are going to examine the above system and derive the conditions for a nontrivial cosymmetry. We identify parameters which give, indeed, a continuous family of equilibria, the calculation of which is presented in Sect. 4. To this end, we introduce a special numerical scheme which is shown to work well for the system (1)–(3).

2. Preliminary Analysis of the System

First, we briefly describe some essential properties of the underlying system. We define a Hilbert space $H = L^2(\Omega) \times L^2(\Omega)$ with the usual scalar product.

In order to enhance the dissipativity of the system we assume that the functions f_u, f_v are such that the following integral equality holds:

$$\int_\Omega (\kappa_v f_u u + \kappa_u f_v v) dx = 0. \tag{5}$$

Then, multiplying the first equation (1) by $\kappa_v u$ and the second one by $\kappa_u v$, summing and integrating over the domain Ω we obtain:

$$\frac{d}{dt} \int_\Omega \left(\kappa_v u^2 + \kappa_u v^2\right) dx = - \int_\Omega \left(\kappa_v \epsilon_u u'^2 + \kappa_u \epsilon_v v'^2\right) dx. \tag{6}$$

One can see that in the case $\kappa_u = 0$ and $\epsilon_u > 0$ the dissipativity of the component $u(x, t)$ is guaranteed, and analogically, for $\kappa_v = 0$ and $\epsilon_v > 0$ we deduce the decrease of $\int_\Omega v^2 dx$.

2.1. Cosymmetry Condition

We are interested in conditions for which the system (1)–(3) has a cosymmetry. We restrict ourselves to the case of a linear cosymmetry in the following form:

$$L(w) = \begin{pmatrix} 0 & \epsilon_v \\ -\epsilon_u & 0 \end{pmatrix} w^\top. \tag{7}$$

In order to cover a reasonable class of nonlinearities, we assume f_u and f_v to be arbitrary bilinear forms in w and w'

$$f_u = \epsilon_u w^\top M^u w', \quad f_v = \epsilon_v w^\top M^v w'. \tag{8}$$

Here the factors ϵ_u and ϵ_v are used for convenience of subsequent presentation. The following Lemma shows the existence of matrices that lead to a cosymmetry of the underlying problem.

Lemma 1. *Let the matrices in* (8) *be of the form*

$$M^u = \begin{pmatrix} \delta - 2\gamma & \beta \\ \alpha & 2\delta - \gamma \end{pmatrix}, \quad M^v = \begin{pmatrix} \alpha - 2\beta & \delta \\ \gamma & \beta - 2\alpha \end{pmatrix}, \tag{9}$$

where the constants α, β, γ, δ are free parameters. Then the system (1)–(3) *has a cosymmetry* (7) *and the condition* (5) *is fulfilled.*

The prove is done by direct verification. Let us multiply the equation (1) by $L(w)$, sum and integrate over the domain Ω. One can easily check that the right-hand side of this equation vanishes, because the following equalities are valid

$$\int_\Omega \epsilon_u \epsilon_v (u''v - v''u)dx = 0, \tag{10}$$

$$\int_\Omega (\epsilon_v \kappa_v v'v - \epsilon_u \kappa_v u'u)dx = 0, \tag{11}$$

$$\int_\Omega (\epsilon_v f_u v - \epsilon_u f_v u)dx = 0. \tag{12}$$

This in turn is obtained using integration by parts, Green's formula and the boundary conditions. As result we have

$$\int_\Omega (\epsilon_v \dot{u} v - \epsilon_u \dot{v} u)dx = 0, \tag{13}$$

which by definition means that (7) is a cosymmetry of the system (1)–(3).

The validity of (5) follows from similar transformations.

So, the system (1)–(3) is a cosymmetric one for any values κ_u, κ_v and the expressions for the nonlinearities based on the four-parametric family of matrices (9). We present below examples of parameter values such that the assumptions of Lemma 1 are satisfied, and compute a number of continuous families consisting of steady states.

Another case of nonlinearities which allows a cosymmetry is given by

$$f_u = \epsilon_u \sum_{k=1}^{K} \mu_k u^{\alpha_k+1} v^{\beta_k}, \quad f_v = \epsilon_v \sum_{k=1}^{K} \mu_k u^{\alpha_k} v^{\beta_k+1}, \tag{14}$$

where for simplicity we will suppose $\alpha_k \geq 0$, $\beta_k \geq 0$, and $\mu_k \in \mathbb{R}$.

As a particular case the nonlinearities (14) may model logistic growth, if we specify

$$f_u = \epsilon_u \mu u \left(1 - \frac{u}{u_\infty}\right), \quad f_v = \epsilon_v \mu v \left(1 - \frac{u}{u_\infty}\right). \tag{15}$$

Here u_∞ plays the role of a limiting value on the concentration $u(x,t)$. It should be mentioned that the logistic law is applied only to the growth of the first population, whereas the other is regulated by closeness to u_∞.

2.2. Stability of Zero Equilibrium

It is easy to analyze the stability of the cosymmetrical equilibrium $w = 0$. Without loss of generality we take $\kappa_u > 0$ and $\kappa_v < 0$. Then, we change variables

$$u = \tilde{u}\sqrt{\kappa_u \epsilon_v}, \quad v = \tilde{v}\sqrt{-\kappa_v \epsilon_u}, \tag{16}$$

and set

$$\lambda = \frac{1}{2}\sqrt{\frac{-\kappa_u \kappa_v}{\epsilon_u \epsilon_v}}. \tag{17}$$

For simplicity we take $\kappa_u = -\kappa_v = \lambda$. After that the corresponding spectral problem is given

$$0 = \tilde{u}'' + 2\lambda\tilde{v}', \quad 0 = \tilde{v}'' - 2\lambda\tilde{u}', \quad \tilde{u}, \tilde{v}|_{\delta\Omega} = 0. \tag{18}$$

We introduce the complex function $W = \tilde{u} + i\tilde{v}$ and rewrite (18) as

$$0 = W'' - 2\lambda i W', \quad W|_{\delta\Omega} = 0, \tag{19}$$

Finally, we substitute $W = U\exp(i\lambda x)$ and deduce

$$0 = U'' + \lambda^2 U, \quad U|_{\delta\Omega} = 0, \tag{20}$$

So, we find that $\lambda_j = j\pi/a$, $j \in \mathbb{Z}$ and all eigenvalues are simple [11].

Because the original problem is equivalent to the eigenvalue problem for the 1d Laplace operator corresponding to pairs of functions, the multiplicity of each eigenvalue doubles. Thus the critical value is $\lambda_{crit} = \pi/a$, and at $\lambda > \lambda_{crit}$ the zero equilibrium becomes unstable.

In the case of nonlinearities given by (15) similar manipulations give the following result

$$\lambda_j = \sqrt{\frac{j^2\pi^2}{a} - \mu}.$$

3. Solution Method

We apply the method of lines to the solution of the system (1)–(3). To discretize the right-hand side we use second-order finite difference operators, paying special attention to the approximation of nonlinear terms. It is a crucial point for systems with cosymmetry: an inappropriate approximation may lead to the loss of the cosymmetry [6]. As result, one may obtain a number of isolated equilibria instead of a continuous family of steady states.

3.1. Semi-discretization

We consider the case of a uniform grid on $\Omega = [0, a]$

$$x_j = jh, \quad j = 0, \dots, N+1, \quad h = a/(N+1). \tag{21}$$

All unknowns are defined at the nodes x_j, and the notations u_j and v_j are used for discrete values. The discretized boundary conditions take the following form:

$$u_0 = v_0 = u_{N+1} = v_{N+1} = 0. \tag{22}$$

For first-order derivatives and the Laplacian we use centered difference operators

$$D_j^1(u) = \frac{u_{j+1} - u_{j-1}}{2h}, \quad D_j^2(u) = \frac{u_{j+1} - 2u_j + u_{j-1}}{h^2}. \tag{23}$$

Then, the discrete analog of the system (1) is given as

$$\dot{u}_j = \epsilon_u D_j^2(u) + \kappa_u D_j^1(v) + f_{u,j}, \quad j = 1, \dots, N, \tag{24}$$

$$\dot{v}_j = \epsilon_v D_j^2(v) + \kappa_v D_j^1(u) + f_{v,j}. \tag{25}$$

The nonlinear terms $f_{u,j}$ and $f_{v,j}$ are defined below. Using the discrete version of Green's formulae and reordering the sum (discrete analog of integration by parts) we conclude that the following equalities are valid

$$h\epsilon_v\epsilon_u \sum_{j=1}^{N} \left(D_j^2(u)v_j - D_j^2(v)u_j \right) = 0, \tag{26}$$

$$\sum_{j=1}^{N} \left(\epsilon_v\kappa_v D_j^1(v)v_j - \epsilon_u\kappa_v D_j^1(u)u_j \right) = 0. \tag{27}$$

The given relations are the analogs of equalities (10) and (11), respectively. Thus, the discrete analog of (7) is a cosymmetry for (24)–(25) iff the following equality is fulfilled

$$h\sum_{j=1}^{N} (\epsilon_v f_{u,j} v_j - \epsilon_u f_{v,j} u_j) = 0. \tag{28}$$

The approximation of the nonlinear terms (14) can be done in a straightforward way. A special approximation has been developed to deal with the nonlinear terms (8) which depend both on the function values and their slopes. We apply for this purpose the direct method of free parameters, doing the calculations by the computer algebra system Maple.

The general second order accurate finite-difference operator on a three nodes stencil may be written as

$$u'v|_{x_j} \approx D_j(u,v) \equiv \sum_{k,l=-1}^{1} p_{kl} u_{j+k} v_{j+l}. \tag{29}$$

Here p_{kl}, $k, l = 1, 2, 3$ are unknown or free parameters.

We compute a Taylor series for the expression $u'v|_{x_j} - D_j(u,v)$ and demand terms up to third order on h to vanish. Thus, we derive a system of six linear equations for the p_{kl} and then solve it. As result we obtain a three-parameter family of operators giving second order accuracy for the term $u'v$.

To establish an additional relation on the unknown parameters we impose the integral equality

$$\int_0^a u'u\phi(u)dx = 0. \tag{30}$$

This is valid, in particular, for homogeneous boundary conditions.

The discrete analog of (30) must be satisfied at least for some functions $\phi(u)$

$$h\sum_{j=1}^{N} D_j(u,u)\phi(u_j) = 0. \tag{31}$$

We get two additional equations on the parameters p_{kl} from this requirement for $\phi(u) = u$.

Eventually, we find a one-parameter formula approximating $u'v$

$$\begin{aligned} D_j(u,v) = \frac{1}{6h}[&(1+\zeta)(u_{j+1} - u_{j-1})v_j - \zeta u_j(v_{j+1} - v_{j-1}) \\ &+ u_{j+1}v_{j+1} - u_{j-1}v_{j-1} + (1-\zeta)(u_{j+1}v_{j-1} - u_{j-1}v_{j+1})]. \end{aligned} \tag{32}$$

Finally, from the cosymmetry relation (28) we deduce that only for $\zeta = 1$ all requirements are met.

We remark that more examples of the application of computer algebra systems to the derivation of numerical schemes are given, for instance, in [12].

Let us rewrite the system of ordinary differential equations (24)–(25) as follows:

$$\dot{Y} = (A + \lambda B)Y + F(Y). \tag{33}$$

Here A is a positive-definite matrix, B is a skew-symmetric matrix, and $F(Y)$ denotes a nonlinear term. The vector Y is given by

$$Y = (u_1, \ldots, u_N, v_1, \ldots, v_N). \tag{34}$$

Note that (33) is a vector equation in $\mathbb{R}^{2N}$, where N is the number of inner grid points in Ω.

Without loss of generality we consider further $\kappa_u = -\kappa_v = \lambda$. The semi-discretization described above may be easily adapted for arbitrary initial-boundary conditions. For brevity, we present here only the case of homogeneous boundary conditions. For this case we establish numerically the stability of the trivial solution $Y = 0$. Parameter values for which instability occurs are found via an eigenproblem for the Jacobian matrix J of the system of ordinary differential equations (33).

It should be noted that preservation of cosymmetry in a finite-difference approximation of a two-dimensional filtration-convection problem on a rectangular domain is considered in [6], [9].

3.2. Computation of the Family of Equilibria

If λ is slightly greater than λ_{crit}, then all points on the family are stable. So, we can start to integrate (33) in an arbitrary nonzero state, and compute the solution. It approaches a limit state which, in general, depends on the initial state. This limit state is not an isolated asymptotically stable one, but one point on a continuous curve of stationary states.

To compute the family of equilibria we apply a technique similar to the approach developed in [5], [7]. This technique is based on the cosymmetric version of the Implicit Function Theorem, cf. [4], which asserts the existence of a continuous family of equilibria, and states that the kernel of the Jacobian is tangential to the family (curve) in phase space.

We trace this curve consisting of equilibria, defined by

$$(A + \lambda B)Y + F(Y) = 0. \tag{35}$$

The Jacobian matrix $J = A + \lambda B + \partial_Y F(Y)$ has a constant rank deficiency of one. We find the family of equilibria (stationary solutions) by following the 1d solution manifold of this algebraic system of equations by a predictor–corrector method. This means, we cancel one equation of the system (35), adding instead one equation of the form:

$$(Y - Y_e) \cdot Y_t = \tau, \tag{36}$$

where Y_e is our actual equilibrium solution and Y_t is a unit basis vector of ker J, τ is the step-size. The thus modified equilibrium condition defines for small enough τ a unique new equilibrium point. A reasonable starting point for Newton's method is $Y_e + \tau Y_t$.

Without going into the details we remark that for higher ratios of λ stability of some stationary solutions may be lost. However, we can still trace the family by the above method, finding an initial guess by increasing λ gradually.

It should be mentioned that if we take either parameters not satisfying the conditions (9) or we use an approximation which destroys the cosymmetry during discretization, we end up with a set of isolated equilibrium states.

4. Numerical Results

Our aim is to verify numerically the existence of a *continuous family* of equilibria for the system (1)–(3), and at the same time to demonstrate the robustness of the described numerical approach.

We take the particular set of parameters

$$\epsilon_u = \epsilon_v = 0.1, \quad \kappa_u = -\kappa_v = \lambda, \quad \alpha = 2, \quad \beta = 4, \quad \gamma = 0, \quad \delta = 0, \tag{37}$$

and vary the parameter λ to observe the transformation of the family. The computations were done on a grid with $N = 40$ nodes. The results on the grid $N = 20$ are practically the same. To display the computed families we will use a projection to the plane with coordinates

$$U = \sum_{j=1}^{N} u_j, \quad V = \sum_{j=1}^{N} v_j. \tag{38}$$

In Fig. 1, we present the families for a number of parameter values λ. It is clearly visible that the family of equilibria expands and shifts with increasing parameter λ, and its form changes. We observe symmetry of all curves with respect to the horizontal line $V = 0$. This is a consequence of a discrete symmetry which exists for the given choice of parameters. In fact, Eq. (1) is invariant under the reflection

$$\{x, u, v\} \mapsto \left\{\frac{a}{2} - x, u, -v\right\}. \tag{39}$$

In Fig. 2, we present the real parts of a number of greatest eigenvalues from the spectra of equilibria. We parameterized the curve by arc coordinate along the family on ([0,1]). One can see that the spectrum varies when the point moves along the family. Hence, the computed families are not the result of the action of a

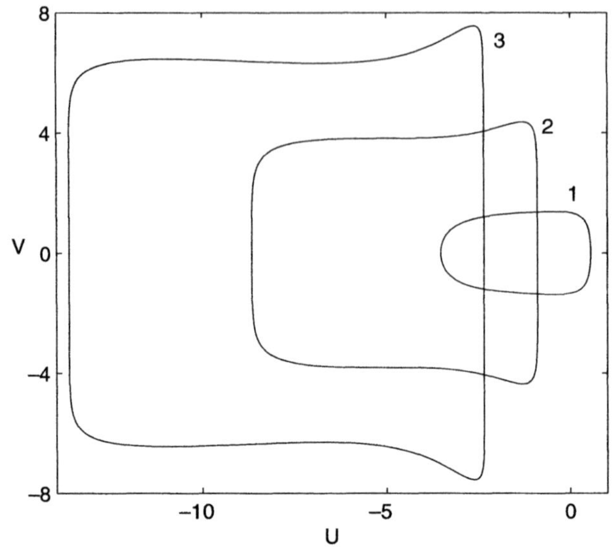

Fig. 1. Families of equilibria for different values λ

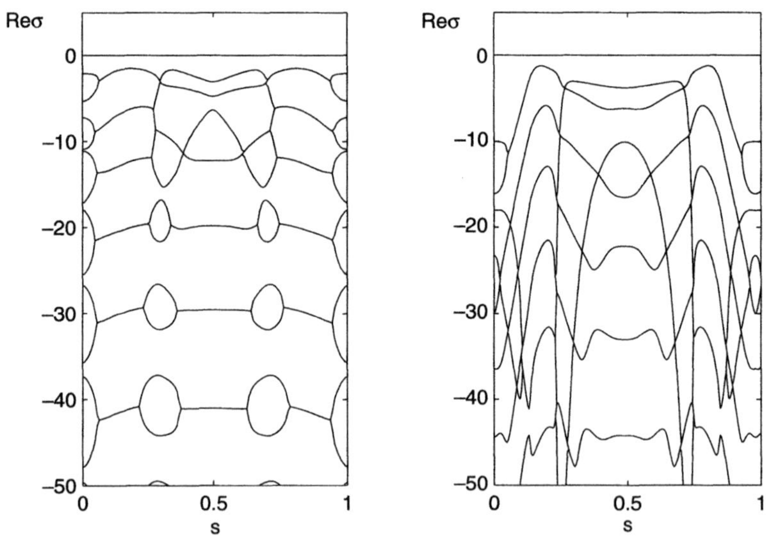

Fig. 2. Real part of spectrum for equilibria on the family; $\lambda = 1$ (left) and $\lambda = 2$ (right)

symmetry group. We also note the complication of the spectrum with increasing λ. All states from the families presented in Fig. 1 are transversally stable with respect to the family. There always exists a zero eigenvalue corresponding to the direction tangential to the family. Thus, the direction along the family of equilibria is a neutral one. The first instability on the family occurs at $\lambda = 7$.

In Figs. 3 and 4, we present the distributions of $u(x)$ and $v(x)$ for a number of stationary points on the family. One can see how the steady state varies. The solution for $u(x,t)$ may be symmetric with respect to the center of the interval $[0,a]$ (even function w.r.t. $a/2$), and the corresponding distribution of $v(x,t)$ is given by a skew-symmetric curve (odd function w.r.t. $a/2$).

Figure 5 demonstrates that the equilibria are really accessible through the solution of initial value problems. We assume the initial conditions

$$u_j = A\sin\left(\frac{\pi x_j}{a}\right), \quad v_j = B\sin\left(\frac{\pi x_j}{a}\right), \quad j = 1,\ldots,N,$$

and solve the system of ordinary differential equations (35) by a Runge-Kutta method on the time span $[0,4]$.

In Figs. 5–7, an initial point is marked by a big dot, a trajectory is displayed by a dotted line and a final position is marked by a circle. Because of the symmetry of

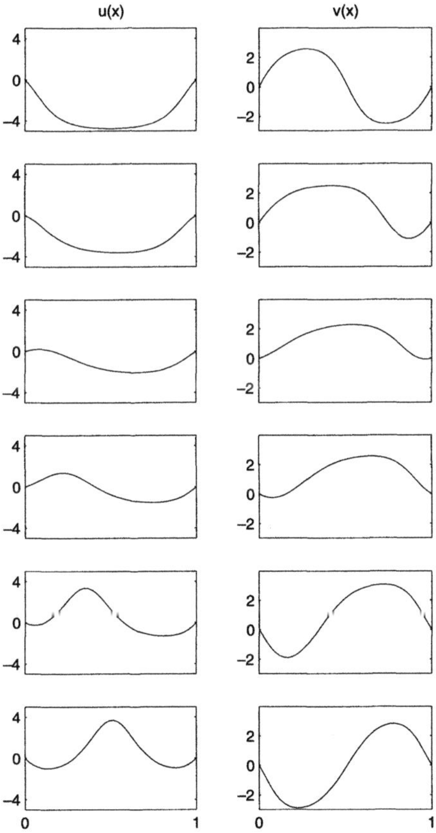

Fig. 3. Selected solutions from the family; $\lambda = 1$

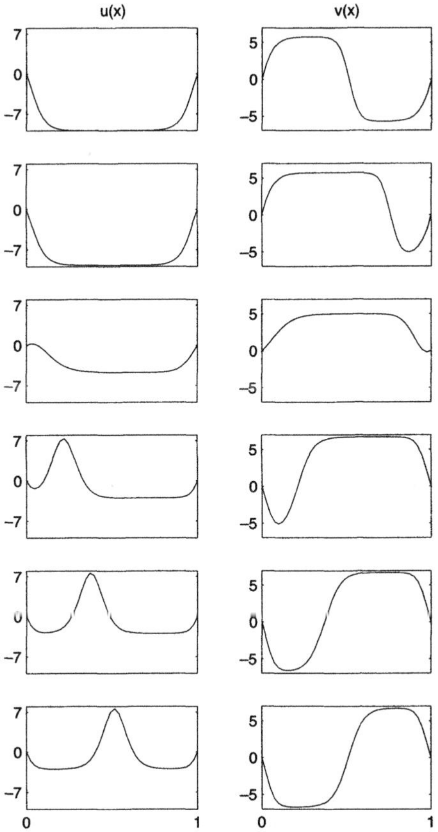

Fig. 4. Selected solutions from the family; $\lambda = 2$

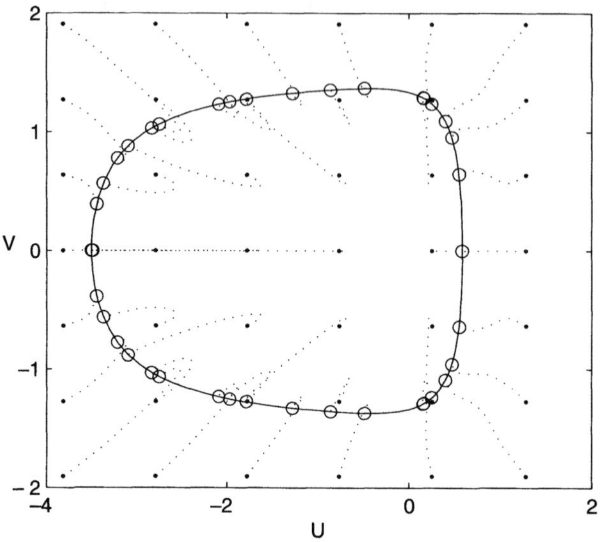

Fig. 5. Attraction to the family of equilibria

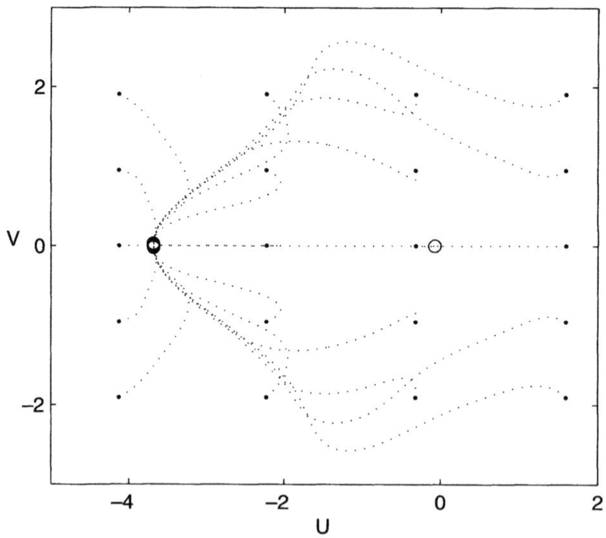

Fig. 6. Destruction of continuity due to loss of cosymmetry; $M_{22}^{v} = -2$

the initial distribution, the resulting phase portrait keeps the symmetry w.r.t. the line $V = 0$.

Note that deviation of parameters away from the requirements of Lemma 1 immediately leads to the destruction of the family – trajectories are attracted to a

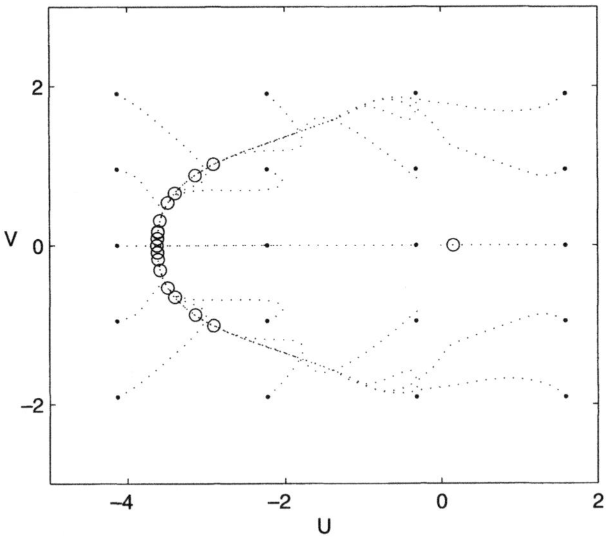

Fig. 7. Destruction of continuity due to loss of cosymmetry; $M_{22}^{v} = -4$

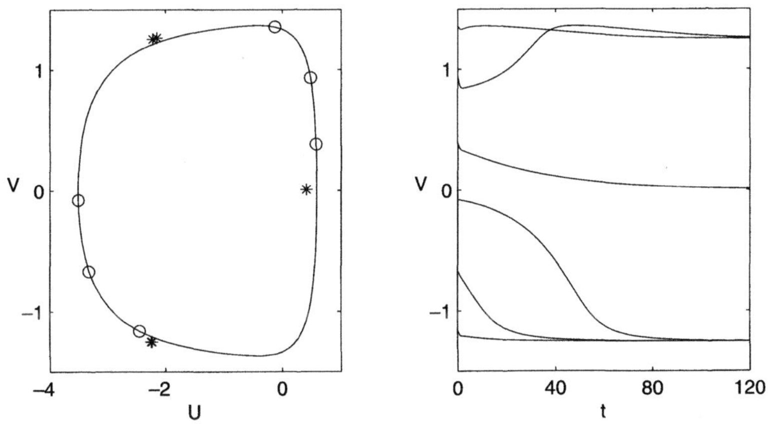

Fig. 8. Destruction of continuity by bad choice of FDM parameter $\zeta = 6$

limited number of isolated equilibrium points. This is illustrated in Figs. 6 and 7, where we use a value of M_{22}^{v} not equal to −6 as Lemma 1 would imply for the given set of parameters.

All trajectories in Fig. 6 ($M_{22}^{v} = -2$) are attracted to two stable isolated equilibria. It is interesting that the transient dynamics reproduces partially the continuous family curve. There is no more a continuous family – but it looks as if the system "remembered" it. We take time span $T = 6$ to get reasonably close to a steady situation in Fig. 6.

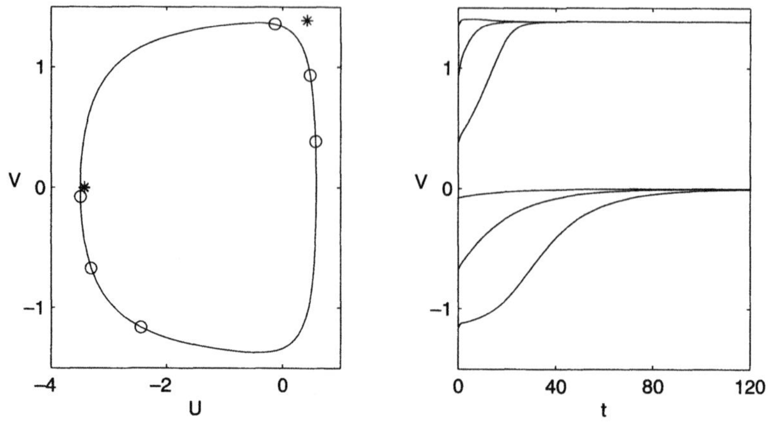

Fig. 9. Destruction of family due to bad FDM parameter $\zeta = -4$

In the case $M_{22}^{v} = -4$ the evolution is very slow, see Fig. 7, and one can observe a number of the points are situated practically on the (lost) "family" curve. Here we see that the time span $T = 9$ was insufficient to complete the transient process.

Figures 6 and 7 demonstrate that study of cosymmetry may be very useful for nearly cosymmetrical cases for which the system has only a finite number of isolated equilibria. It provides a better understanding of the dynamics and explains its essential features.

Another reason for the loss of continuity of the family of equilibria may be an inappropriate finite-difference approximation. To demonstrate that we put $\zeta = 6$ instead of correct value $\zeta = 1$ which preserves the cosymmetry. We take a number points from the family as initial conditions and integrate the system on the time interval $[0, 120]$ up to relaxation.

In the left part of Fig. 8, the initial and final states are marked by circles and stars, respectively. The right part of Fig. 8 gives the dependence of the variable V on t. Note that the transient process takes a long time, and moreover, that the trajectories are located close to the family.

Our computations display three stable equilibria in the case of non-adequate approximation of nonlinear terms. It is simple to check that the linear function (7) is not a cosymmetry now, although it vanishes on isolated equilibria. We also detect that there exist three unstable equilibria.

The equilibrium with $V = 0$ has a symmetric distribution on $u(x, t)$ and an anti-symmetric one on $v(x, t)$ as shown in the last row in Fig. 3. Two other equilibria form a pair without symmetry, but one of them transforms to the other under reflection w.r.t. $a/2$ (mirror symmetric pair).

This way, the continuous family of equilibria is transformed to a finite number of isolated equilibria. The number and location of the isolated states depend on the deviation from the parameter $\zeta = 1$. We take the same initial points and $\zeta = -4$ and repeat the direct integration. One can see that the location of the equilibria is transformed in comparison with Fig. 8. In this case the destruction of the continuous family also leads to the appearance of three stable and three unstable equilibria lying in the vicinity of the curve corresponding to the lost family.

It should be mentioned that for ζ close to 1 we need a long-time integration to reach a stable equilibrium.

For the application in population kinetics, we note that a trivial equilibrium would assume constant population densities along the whole space. The results presented in this section show that for the special choice of parameters, stable distributions may occur while members of both groups move all the time due to diffusion – in order to level out the distributions. These diffusive fluxes, however, are balanced by the growth/decay caused by the flux of the respective other group, cf. Figs. 3 and 4.

5. Conclusion

Known examples of cosymmetric systems [1]–[10] include the planar problem of filtration of an incompressible fluid in a saturated porous medium, the abstract Darcy model, and some applications to analytical mechanics.

Here we considered a population model with two species of quite different behavior. A system of two parabolic differential equations was formulated and continuous non-symmetric families of equilibrium states were computed. We intend to extend the proposed model to the case of two and three space dimensions.

Loss of cosymmetry either from inappropriate parameter choice or bad discretization may lead to dynamics without a continuous family of equilibria. However, in a certain sense, the behavior of the systems stays close to one with cosymmetry. Thus, the investigation of the family of equilibria brings insight to the dynamics of the system.

Acknowledgements

V. Ts. thanks Rostock University for hospitality during visits in 2001–2002, the Russian Foundation for Basic Research (grant no. 01-01-22002) and the program "Russian Universities – Fundamental Research" (project no. UR.04.01.063) for partial support.

References

[1] Yudovich, V. I.: Cosymmetry, degeneracy of the solutions of operator equations, and the onset of filtrational convection. Matem. Zametki *49*, 142–160 (1991).

[2] Yudovich, V. I.: Secondary cycle of equilibria in a system with cosymmetry, its creation by bifurcation and impossibility of symmetric treatment of it. Chaos *5*, 402–441 (1995).

[3] Bratsun, D. A., Lyubimov, D. V., Roux, B.: Co-symmetry breakdown in problems of thermal convection in porous medium. Physica D *82*, 398–417 (1995).

[4] Yudovich, V. I.: Implicit function theorem for cosymmetric equations. Matem. Zametki *60*, 313–317 (1996).

[5] Govorukhin, V. G.: Numerical simulation of the loss of stability for secondary steady regimes in the Darcy plane-convection problem. Doklady Akademii Nauk *363*, 806–808 (1998).
[6] Karasözen, B., Tsybulin, V. G.: Finite-difference approximations and cosymmetry conservation in filtration convection problem. Physics Letters *A 262*, 321–329 (1999).
[7] Govorukhin, V. G.: Calculation of one-parameter families of stationary regimes in a cosymmetric case and analysis of plane filtrational convection problem. In: Continuation methods in fluid dynamics (Aussois, ed., 1998), Notes Numer. Fluid Mech. *74*, 133–144. Braunschweig: Vieweg 2000.
[8] Kurakin, L. G., Yudovich, V. I.: Bifurcations accompanying monotonic instability of an equilibrium of a cosymmetric dynamical system. Chaos *10*, 311–330 (2000).
[9] Karasözen, B., Tsybulin, V. G.: Conservative finite difference schemes for cosymmetric systems. Proc. 4th Conf. on Computer Algebra in Scientific Computing. Springer 2001, pp 363–375.
[10] Yudovich, V. I.: Cosymmetry and conservative systems. Preprint VINITI, 2001 (in Russian).
[11] Courant, R., Hilbert, D.: Methods of mathematical physics, Vol. 1. New York: Intersience 1953.
[12] Ganzha, V. G., Vorozhtsov, E. V.: Numerical solutions for partial differential equations. Problem solving using mathematica. New York London: CRC Press, Boca Raton 1996.

K. Frischmuth
Department of Mathematics
University of Rostock
Rostock
Germany
e-mail: kurt.frischmuth@mathematik.uni-rostock.de

V. G. Tsybulin
Department of Mathematics and Mechanics
Rostov State University
344090 Rostov on Don
Russia
e-mail: tsybulin@math.rsu.ru

Computing [Suppl] 16, 83–105 (2002)

Computing

Quadratic Convergence of Scaled Iterates by Kogbetliantz Method

V. Hari and **J. Matejaš**

Abstract

A sharp quadratic convergence bound of scaled iterates for the serial Kogbetliantz method is derived. Iterates are symmetrically scaled by diagonal matrices so that diagonal elements are ones. The result is obtained for a scaled diagonally dominant complex triangular matrix with multiple singular values. The estimate depends on the relative separation of the singular values.

AMS Subject Classification: 65F15, 65G05.

Keywords: Kogbetliantz method, scaled matrices, quadratic convergence.

1. Introduction

Kogbetliantz method for computing SVD of triangular matrices is a reliable simple method well suited for ordinary [9], [3] and parallel [9], [13] computers. Our numerical tests give directions to conclusion that it is relatively accurate at least for matrices of the form DA, AD and DAD where D is diagonal and A is well conditioned. Using recent estimates [19] on the relative distance between diagonal elements of a scaled diagonally dominant triangular matrix and the corresponding singular values, one can conclude that the off-norm of the symmetrically scaled iterate is a good measure for advancing of the algorithm. The phenomenon is quite similar to the known one (see [4]) for the standard Jacobi method for Hermitian positive definite matrices. The stopping criterion based on the new measure can ensure the relative accuracy of the computed singular values and sometimes (e.g. graded matrices) it can terminate the process sooner than the classical stopping criterion.

These observations have motivated a research of scaled iterates by diagonalization methods (see [16], [17], [18], [20]) and this paper is part of that program. In [20] we have obtained usable estimates of scaled iterates by Kogbetliantz method and have made a sketch of the quadratic convergence proof in the case of simple singular values. Here we provide a detailed quadratic convergence proof in the general situation of multiple singular values. The proof is valid for the case of a complex triangular initial matrix and the complex Kogbetliantz algorithm. Here we use the technique from [18], [16], hence the reader is suggested to be acquainted with [17], [18] (or [16]) and [20].

The paper is organized as follows. In Sect. 1, we present recent results on the structure of scaled diagonally dominant triangular matrices which use relative gaps in the set of singular values. In Sect. 2, we shortly describe the Kogbetliantz method and in Sect. 3, we provide some preparatory results. In Sect. 4, we formulate the asymptotic assumptions and provide more auxiliary results. In Sect. 5, we prove the main result, Theorem 5.1 and in Sect. 6, we provide some numerical examples.

2. Structure of Scaled Almost Diagonal Matrices

Let $\mathbf{C}^{n\times n}$ denote the set of complex n by n matrices and let $B \in \mathbf{C}^{n\times n}$. Let

$$\sigma_1 = \sigma_2 = \cdots = \sigma_{s_1} > \sigma_{s_1+1} = \cdots = \sigma_{s_2} > \cdots > \sigma_{s_{p-1}+1} = \cdots = \sigma_{s_p} \tag{1.1}$$

be the singular values of B. Then p is the number of distinct singular values of B and for each i, $1 \le i \le p$, $n_i = s_i - s_{i-1}$ ($s_0 = 0$, $s_p = n$) is the multiplicity of σ_{s_i}. Let us define the sets of indices

$$\mathcal{N}_i = \{t \in \mathbf{N} :\ s_{i-1} + 1 \le t \le s_i\}, \quad 1 \le i \le p. \tag{1.2}$$

According to the partition $n = n_1 + \cdots + n_p$ we define for every $X \in \mathbf{C}^{n\times n}$ the block partition

$$X = \begin{bmatrix} X_{11} & \dots & X_{1p} \\ \vdots & & \vdots \\ X_{p1} & \dots & X_{pp} \end{bmatrix}, \qquad X_{ij} \in \mathbf{C}^{n_i\times n_j}, \quad 1 \le i,j \le p.$$

In the sequel we use the following notation: $\Omega(X) = X - \mathrm{diag}(X)$ is the off-diagonal part of X, $\pi(X) = \mathrm{diag}(X_{11}, \dots, X_{pp})$ is the block-diagonal and $\tau(X) = X - \pi(X)$ is the off-block-diagonal part of X. The i'th diagonal block and the corresponding off-diagonal block-row of X are denoted by

$$\pi_i(X) = X_{ii}, \quad \tau_i(X) = [X_{i1} \dots X_{i,i-1}\ X_{i,i+1} \dots X_{ip}], \quad 1 \le i \le p.$$

By $\|X\|_2$ and $\|X\|$ are denoted the spectral and the Frobenius norm of X respectively. For B with nonzero diagonal, the matrix

$$B_S = |\mathrm{diag}(B)|^{-1/2} B |\mathrm{diag}(B)|^{-1/2} \tag{1.3}$$

is referred to as the *scaled matrix* (we use here only symmetric scaling). Note that all diagonal elements of B_S have absolute value one.

If $\|\Omega(B_S)\| \le \alpha$ with $\alpha < 1$ then B is *α-scaled diagonally dominant* (α-s.d.d.) matrix with respect to norm $\|\cdot\|$ (see [1]).

The *relative gap function* rg of two real arguments is defined by (see [12])

$$\mathrm{rg}(a,b) = \begin{cases} \frac{|a-b|}{|a|+|b|}, & |a| + |b| > 0 \\ 0, & a = b = 0. \end{cases} \tag{1.4}$$

Relative gaps in the set of singular values of B are numbers

$$\theta_i = \min_{\substack{1 \le j \le p \\ j \ne i}} \mathrm{rg}(\sigma_{s_i}, \sigma_{s_j}), \quad 1 \le i \le p, \tag{1.5}$$

and the *minimum relative gap* is

$$\theta = \min_{1 \le i \le p} \theta_i. \tag{1.6}$$

Here we have used (1.1). Note that $\theta_i \le 1$ for all $1 \le i \le p$.

The structure of an α-s.d.d. matrix with multiple singular values is given in [19]. Here we consider only triangular and *essentially triangular* matrices (satisfying $b_{rt}b_{tr} = 0$, $r \ne t$). Therefore, we recall [19, Corollary 6] which reveals the special structure of an α-s.d.d. essentially triangular matrix[1].

Theorem 1.1. *Let $B \in \mathbf{C}^{n\times n}$ be essentially triangular matrix with positive diagonal elements. Let*

(a) $(\forall q \in \mathcal{N}_i)(\forall r \in \mathcal{N}_j) \quad i < j \Rightarrow b_{qq} > b_{rr}$,

(b) $\|\Omega(B_S)\|_2 < \frac{\theta}{3}$,

where $\mathcal{N}_i$, B_S and θ are defined by the relations (1.2), (1.3) *and* (1.6), *respectively. Then*

(i) $\sum_{t=s_{i-1}+1}^{s_i} |1 - \frac{\sigma_{s_i}}{b_{tt}}|^2 + \frac{1}{2}\|\Omega(\pi_i(B_S))\|_F^2 \le \frac{4}{\theta_i^2}$
$\left[\|\Omega(\pi_i(B_S))\|_F^2 + \|\tau_i(B_S)\|_F^2 + \|\tau_i(B_S{}^*)\|_F^2\right]^2, \quad 1 \le i \le p,$

(ii) $\sum_{t=1}^{n} |1 - \frac{\sigma_t}{b_{tt}}|^2 + \frac{1}{2}\|\Omega(\pi(B_S))\|_F^2 \le \frac{4}{\theta^2}\left[\|\Omega(B_S)\|_F^2 + \|\tau(B_S)\|_F^2\right] \cdot \|\Omega(B_S)\|_F^2$.

We see that for an α-s.d.d. matrix with $\alpha < \theta/3$, the off-diagonal elements of diagonal block $\pi_i(B_S)$ (which is affiliated with σ_{s_i}) are quadratically small with respect to the norm of the off-block-diagonal part of the matrix, $\|\tau(B_S)\|_F$. This structure has impact on the rate of convergence of scaled iterates by the Kogbetliantz method.

3. Kogbetliantz Method

Here we give a brief description of the serial Kogbetliantz method (see [14], [15], [7], [22]) for complex triangular matrices (cf. [9], [10]) with positive diagonal elements. Kogbetliantz method for complex triangular matrices has a nice property to maintain the reality (positivity) of the initially real (positive) diagonal elements. Since we consider scaled iterates, we shall assume positive diagonal elements which satisfy the condition (a) of Theorem 1.1.

[1] The original assumption $b_{11} \ge \cdots \ge b_{nn}$ is replaced here by the weaker one (a).

In general, nonnegative diagonal elements can easily be obtained by premultiplying a complex triangular matrix with a unitary diagonal matrix. Note that, as part of a SVD computation, QR factorization with column pivoting is often used as a pre-processing step (see also [5]). And it leads to a triangular matrix with real diagonal satisfying (up to signs) the condition (a) of Theorem 1.1. Note also that zero singular values of a triangular matrix can be deflated using Givens rotations.

The method we work with is actually the one from [20], but for the completeness of this paper we shortly describe it here.

Starting from a triangular matrix $B = (b_{lm})$, Kogbetliantz method generates a sequence of matrices $B^{(k)} = (b_{lm}^{(k)})$, $k \geq 0$, by the rule

$$B^{(k+1)} = (U^{(k)})^* B^{(k)} V^{(k)}, \qquad k \geq 0; \;\; B^{(0)} = B, \tag{2.1}$$

where $U^{(k)}$ and $V^{(k)}$, $k \geq 0$ are unitary plane matrices whose essential four elements are

$$(U^{(k)})_{ii} = (U^{(k)})_{jj} = \cos\varphi^{(k)}, \quad (U^{(k)})_{ij} = -\overline{(U^{(k)})_{ji}} = e^{\imath\omega_k}\sin\varphi^{(k)},$$

$$(V^{(k)})_{ii} = (V^{(k)})_{jj} = \cos\psi^{(k)}, \quad (V^{(k)})_{ij} = -\overline{(V^{(k)})_{ji}} = e^{\imath\vartheta_k}\sin\psi^{(k)}.$$

For all other elements, we have $(U^{(k)})_{lm} = (V^{(k)})_{lm} = \delta_{lm}$. Here, $\imath$ is the imaginary unit, $\bar{z}$ is the complex conjugate z and δ_{lm} is the Kronecker's delta. The 2×2 matrices obtained on the intersections of rows and columns i and j of $B^{(k)}$, $U^{(k)}$, $V^{(k)}$ are called *pivot submatrices* of these matrices. The pair of indices $(i, j) = (i(k), j(k))$ is called *pivot pair* and $b_{ij}^{(k)}$ is the *pivot element*. The equality in (2.1) defines the k'th *step* or the k'th *iteration* of the method. In the sequel, i and j will be reserved for pivot indices.

For $b_{ij}^{(k)} \neq 0$ the angles $\varphi^{(k)}$, $\psi^{(k)}$, ω_k and ϑ_k are determinated by the requirement

$$b_{ij}^{(k+1)} = b_{ji}^{(k+1)} = 0, \tag{2.2}$$

which ensures that $\|\Omega(B^{(k+1)})\|^2 = \|\Omega(B^{(k)})\|^2 - |b_{ij}^{(k)}|^2$, $k \geq 0$, holds. The relation (2.2) yields several sets of formulas for required angles, depending on whether $U^{(k)}$ or $V^{(k)}$ is calculated first (see [7], [22], [3], [10], [11]). Here we consider only the serial, that is, the column- and the row-cyclic pivot strategies. In the first (second) case, the off-diagonal elements are annihilated in column- (row-) wise fashion. Serial Kogbetliantz methods do not essentially ruin the triangular form. If B is upper-triangular, then within an odd (even) cycle the nontrivial off-diagonal elements push downwards (upwards) so that after each odd (even) cycle the current matrix is lower- (upper-) triangular. Similar situation is when B is lower-triangular. Within every cycle each current matrix is permutationally similar to a matrix in triangular form (PST), hence it is also essentially triangular (see [9]). Therefore, the whole process can be performed on an upper-triangular array (see [10], [11]). At step k, we can associate the upper-triangular matrix $G^{(k)} = (g_{lm}^{(k)})$ with the content of that array. Then, we have $G^{(k)} + G^{(k)T} = B^{(k)} + B^{(k)T}$, $k \geq 0$. In the following, as we mentioned earlier, we assume positive diagonal elements. We shall express the angle formulas in terms of the elements of $G^{(k)}$. If the right-hand transformation $V^{(k)}$ is computed first, the relation (2.2) implies

$$e^{\imath\vartheta_k} \tan 2\psi^{(k)} = \frac{2g_{ii}^{(k)} g_{ij}^{(k)}}{(g_{jj}^{(k)})^2 - (g_{ii}^{(k)})^2 + |g_{ij}^{(k)}|^2},$$

$$e^{\imath\omega_k} \tan \varphi^{(k)} = e^{\imath\vartheta_k} \frac{|g_{ij}^{(k)}| + g_{ii}^{(k)} \tan \psi^{(k)}}{g_{jj}^{(k)}} = \frac{g_{jj}^{(k)} e^{\imath\vartheta_k} \tan \psi^{(k)}}{g_{ii}^{(k)} - |g_{ij}^{(k)}| \tan \psi^{(k)}}.$$

Hence, we obtain the following angle formulas (cf. [22], [9], [10], [11]) for the first sweep:

$$\left.\begin{aligned}
e^{\imath\vartheta_k} &= \frac{g_{ij}^{(k)}}{|g_{ij}^{(k)}|}, \qquad \text{i.e. } \vartheta_k = \arg(g_{ij}^{(k)}) \\
\tan 2\psi^{(k)} &= \frac{2g_{ii}^{(k)} |g_{ij}^{(k)}|}{(g_{jj}^{(k)})^2 - (g_{ii}^{(k)})^2 + |g_{ij}^{(k)}|^2}, \qquad \psi^{(k)} \in \left[-\frac{\pi}{4}, \frac{\pi}{4}\right], \\
z_k &= \frac{g_{ij}^{(k)} + g_{ii}^{(k)} e^{\imath\vartheta_k} \tan \psi^{(k)}}{g_{jj}^{(k)}} = e^{\imath\vartheta_k} \frac{|g_{ij}^{(k)}| + g_{ii}^{(k)} \tan \psi^{(k)}}{g_{jj}^{(k)}} \\
&= \frac{g_{jj}^{(k)} e^{\imath\vartheta_k} \tan \psi^{(k)}}{g_{ii}^{(k)} - |g_{ij}^{(k)}| \tan \psi^{(k)}}, \\
e^{\imath\omega_k} &= \frac{z_k}{|z_k|}, \qquad \text{i.e. } \omega_k = \arg(z_k), \\
\tan \varphi^{(k)} &= |z_k|, \qquad \varphi^{(k)} \in \left[0, \frac{\pi}{2}\right].
\end{aligned}\right\} \tag{2.3}$$

Using convenient notation: $c_\varphi^{(k)} = \cos \varphi^{(k)}$, $s_\varphi^{(k)} = \sin \varphi^{(k)}$, $c_\psi^{(k)} = \cos \psi^{(k)}$, $s_\psi^{(k)} = \sin \psi^{(k)}$, the transformation formulas for the elements of $G^{(k)}$ are given by the relation (2.4). The transformation $G^{(k)} \to G^{(k+1)}$, defined by the relations (2.3) and (2.4) is referred to as *Kogbetliantz transformation* or *Kogbetliantz step* on the associated matrix $G^{(k)}$, with pivot pair (i, j). Shorter, we call it (associated) Kogbetliantz transformation or step. The algorithm (and the whole iterative process) on matrices $G^{(k)}$ is called associated Kogbetliantz algorithm (process) or shorter Kogbetliantz algorithm (process) on G.

$$\left.\begin{aligned} g_{li}^{(k+1)} &= c_{\varphi}^{(k)} g_{li}^{(k)} - e^{\imath\omega_k} s_{\varphi}^{(k)} g_{lj}^{(k)} \\ g_{lj}^{(k+1)} &= e^{-\imath\omega_k} s_{\varphi}^{(k)} g_{li}^{(k)} + c_{\varphi}^{(k)} g_{lj}^{(k)} \end{aligned}\right\} \quad 1 \le l \le i-1$$

$$\left.\begin{aligned} g_{il}^{(k+1)} &= c_{\psi}^{(k)} g_{il}^{(k)} - e^{-\imath\vartheta_k} s_{\psi}^{(k)} g_{lj}^{(k)} \\ g_{lj}^{(k+1)} &= e^{\imath\vartheta_k} s_{\psi}^{(k)} g_{il}^{(k)} + c_{\psi}^{(k)} g_{lj}^{(k)} \end{aligned}\right\} \quad i+1 \le l \le j-1$$

$$\left.\begin{aligned} g_{il}^{(k+1)} &= c_{\varphi}^{(k)} g_{il}^{(k)} - e^{\imath\omega_k} s_{\varphi}^{(k)} g_{jl}^{(k)} \\ g_{jl}^{(k+1)} &= e^{-\imath\omega_k} s_{\varphi}^{(k)} g_{il}^{(k)} + c_{\varphi}^{(k)} g_{jl}^{(k)} \end{aligned}\right\} \quad j+1 \le l \le n \tag{2.4}$$

$$g_{ii}^{(k+1)} = \frac{c_{\varphi}^{(k)}}{c_{\psi}^{(k)}} g_{ii}^{(k)} = \frac{|s_{\psi}^{(k)}|}{s_{\varphi}^{(k)}} g_{jj}^{(k)} = \frac{c_{\psi}^{(k)}}{c_{\varphi}^{(k)}} g_{ii}^{(k)} - \frac{s_{\psi}^{(k)}}{c_{\varphi}^{(k)}} |g_{ij}^{(k)}|$$

$$g_{jj}^{(k+1)} = \frac{c_{\psi}^{(k)}}{c_{\varphi}^{(k)}} g_{jj}^{(k)} = \frac{s_{\varphi}^{(k)}}{|s_{\psi}^{(k)}|} g_{ii}^{(k)} = \frac{c_{\varphi}^{(k)}}{c_{\psi}^{(k)}} g_{jj}^{(k)} + \operatorname{sgn}(s_{\psi}^{(k)}) \frac{s_{\varphi}^{(k)}}{c_{\psi}^{(k)}} |g_{ij}^{(k)}|$$

$$g_{ij}^{(k+1)} = 0.$$

Starting with upper-triangular B, the relation (2.4) holds for odd cycles only. For even cycles $\varphi^{(k)}, \omega_k$ and $\psi^{(k)}, \vartheta_k$ interchange their places. We can consider a method in which $U^{(k)}$ and $V^{(k)}$ interchange their places in subsequent cycles and thus ensure that (2.4) holds for every cycle.[2] But we need not be concerned about that since we shall use estimates from [20] which are symmetric in the arguments $\varphi^{(k)}$ and $\psi^{(k)}$. For this reason we do not consider the angle formulas when $U^{(k)}$ is calculated first.

Let us now consider the scaled iterates of the matrices $G^{(k)}$,

$$G_S^{(k)} = |\operatorname{diag}(G^{(k)})|^{-1/2} G^{(k)} |\operatorname{diag}(G^{(k)})|^{-1/2}, \quad k \ge 0. \tag{2.5}$$

The matrix $G_S^{(k)}$ is associated with the content of the upper-triangular array when $B_S^{(k)}$ is compactly stored in it. Next, we define

$$A^{(k)} = \Omega(G_S^{(k)} + G_S^{(k)T}) = \Omega(G_S^{(k)}) + \Omega(G_S^{(k)})^T, \quad k \ge 0. \tag{2.6}$$

Since each $B_S^{(k)}$ is essentially triangular, the relations (2.5) and (2.6) imply $A^{(k)} = \Omega(B_S^{(k)}) + \Omega(B_S^{(k)})^T$, $k \ge 0$. Note that $A^{(k)} = (a_{lm}^{(k)})$ is a (complex) symmetric matrix with zero diagonal, whose off-diagonal elements are given by

$$a_{lm}^{(k)} = a_{ml}^{(k)} = \frac{g_{lm}^{(k)}}{\sqrt{g_{ll}^{(k)} g_{mm}^{(k)}}} = \frac{b_{lm}^{(k)} + b_{ml}^{(k)}}{\sqrt{b_{ll}^{(k)} b_{mm}^{(k)}}}, \quad l \ne m, \ k \ge 0. \tag{2.7}$$

[2]The interchanges reflect only on singular vector updates.

The Frobenius norm of $\Omega(G_S^{(k)})$ shall be denoted by α_k,

$$\alpha_k = \|\Omega(G_S^{(k)})\| = \|\Omega(B_S^{(k)})\| = \frac{\sqrt{2}}{2}\|A^{(k)}\|, \quad k \geq 0. \tag{2.8}$$

This paper estimates the quadratic reduction of α_k within one cycle. To this end, we can presume that $k = 0$.

4. Preliminary Results

We recall several auxiliary results from [20] concerning nonorthogonal transformation $G_S^{(k)} \to G_S^{(k+1)}$ which do not depend on multiplicities of singular values. We assume that the initial matrix has positive diagonal elements.

Lemma 3.1. *Suppose the sequence $G^{(k)} = (g_{lm}^{(k)})$, $k \geq 0$ is obtained by the associated Kogbetliantz process, and suppose that $G^{(0)}$ has positive diagonal elements. For a fixed k, let us denote $G = (g_{lm}) = G^{(k)}$ and $\widetilde{G} = (\widetilde{g}_{lm}) = G^{(k+1)}$. Thus, $\widetilde{G}$ is obtained from G by a single Kogbetliantz transformation which annihilates the element g_{ij}. Let $A = (a_{lm})$ and $\widetilde{A} = (\widetilde{a}_{lm})$ be defined by $A = \Omega(G_S) + \Omega(G_S)^T$, $\widetilde{A} = \Omega(\widetilde{G}_S) + \Omega(\widetilde{G}_S)^T$, respectively.*

(i) *If $|a_{ij}| < 1$, then $|\widetilde{a}_{li}|^2 + |\widetilde{a}_{lj}|^2 \leq \frac{1+|a_{ij}|}{1-|a_{ij}|}(|a_{li}|^2 + |a_{lj}|^2)$, $\quad l \neq i,j$.*

(ii) *If $|a_{ij}| < 1$ and $\|\Omega(\widetilde{G}_S)\| > \|\Omega(G_S)\|$, then*

(a) $\|\Omega(\widetilde{G}_S)\|^2 - \|\Omega(G_S)\|^2 \leq |a_{ij}| \cdot \frac{2\|\Omega(G_S)\|^2 - |a_{ij}|}{1-|a_{ij}|}$ *and*

(b) $|a_{ij}| \leq 2\|\Omega(G_S)\|^2$.

Proof: See the proof of [20, Lemma 1 and Lemma 2]. □

The next lemma bounds the growth of the off-norm of scaled matrices which can temporarily increase during the process.

Lemma 3.2. *Let $G^{(0)} = G, G^{(1)}, \ldots, G^{(N)}$ be obtained by applying N Kogbetliantz steps to G and let α_k, $0 \leq k \leq N$ be defined by the relation (2.8). If*

$$\alpha_0 \leq \frac{1}{10n}, \quad n \geq 3,$$

then

$$\alpha_k^2 \leq c_k \alpha_0^2, \quad c_k = \left(1 + \frac{0.011}{n^2}\right)^k < 1.006 \quad 0 \leq k \leq N.$$

Proof: See the proof of [20, Lemma 3]. □

The preceeding lemmas hold for both the column- and the row-cyclic strategies. The next lemma estimates the growth of the affected part of the r-th column after $s(s-1)/2$ Kogbetliantz steps under the column-cyclic strategy. Let

$$\eta_{sr}^{(k)} = [a_{1r}^{(k)}, \ldots, a_{sr}^{(k)}]^T, \quad 1 \leq s < r \leq n, \ 0 \leq k \leq N \text{ and}$$

$$Q_s = 1 + 2 + \cdots + (s-1).$$

Lemma 3.3. *Let $G^{(0)} = G, G^{(1)}, \ldots, G^{(N)}$ be obtained by applying N Kogbetliantz steps to G under the column-cyclic strategy.*

$$\textit{If } \alpha_0 \leq \frac{1}{10n} \quad \textit{then} \quad \|\eta_{sr}^{(Q_s)}\|^2 \leq K_s \|\eta_{sr}^{(0)}\|^2,$$

where

$$K_s = [(1+\alpha)/(1-\alpha)]^{-(2s-3)} < 1.565 \quad \textit{and} \quad \alpha = \sqrt{1.006}\alpha_0. \tag{3.1}$$

Proof: See the proof of [17, 18, 20, Lemma 4]. Let

$$a = \max\left\{ \frac{|g_{i(k)j(k)}^{(k)}|}{\sqrt{g_{i(k)i(k)}^{(k)} g_{j(k)j(k)}^{(k)}}}, \quad 0 \leq k < Q_s \right\}.$$

By Lemma 3.2, $a \leq \sqrt{1.006}\alpha_0 = \alpha < 1$. According to Lemma 3.1(i) instead of the term $1/(1-a)$ in [18, Lemma 4], we now have $(1+a)/(1-a)$. Following the lines in the proof of [18, Lemma 4], we obtain

$$\|\eta_{sr}^{(Q_s)}\|^2 \leq \left(\frac{1+a}{1-a}\right)^{2s-3} \|\eta_{sr}^{(0)}\|^2. \tag{3.2}$$

By the assumption on α_0, we have $n \cdot 2a/(1-a) < 1$. Therefore, we can use the inequality $(1+x)^k \leq 1/(1-kx)$. We obtain

$$\left(\frac{1+a}{1-a}\right)^{2s-3} \leq \left(1 + \frac{2a}{1-a}\right)^{2(n-1)} \leq \left(1 - \frac{2(n-1)a}{1-a}\right)^{-2}$$
$$= \left(1 + \frac{2(n-1)a}{1+a-2na}\right)^2 \leq \left(1 + \frac{2na}{1-2na}\right)^2 < 1.565. \tag{3.3}$$

□

5. Asymptotic Assumptions

Here we define asymptotic assumptions for $B = (b_{lm})$ and prove some auxiliary results.

(B1) $B \in \mathbf{C}^{n\times n}$, $n \geq 3$ is upper triangular nonsingular matrix, satisfying

$$\alpha_0 \leq \frac{1}{10}\min\left\{\frac{1}{n}, \theta\right\},$$

where α_0 and θ are defined by the relations (2.8) and (1.6), respectively.

(B2) Diagonal elements of B are positive and satisfy

$$(\forall l \in \mathcal{N}_q)\ (\forall m \in \mathcal{N}_r)\ \ q < r \ \ \Rightarrow b_{ll} > b_{mm},$$

where the sets $\mathcal{N}_r$, $1 \leq r \leq p$ are defined by the relation (1.2).

The singular values of B are given by the relation (1.1) and their relative gaps by the relations (1.5) and (1.6). Since B is nonsingular all relative gaps are smaller than one. It is clear that the Assumptions (B1) and (B2) hold also for the matrix G $(= B)$.

Lemma 4.1. *Let $G = G^{(0)}$ satisfy the asymptotic Assumptions* (B1) *and* (B2). *Let the sequence $G^{(k)} = (g_{lm}^{(k)})$, $k = 0, 1, \ldots N$ be generated by the associated Kogbetliantz algorithm. Then for $0 \leq k \leq N$ holds*

(i) $\left(1 - \frac{\theta}{49}\right)\sigma_{s_r} < g_{tt}^{(k)} < \left(1 + \frac{\theta}{49}\right)\sigma_{s_r}, \quad t \in \mathcal{N}_r,\ 1 \leq r \leq p,$

(ii) $rg(g_{ll}^{(k)}, g_{mm}^{(k)}) > \frac{24}{25}\theta, \quad l \in \mathcal{N}_q,\ m \in \mathcal{N}_r,\ q \neq r,$

(iii) $|\tan 2\psi^{(k)}| \leq \frac{2|g_{ij}^{(k)}|}{g_{ii}^{(k)} - g_{jj}^{(k)}}, \quad i(k) \in \mathcal{N}_q,\ j(k) \in \mathcal{N}_r,\ q < r,$

(iv) $\max\{|\tan \psi^{(k)}|, |\tan \varphi^{(k)}|\} \leq \frac{|g_{ij}^{(k)}|}{g_{ii}^{(k)} - g_{jj}^{(k)}} \leq \frac{25}{48}\frac{|a_{ij}^{(k)}|}{\theta}, \ i(k) \in \mathcal{N}_q,\ j(k) \in \mathcal{N}_r,\ q < r,$

where $rg(\cdot,\cdot)$ and $a_{ij}^{(k)}$ are defined by the relations (1.4) *and* (2.7), *respectively.*

Proof: (i) We follow the proof of [18, Lemma 5(i)]. Note that (B1) and Lemma 3.2 ensure that condition (b) of Theorem 1.1 holds for all $0 \leq k \leq N$. (B2) ensures that condition (a) of Theorem 1.1 holds for $k = 0$. We use the induction over k. By Theorem 1.1(i) we have for $t \in \mathcal{N}_r$, $1 \leq r \leq p$, and some k,

$$1 - \frac{2\alpha_k^2}{\theta_r} \leq \frac{\sigma_{s_r}}{g_{tt}^{(k)}} \leq 1 + \frac{2\alpha_k^2}{\theta_r},$$

or equivalently

$$\sigma_{s_r}\left(1 - \frac{\frac{2\alpha_k^2}{\theta_r}}{1 + \frac{2\alpha_k^2}{\theta_r}}\right) \leq g_{tt}^{(k)} \leq \sigma_{s_r}\left(1 + \frac{\frac{2\alpha_k^2}{\theta_r}}{1 - \frac{2\alpha_k^2}{\theta_r}}\right).$$

Using Lemma 3.2 and the Assumption (B1) we obtain

$$\sigma_{s_r}\left(1 - \frac{2\cdot 1.006}{100}\theta\right) \leq g_{tt}^{(k)} \leq \sigma_{s_r}\left(1 + \frac{\frac{2\cdot 1.006}{100}}{1 - 2\cdot 1.006 \cdot \frac{1}{10}\cdot\frac{1}{30}}\theta\right),$$

which yields $g_{tt}^{(k)} \in \mathscr{D}_r$, $t \in \mathscr{N}_r$ where $\mathscr{D}_\tau = \{x:\ (1-\theta/49)\sigma_{s_\tau} < x < (1+\theta/49)\sigma_{s_\tau}\}$, $1 \le \tau \le p$. Using the definition (1.6) of θ we obtain for $r \neq q$,

$$\left(1-\frac{\theta}{49}\right)\max\{\sigma_{s_q},\sigma_{s_r}\} - \left(1+\frac{\theta}{49}\right)\min\{\sigma_{s_q},\sigma_{s_r}\}$$
$$\ge |\sigma_{s_q}-\sigma_{s_r}| - \frac{\theta}{49}(\sigma_{s_q}+\sigma_{s_r}) \ge \theta(\sigma_{s_q}+\sigma_{s_r}) - \frac{\theta}{49}(\sigma_{s_q}+\sigma_{s_r})$$
$$= (\sigma_{s_q}+\sigma_{s_r})\cdot\left(1-\frac{1}{49}\right)\theta = \frac{48}{49}\theta(\sigma_{s_q}+\sigma_{s_r}).$$

which yields

$$|x-y| \ge \tfrac{48}{49}\theta(\sigma_{s_q}+\sigma_{s_r}) \quad \text{whenever } x \in \mathscr{D}_q,\ y \in \mathscr{D}_r,\ q \neq r.$$

Using the obtained relations, in the same way as in the proof of [18, Lemma 5(i)], we obtain for $i = i(k) \in \mathscr{N}_q$, $j = j(k) \in \mathscr{N}_r$, $q < r$,

$$|g_{ii}^{(k+1)} - g_{ii}^{(k)}| \le |g_{ij}^{(k)}| = |a_{ij}^{(k)}|\cdot\sqrt{g_{ii}^{(k)}g_{jj}^{(k)}}$$
$$\le \sqrt{1.006}\alpha_0\cdot\left(1+\frac{\theta}{49}\right)\cdot\sqrt{\sigma_{s_r}\sigma_{s_q}}$$
$$\le \sqrt{1.006}\cdot\frac{\theta}{10}\cdot\frac{50}{49}\cdot\frac{1}{2}(\sigma_{s_r}+\sigma_{s_q})$$
$$\le \frac{3}{49}\cdot(\sigma_{s_r}+\sigma_{s_q}).$$

This yields implications $g_{ii}^{(k)} \in \mathscr{D}_q \Rightarrow g_{ii}^{(k+1)} \in \mathscr{D}_q$ and $g_{jj}^{(k)} \in \mathscr{D}_r \Rightarrow g_{jj}^{(k+1)} \in \mathscr{D}_r$. The rest of the proof of (i) follows the proof of [18, Lemma 5(i)].

(ii) The proof follows the lines of the proof of [18, Lemma 5(ii)] and uses (i).

(iii) To simplify notation we omit the superscript k. By (ii), Lemma 3.2 and the Assumption (B1), we have

$$\frac{\frac{|g_{ij}|^2}{g_{jj}}}{g_{ii}-g_{jj}} = \frac{\frac{|g_{ij}|^2}{g_{ii}g_{jj}}}{\frac{g_{ii}-g_{jj}}{g_{ii}+g_{jj}}\cdot\frac{g_{ii}+g_{jj}}{g_{ii}}} = \frac{|a_{ij}|^2}{rg(g_{ii},g_{jj})\cdot\left(1+\frac{g_{jj}}{g_{ii}}\right)} \le \frac{|a_{ij}|^2}{\frac{24}{25}\theta\cdot 1} < 1,$$

hence $|g_{ij}|^2/(g_{ii}-g_{jj}) < g_{jj}$ and (see the relation (2.3))

$$\tan 2\psi = \frac{2|g_{ij}|}{g_{jj}-g_{ii}}\cdot\frac{g_{ii}}{\frac{g_{jj}^2-g_{ii}^2+|g_{ij}|^2}{g_{jj}-g_{ii}}} = \frac{2|g_{ij}|}{g_{jj}-g_{ii}}\cdot\frac{g_{ii}}{g_{ii}+\left(g_{jj}-\frac{|g_{ij}|^2}{g_{ii}-g_{jj}}\right)},$$

which yields the assertion. Note that the Assumption (B2) yields $g_{ii} > g_{jj}$.

(iv) Using the assertion (iii), the first inequality for the angle ψ follows from $|\tan\psi| \leq \frac{1}{2}|\tan 2\psi|$. For the angle φ we use (ii) and the Assumption (B1),

$$\frac{|g_{ij}\tan\psi|}{g_{ii}-g_{jj}} \leq \frac{|g_{ij}|^2}{(g_{ii}-g_{jj})^2} = \frac{|a_{ij}|^2}{(\frac{g_{ii}-g_{jj}}{g_{ii}+g_{jj}})^2 \cdot \frac{(g_{ii}+g_{jj})^2}{g_{ii}g_{jj}}} \leq \frac{|a_{ij}|^2}{(\frac{24}{25}\theta)^2 \cdot 4} \leq \frac{|a_{ij}|^2}{\theta^2} < 1.$$

Hence

$$g_{jj} < g_{ii} - |g_{ij}\tan\psi| \leq g_{ii} - |g_{ij}|\tan\psi$$

and therefore

$$|\tan\varphi| \leq \left|\frac{g_{jj}}{g_{ii}-|g_{ij}|\tan\psi}\right| \cdot |\tan\psi| \leq |\tan\psi|,$$

which is the first inequality for the angle φ. In the rest of the proof we use $|g_{ij}|/(g_{ii}-g_{jj}) \leq |a_{ij}|/(2rg(g_{ii},g_{jj}))$ and the assertion (ii). □

6. The Main Theorem

Here we state and prove the quadratic convergence result.

Theorem 5.1. *Let $B \in \mathbf{C}^{n\times n}$ satisfy the asymptotic Assumptions* (B1) *and* (B2). *Let the sequence $B^{(0)} = B, B^{(1)}, \ldots, B^{(N)}$ be generated by the serial Kogbetliantz method. Then*

$$\alpha_N \leq \frac{13}{2} \cdot \frac{\alpha_0^2}{\theta},$$

where α_0, α_N and θ are defined by the relations (2.8) *and* (1.6), *respectively.*

This result is quite analogous to the one for the Hermitian Jacobi method (see [18], [17], [16]). Note that the classical result from [11] (see also [2], [8], [21], [6])

$$\|\Omega(B^{(N+1)})\|_F \leq 4\frac{\|\Omega(B)\|_F^2}{\delta} \quad \text{provided that } \Omega(B) < \frac{\delta}{3}, \tag{5.1}$$

where δ is minimum absolute gap in the set $\{-\sigma_{s_1}, \ldots, -\sigma_{s_p}, \sigma_{s_p}, \ldots, \sigma_{s_1}\}$, is either not valid or in praxis not applicable for tiny δ. In particular, if σ_{s_p} is tiny, then δ ($\leq 2\sigma_{s_p}$) is also tiny, while θ need not be tiny. Hence, the new result is especially applicable in cases when singular values cluster around zero.

Proof of Theorem 5.1

Because the row- and column-cyclic strategies are equivalent it suffices to prove the theorem for the column-cyclic strategy. The proof is based on the proof of [18, Theorem 6]. It uses the same technique: induction over the set $\{1,2,\ldots,p\}$.

Let $A^{(k)}=(a_{ij}^{(k)})$, $k\geq 0$ be defined by the relation (2.6). For $1\leq t\leq n$ we define the following submatrices of $A^{(k)}$, $k\geq 0$ (we recall notation from [18]),

$$\begin{aligned} M_t^{(k)} &= (a_{ij}^{(k)}), \quad 1\leq i,j\leq t, \quad M_t=M_t^{(0)}, \\ N_t^{(k)} &= (a_{ij}^{(k)}), \quad 1\leq i\leq t,\ 1\leq j\leq n, \quad N_t=N_t^{(0)}, \\ T_t^{(k)} &= (\tau_{ij}^{(k)}), \quad 1\leq i,j\leq t, \quad T_t=T_t^{(0)} \end{aligned}$$

where

$$\tau_{ij}^{(k)}=\begin{cases} a_{ij}^{(k)}, & 1\leq i\leq s_{r-1}<j\leq \min\{t,s_r\}, \quad 2\leq r\leq p \\ 0, & \text{otherwise} \end{cases}$$

and certain blocks of $A^{(k)}$, $k\geq 0$ (see the figure below):

$$\begin{aligned} F_r^{(k)} &= (a_{ij}^{(k)}), \quad 1\leq i\leq s_{r-1}<j\leq s_r, \quad F_r=F_r^{(0)}, \\ \overline{F}_r^{(k)} &= (a_{ij}^{(k)}), \quad s_{r-1}<j\leq s_r<i\leq n, \quad \overline{F}_r=\overline{F}_r^{(0)}, \\ G_r^{(k)} &= (a_{ij}^{(k)}), \quad 1\leq i\leq s_r, \quad \max\{i,s_{r-1}\}<j\leq s_r, \quad G_r=G_r^{(0)}. \end{aligned}$$

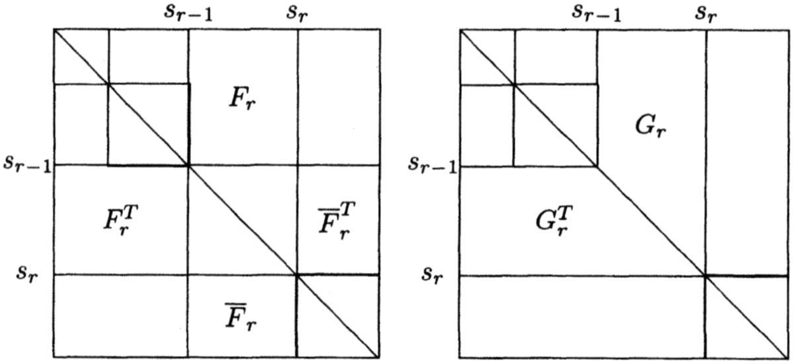

Using induction with respect to r we shall prove the key relation

$$\|T_{s_r}^{(Q_{s_r})}\|\leq C_r\cdot\frac{\|N_{s_r}\|^2}{\theta}, \quad 1\leq r\leq p, \tag{5.2}$$

where $Q_{s_r}=1+2+\cdots+(s_r-1)$, and

$$C_r=2.2\prod_{i=1}^{r}\left(1+50\frac{\|G_i\|^2}{\theta^2}\right)^{1/2}, \quad 1\leq r\leq p\,. \tag{5.3}$$

Since $\sum_{i=1}^{r} \|G_i\|^2/\theta^2 \leq \alpha_0^2/\theta^2$, the Assumption (B1) implies

$$C_r^2 \leq \frac{2.2^2}{1 - 50 \sum_{i=1}^{r} \frac{\|G_i\|^2}{\theta^2}} \leq \frac{2.2^2}{1 - 50 \cdot \frac{1}{100}} = 9.68, \quad 1 \leq r \leq p \, . \tag{5.4}$$

The proof of the relation (5.2) is divided into three parts. We shall follow the proof of [18, Theorem 6] and briefly elaborate the proof of Theorem 5.1.

The induction base (Part I): We prove that the relation (5.2) holds for $r = 1$. It is obvious because $T_{s_1}^{(Q_{s_1})}$ is null-matrix.

The induction hypothesis: We assume that the relation (5.2) holds for some $r - 1$, $1 < r \leq p$, i.e.

$$\|T_{s_{r-1}}^{(I)}\| \leq C_{r-1} \cdot \frac{\|N_{s_{r-1}}\|^2}{\theta}, \quad I := Q_{s_{r-1}} = 1 + 2 + \cdots + (s_{r-1} - 1). \tag{5.5}$$

The induction step (Parts II and III):

Part II: Annihilation of the elements in block F_r by columns. We shall prove

$$\|T_{s_{r-1}}^{(II)}\| \leq C_{r-1} \cdot \frac{\|N_{s_r}\|^2}{\theta}, \tag{5.6}$$

$$\|F_r^{(II)}\| \leq 13.5 \frac{\|N_{s_r}\|^2}{\theta^2} \|G_r\| \, , \tag{5.7}$$

where $II := Q_{s_{r-1}} + s_{r-1}(s_r - s_{r-1})$.

Proof of the relation (5.6)

We shall use the superscript w_m defined by

$$w_t = Q_{s_{r-1}} + (t - 1 - s_{r-1})s_{r-1}, \quad s_{r-1} + 1 \leq t \leq s_r. \tag{5.8}$$

Lemma 5.2. *Let B satisfy the assumptions of Theorem 5.1. Let w_m be defined by the relation (5.8). Then*

(i) $a_{km}^{(w_m+k)} = 0$, $1 \leq k \leq s_{r-1}$,

(ii) $|a_{lm}^{(w_m+k)}| \leq \sqrt{1.006}(|a_{lm}^{(w_m)}| + \frac{25}{240} \sum_{t=1}^{k} |a_{lt}^{(w_m)} a_{tm}^{(w_m+t-1)}|)$, $1 \leq k \leq s_{r-1}$, $k < l \leq m - 1$,

(iii) $|a_{lm}^{(w_m+k)}| \leq \frac{1.045}{\theta} \sum_{t=l+1}^{k} |a_{tm}^{(w_m+t-1)}| (|a_{lt}^{(w_m)}| + \frac{0.27}{\theta} |a_{lm}^{(w_m+l-1)} a_{tm}^{(w_m+l-1)}|)$, $1 \leq l < k \leq s_{r-1}$,

(iv) $|a_{lt}^{(w_m+s_{r-1})}| \leq |a_{lt}^{(w_m)}| + \frac{25}{48\theta}|a_{lm}^{(w_m+l-1)} a_{tm}^{(w_m+l-1)}|, \quad 1 \leq l \neq t \leq s_{r-1},$

(v) $|a_{lt}^{(w_m+s_{r-1})}| \leq |a_{lt}^{(w_m)}| + \frac{25}{48\theta}|a_{lm}^{(w_m+l-1)} a_{tm}^{(w_m+l-1)}|, \; 1 \leq l \leq s_{r-1} < t < m.$

Proof: In the proof we shall omit the superscript w_m. The assertion (i) is obvious because $a_{km}^{(k)}$ is annihilated pivot element. (ii) Using Lemma 4.1 (ii), (iv) the proof becomes analogous to the proof of [18, Lemma 8 (ii)]. The only difference arises in deriving the bound for $g_{mm}^{(t-1)}/g_{mm}^{(k)}$, $1 \leq t \leq k \leq s_{r-1}$. The bound will be obtained in the following way. Using Lemma 4.1 (ii), we have

$$\frac{|g_{km}^{(k-1)}|^2}{(g_{kk}^{(k-1)})^2 - (g_{mm}^{(k-1)})^2} = \frac{|a_{km}^{(k-1)}|^2}{\underbrace{\frac{(g_{kk}^{(k-1)})^2 - (g_{mm}^{(k-1)})^2}{(g_{kk}^{(k-1)} + g_{mm}^{(k-1)})^2}}_{\geq 24\theta/25 > \theta/2} \cdot \underbrace{\frac{(g_{kk}^{(k-1)} + g_{mm}^{(k-1)})^2}{g_{kk}^{(k-1)} g_{mm}^{(k-1)}}}_{\geq 4}}$$

$$\leq \frac{1}{2} \cdot \frac{|a_{km}^{(k-1)}|^2}{\theta} < 1 .$$

From the above relation we have

$$(g_{mm}^{(k-1)})^2 - (g_{kk}^{(k-1)})^2 + |g_{km}^{(k-1)}|^2 < 0,$$

which together with the relation (2.3) yields

$$|\tan\varphi^{(k-1)}| \leq \frac{g_{mm}^{(k-1)}}{g_{kk}^{(k-1)}} |\tan\psi^{(k-1)}| < |\tan\psi^{(k-1)}| \leq 1 \tag{5.9}$$

and

$$|\tan\psi^{(k-1)}| \leq \frac{g_{mm}^{(k-1)}}{g_{kk}^{(k-1)}} |\tan\varphi^{(k-1)}| + \frac{|g_{km}^{(k-1)}|}{g_{kk}^{(k-1)}} \leq |\tan\varphi^{(k-1)}| + |a_{km}^{(k-1)}|. \tag{5.10}$$

Using now the relations (2.4), (5.9), (5.10) and Lemma 4.1 (iv), we have

$$\begin{aligned} \frac{g_{mm}^{(k-1)}}{g_{mm}^{(k)}} &= \frac{\cos\varphi^{(k-1)}}{\cos\psi^{(k-1)}} = \sqrt{1 + \frac{\tan^2\psi^{(k-1)} - \tan^2\varphi^{(k-1)}}{1 + \tan^2\varphi^{(k-1)}}} \\ &< \sqrt{1 + \tan^2\psi^{(k-1)} - \tan^2\varphi^{(k-1)}} \\ &\leq 1 + \frac{1}{2}|\tan\psi^{(k-1)} + \tan\varphi^{(k-1)}| \cdot \left|\tan\psi^{(k-1)} - \tan\varphi^{(k-1)}\right| \\ &\leq 1 + \frac{1}{2} \cdot 2\left|\tan\psi^{(k-1)}\right| \cdot \left|a_{km}^{(k-1)}\right| \\ &\leq 1 + \frac{25}{48} \cdot \frac{\left|a_{km}^{(k-1)}\right|^2}{\theta} \leq 1 + \frac{25}{48} \cdot \frac{\alpha^2}{\theta}, \qquad 1 \leq k \leq s_{r-1}. \end{aligned}$$

Using the assumption (B1) and the relation (3.1), we obtain

$$\frac{g_{mm}^{(t-1)}}{g_{mm}^{(k)}} \leq \left(1+\frac{25\alpha^2}{48\theta}\right)^{k-t+1} < \left(1+\frac{25\alpha^2}{48\theta}\right)^{n} < \frac{1}{1-\frac{25}{48}n\alpha\cdot\frac{\alpha}{\theta}}$$
$$\leq \frac{1}{1-\frac{25\cdot 1.006}{48\cdot 100}} < 1.006, \quad 1 \leq t \leq k \leq s_{r-1}$$

which is the desired bound.

Using Lemma 4.1(ii) and Lemma 4.1(iv), and following the proof of [18, Lemma 8 (iii)], [18, Lemma 8 (iv)] and [18, Lemma 8 (v)], we obtain (iii), (iv) and (v), respectively. □

Using Lemma 5.2, we can estimate the elements of the m-th column prior and after annihilations. Let

$$\eta_m^{(k)} = [a_{1m}^{(k)}, \ldots, a_{s_{r-1},m}^{(k)}], \quad s_{r-1}+1 \leq m \leq s_r, \ 0 \leq k \leq N, \tag{5.11}$$

$$\rho_m = [a_{1m}, a_{2m}, \ldots, a_{m-1,m}], \quad s_{r-1}+1 \leq m \leq s_r. \tag{5.12}$$

Lemma 5.3. *Let B satisfy the assumptions of Theorem 5.1. Let w_m, $\eta_m^{(w_m)}$ and $M_{s_{r-1}}^{(w_m)}$ be defined by the relations (5.8), (5.11) and (5.12), respectively. Then*

(i) $\sum_{l=1}^{s_{r-1}} |a_{lm}^{(w_m+q_l)}|^2 \leq \mu_m^2 \|\eta_m^{(w_m)}\|^2, \quad 0 \leq q_l < l,$

(ii) $\sum_{l=1}^{s_{r-1}} |a_{lm}^{(w_m+q_l)}|^2 \leq \frac{0.55\mu_m^2}{\theta^2} \|\eta_m^{(w_m)}\|^2, \quad l \leq q_l \leq s_{r-1},$

where

$$\mu_m = \frac{\sqrt{1.006}}{1-\frac{0.74}{\theta}\|M_{s_{r-1}}^{(w_m)}\|}. \tag{5.13}$$

Proof: Using the assertions (ii) and (iii) of Lemma 5.2, we can prove the assertions (i) and (ii) in the same way as it is done in [18, Lemma 9]. □

Lemma 5.4. *Let B satisfy the assumptions of Theorem 5.1. Then for $s_{r-1}+1 \leq m \leq s_r$ hold*

(i) $\|M_{s_{r-1}}^{(w_m+s_{r-1})}\| \leq \|M_{s_{r-1}}^{(w_m)}\| + \frac{25\sqrt{2}\mu_m^2\|\eta_m^{(w_m)}\|^2}{48\theta} \cdot [1+\frac{\sqrt{0.55}}{\theta}(\|M_{s_{r-1}}^{(w_m)}\| + \frac{0.39\mu_m^2}{\theta}\|\eta_m^{(w_m)}\|^2)],$

(ii) $\|T_{s_{r-1}}^{(w_m+s_{r-1})}\| \leq \|T_{s_{r-1}}^{(w_m)}\| + \frac{25\mu_m^2\|\eta_m^{(w_m)}\|^2}{48\theta} \cdot [1+\frac{\sqrt{0.55}}{\theta}(\|M_{s_{r-1}}^{(w_m)}\| + \frac{0.39\mu_m^2}{\theta}\|\eta_m^{(w_m)}\|^2)],$

where w_m, $\eta_m^{(k)}$ and μ_m are defined by the relations (5.8), (5.11) *and* (5.13), *respectively.*

Proof: Using Lemma 5.2(iv) and Lemma 5.3, we just follow the lines in the proof of [18, Lemma 10]. □

Lemma 5.5. *Let B satisfy the assumptions of Theorem* 5.1. *Let $s_{r-1}+1 \le q < m \le s_r$ and let w_m, $\eta_m^{(k)}$ and μ_m be defined by the relations* (5.8), (5.11) *and* (5.13). *Then*

(i) $|a_{qm}^{(w_m+l)}| \le \sqrt{1.006}(|a_{qm}^{(w_m)}| + \frac{25\mu_m}{24\theta}\|\eta_q^{(w_m)}\| \cdot \|\eta_m^{(w_m)}\|)$,

(ii) $|a_{qm}^{(w_m)}| = |a_{qm}^{(w_{q+1})}| \le \sqrt{1.006}(|a_{qm}^{(w_q)}| + \frac{25\mu_q}{24\theta}\|\eta_q^{(w_q)}\| \cdot \|\eta_m^{(w_q)}\|)$.

Proof: Using Lemma 5.2(ii) and Lemma 5.3(i), the proof is analogous to the proof of [18, Lemma 11]. □

Lemma 5.6. *Let B satisfy the assumptions of Theorem* 5.1. *If the hypothesis* (5.5) *holds, then*

$$\|M_{s_{r-1}}^{(I)}\| \le 9.57\frac{\|N_{s_{r-1}}\|^2}{\theta}.$$

Proof: Following the proof of [18, Lemma 12], we apply Theorem 1.1(i) to bound $\sum_{t=1}^{r-1}\|A_{tt}^{(I)}\|^2$. Since by Lemma 3.2, the matrices $B^{(k)}$ and $G^{(k)}$ are α-s.d.d. ($\alpha = \sqrt{1.006}\alpha_0$) for all $0 \le k \le N$, we have for $1 \le t \le p$, $0 \le k \le N$

$$\frac{2}{\theta}\|A_{tt}^{(k)}\| \le \frac{4}{\theta^2}\left[\|A_{tt}^{(k)}\|^2 + 2\|F_t^{(k)}\|^2 + 2\|\overline{F}_t^{(k)}\|^2\right] \le \frac{8}{\theta^2}\alpha^2 . \tag{5.14}$$

Using Lemma 3.2 and the Assumption (B1), we obtain for $0 \le k \le N$,

$$\|A_{tt}^{(k)}\| \le \frac{1}{1-\frac{2}{\theta}\|A_{tt}^{(k)}\|}\cdot\frac{4}{\theta}\left[\|F_t^{(k)}\|^2 + \|\overline{F}_t^{(k)}\|^2\right] \le 1.088\cdot\frac{4}{\theta}\left[\|F_t^{(k)}\|^2 + \|\overline{F}_t^{(k)}\|^2\right]. \tag{5.15}$$

Using the induction hypothesis (the relation (5.5)), Lemma 3.3 and the relation

$$\frac{\|N_{s_{r-1}}\|^2}{\theta^2} \le \frac{2\alpha_0^2}{\theta^2}, \tag{5.16}$$

we obtain (cf. the proof of [18, Lemma 12])

$$\|M_{s_{r-1}}\|^2 \le \left[2C_{r-1}^2 + 1.088^2\cdot 16(4\frac{\alpha_0^2}{\theta^2}C_{r-1}^2 + 1.565)^2\right]\cdot\frac{\|N_{s_{r-1}}\|^4}{\theta^2}$$

which, together with the relation (5.4), implies the assertion. □

Lemma 5.7. *Let B satisfy the assumptions of Theorem* 5.1. *Let w_m and μ_m be defined by the relations* (5.8) *and* (5.13), *respectively. If the hypothesis* (5.5) *holds, then for $s_{r-1}+1 \le m \le s_r$, we have*

(i) $\mu_m \leq \mu = 1.169$,

(ii) $\|M_{s_{r-1}}^{(w_m+1)}\| \leq 9.57 \frac{\|N_m\|^2}{\theta}$,

(iii) $\|T_{s_{r-1}}^{(w_m+1)}\| \leq C_{r-1} \frac{\|N_m\|^2}{\theta}$.

Proof: See the proof of [18, Lemma 13]. Induction with respect to $m, s_{r-1} + 1 \leq m \leq s_r$ is used.

(i) The relations (5.13), (5.16), Lemma 5.6 and the Assumption (B1) are used.

(ii) The assertion (i), the lemmas 5.4(i), 3.3, 5.6, the Assumption (B1) and the relation (5.16) are used.

(iii) Lemma 5.4 (ii) and the relations (5.5) and (5.4) are used. □

The assertion (iii) of Lemma 5.7 implies (by taking $m = s_r$) the relation (5.6).

Proof of the relation (5.7)

Lemma 5.8. *Let B satisfy the assumptions of Theorem* 5.1. *Let* $s_{r-1} + 1 \leq q \leq s_r$ *and let* w_m, $\eta_m^{(k)}$ *and* ρ_m *be defined by the relations* (5.8), (5.11) *and* (5.12), *respectively. If the hypothesis* (5.5) *holds, then*

$$\|\eta_q^{(w_{s_r}+1)}\| \leq 1.012\|\eta_q^{(w_q+1)}\| + \frac{0.774}{\theta} \sum_{m=q+1}^{s_r} |a_{qm}^{(w_q+1)}| \cdot \|\rho_m\|.$$

Proof: Following the lines in the proof of [18, Lemma 14] and using Lemma 5.2 (v), Lemma 5.3 (i), Lemma 5.5 (i), Lemma 5.7 (i) and Lemma 3.3, we obtain

$$\|\eta_q^{(w_m+1)}\| \leq z_m\|\eta_q^{(w_m)}\| + z|a_{qm}^{(w_q+1)}| \cdot \|\rho_m\|, \quad q+1 \leq m \leq s_r,$$

where

$$z_m = 1 + \frac{25 \cdot 25 \cdot \sqrt{1.006} \cdot \mu^2 \cdot 1.565}{48 \cdot 24} \cdot \frac{\|\rho_m\|^2}{\theta^2} \leq 1 + 1.164 \frac{\|\rho_m\|^2}{\theta^2},$$

$$z = \frac{25\mu\sqrt{1.006} \cdot 1.565}{48\theta} \leq \frac{0.764}{\theta}.$$

The rest of the proof is analogous to the proof of [18, Lemma 14] with the only exception that $\sum_{m=1}^{n} \|\rho_m\|^2$ is now bounded by α_0^2/θ^2 (instead of $\alpha_0^2/(2\gamma^2)$). □

Lemma 5.9. *Let B satisfy the assumptions of Theorem* 5.1. *If the hypothesis* (5.5) *holds, then the relation* (5.7) *holds, too.*

Proof: See the proof of [18, Lemma 15]. Using the relation (5.14), we have

$$\|A_{rr}\| \leq \frac{4}{\theta}\|N_{s_r}\|^2. \tag{5.17}$$

Now, using Lemma 5.3(ii), Lemma 5.7(i),(ii) and Lemma 3.3, we obtain

$$\begin{aligned}\|\eta_q^{(w_{q+1})}\|^2 &\leq \frac{0.55\cdot 1.169^2}{\theta^2}\cdot 1.565\|\rho_q\|^2\left(9.57\frac{\|N_q\|^2}{\theta}\right)^2 \\ &\leq 107.73\cdot\frac{\|\rho_q\|^2}{\theta^2}\cdot\frac{\|N_{s_r}\|^4}{\theta^2}, \quad s_{r-1}+1\leq q\leq s_r.\end{aligned}$$

Using Lemma 5.5(ii), Lemma 5.7(i) and Lemma 3.3, we have

$$\sum_{q=s_{r-1}+1}^{s_r}\sum_{m=q+1}^{s_r}|a_{qm}^{(w_{q+1})}|^2 \leq \sqrt{1.006}\left(\frac{\|A_{rr}\|}{\sqrt{2}}+\frac{25\mu\cdot 1.565}{48\theta}\|G_r\|^2\right)^2.$$

Combining the obtained expressions with Lemma 5.8 and the relation (5.17) as in the proof of [18, Lemma 15], we obtain the desired assertion. □

Part III: Annihilation of the elements in block A_{rr} by columns.

We shall prove

$$T_{s_{r-1}}^{(III)} = T_{s_{r-1}}^{(II)}, \tag{5.18}$$

$$\|F_r^{(III)}\| \leq 1.07\|F_r^{(II)}\|, \tag{5.19}$$

where $III := Q_{s_r} = 1+2+\cdots+(s_r-1)$.

The relation (5.18) is obvios because the matrix $T_{s_{r-1}}^{(II)}$ does not change at this stage. For $s_{r-1}+2\leq m\leq s_r$ we define index

$$v_m = Q_{s_{r-1}} + (s_r - s_{r-1})s_{r-1} + (m - s_{r-1} - 1)(m - s_{r-1} - 2)/2. \tag{5.20}$$

The index v_m corresponds to the stage just before the annihilations in the mth column within A_{rr} take place.

Lemma 5.10. *Let B satisfy the assumptions of Theorem* 5.1. *Then*

$$\|F_r^{(v_{s_r+1})}\| \leq 1.07\cdot\|F_r^{(v_{s_{r-1}+2})}\|,$$

where v_m is defined by the relation (5.20).

Proof: See the proof of [18, Lemma 16]. Using the relation (5.14), we have

$$a \leq \max_{k,m}\frac{\|A_{rr}^{(v_m+k)}\|}{\sqrt{2}} \leq \frac{1}{\sqrt{2}}\cdot\frac{4}{\theta}\cdot\alpha^2.$$

Using the relations (3.2) and (3.3), we obtain

$$\left(\frac{1+a}{1-a}\right)^{2s-3} \leq \left(1+\frac{\sqrt{2}\cdot 4\cdot n\alpha\cdot \alpha/\theta}{1-\sqrt{2}\cdot 4\cdot n\alpha\cdot\alpha/\theta}\right)^2$$
$$< \left(1+\frac{\sqrt{2}\cdot 4\cdot 1.006\cdot 0.01}{1-\sqrt{2}\cdot 4\cdot 1.006\cdot 0.01}\right)^2 < 1.07^2,$$

which together with (3.1) and the assumption (B1) proves the lemma. □

Note that Lemma 5.10 proves the relation (5.19). This completes Part III.

Using now the relations (5.18), (5.19), (5.6), (5.7) and (5.3), the induction step is quite analogous to that in [18].

Completing the proof of Theorem 5.1

Using the relation (5.15), we have

$$\sum_{t=1}^{p} \|A_{tt}^{(N)}\|^2 \leq 1.088^2\cdot\frac{16}{\theta^2}\sum_{t=1}^{p}[\|F_t^{(N)}\|^2+\|\overline{F}_t^{(N)}\|^2]^2$$
$$\leq 1.088^2\cdot\frac{16}{\theta^2}\cdot\|T_{s_p}^{(N)}\|^2\sum_{t=1}^{p}[\|F_t^{(N)}\|^2+\|\overline{F}_t^{(N)}\|^2]^2$$
$$\leq 1.088^2\cdot\frac{16}{\theta^2}\cdot 2\|T_{s_p}^{(N)}\|^4 \;<\; 37.88\cdot\frac{\|T_{s_p}^{(N)}\|^4}{\theta^2}.$$

Finally, using the relation (5.4), one obtains

$$\alpha_N^2 = \|T_{s_p}^{(N)}\|^2+\frac{1}{2}\sum_{t=1}^{p}\|A_{tt}^{(N)}\|^2 \leq \|T_{s_p}^{(N)}\|^2+\frac{1}{2}\cdot 37.88\cdot\frac{\|T_{s_p}^{(N)}\|^4}{\theta^2}$$
$$\leq \|T_{s_p}^{(N)}\|^2\left(1+18.94\cdot\frac{\|T_{s_p}^{(N)}\|^2}{\theta^2}\right) \leq C_p^2\,\frac{\|N_{s_p}\|^4}{\theta^2}\left(1+18.94\cdot C_p^2\cdot\frac{\|N_{s_p}\|^4}{\theta^4}\right)$$
$$\leq 9.68\cdot\frac{(2\alpha_0^2)^2}{\theta^2}\left(1+18.94\cdot 9.68\cdot\frac{(2\alpha_0^2)^2}{\theta^4}\right) \leq 38.72(1+183.34\cdot 4\cdot 0.1^4)\cdot\frac{\alpha_0^4}{\theta^2}$$
$$< \left(\frac{13}{2}\cdot\frac{\alpha_0^2}{\theta}\right)^2,$$

which completes the proof of Theorem 5.1. □

7. Some Numerical Examples

We have performed some numerical tests. Test matrices have been generated in several ways and here we describe the simplest one. For given n, the upper-triangular matrix $R(n) = (r_{ij}^{(n)})$ is generated by "wrapping" the sequence of natural numbers: $1, 2, 3, \ldots$ onto the superdiagonals of $R(n)$, starting with the last one.

Thus, $r_{1n}^{(n)}=1$, $r_{1,n-1}^{(n)}=2$, $r_{2n}^{(n)}=3,\ldots$, $r_{11}^{(n)}=M_{n-1}+1,\ldots$, $r_{nn}^{(n)}=M_n$, where $M_n=n(n+1)/2$. Then $R(n)$ is scaled as $T(n)=D(n)R(n)$, $T(n)=R(n)D(n)$ or $T(n)=[D(n)]^{\frac{1}{2}}R(n)[D(n)]^{\frac{1}{2}}$ where diagonal matrix $D(n)$ is defined by the vector $d(n)$. Then $B=T(n)$ is our initial matrix. The role of the transformation $R(n)\mapsto T(n)$ is to make $T(n)$ scaled almost diagonal which means that the minimum absolute gap δ will be in most cases much smaller than the minimum relative gap θ. The vector $d(n)$ is generated using four integer parameters: M, I1, I2, I3. The following FORTRAN 77 code has been used in obtaining $d(n)$

```
DO I = 1,M
  D(I) = 1D1 * *(DBLE(I1) + DBLE((I2 - I1) * (I - 1))/DBLE(M - 1))
ENDDO
DO I = M + 1,N
  D(I) = 1D1 * *(DBLE(I2) + DBLE((I3 - I2) * (I - M))/DBLE(N - M))
ENDDO
```

The parameters I1, I2 and I3 are integers between −20 and 8 and and M is between 1 and N, N is the matrix order. The parameters I1, I2, I3 have been varying by three nested loops, M has been set to N/2 and N has been varying by several values between 30 and 300.

The computation was done using double precision arithmetic. Visual Fortran Standard Edition 5.0D compiler of Digital corp. has been used.

We have considered the behavior of $\|\Omega(B^{(rN)})\|_F$ and $\|\Omega(B_S^{(rN)})\|_F$, $r=1,2,\ldots$ and how well these measures qualify as predictors on the number of cycles needed to reach the stopping criterion. As stopping criterion, we have chosen the one of Rutishauser [23]. However, instead of using sophisticated Rutishauser's transformation formulas, we have used the LAPACK routine DLASV2 for computing the transformation parameters and the BLAS 1 routine DROT for performing transformation.

In all performed diagonalizations, the quantity $\|\Omega(B_S^{(rN)})\|_F$ has decreased as expected, smoothly quadratically, while $\|\Omega(B^{(rN)})\|_F$ has often behaved unexpectedly.

Say, for N= 30, I1= 4, I2=-10, I3=-10, we have obtained $d(30)=(10^4, 10^3, 10^2, 10, 1, 10^{-1}, 10^{-2}, 10^{-3}, 10^{-4}, 10^{-5}, 10^{-6}, 10^{-7}, 10^{-8}, \ldots, 10^{-8})$ and the starting upper-triangular matrix $B=R(30)D(30)$ whose three largest and smallest diagonals and singular values are displayed in Table 1.

Table 1

i	b_{ii}	σ_{s_i}
1	$0.436E+07$	$\approx 0.4379\text{E}+07$
2	$0.437E+06$	$\approx 0.4370\text{E}+06$
3	$0.438E+05$	$\approx 0.4380\text{E}+05$
28	$0.463E-07$	$\approx 0.2463\text{E}-07$
29	$0.464E-07$	$\approx 0.2407\text{E}-07$
30	$0.465E-07$	$\approx 0.2367\text{E}-07$

Table 2

r	$\|\Omega(B^{(rN)})\|_F$	$\|\Omega(B_S^{(rN)})\|_F$
1	$0.358E+03$	$0.224E+01$
2	$0.781E-07$	$0.499E+00$
3	$0.313E-08$	$0.533E-01$
4	$0.364E-10$	$0.772E-03$
5	$0.682E-14$	$0.151E-06$
6	$0.234E-21$	$0.488E-14$
7	$0.173E-36$	$0.322E-29$

Table 3. B= $D^{\frac{1}{2}}R(N)D^{\frac{1}{2}}$
N = 30, I1 = 8, I2 = −10 I3 = −10

r	$\|\Omega(B^{(rN)})\|_F$	$\|\Omega(B_S^{(rN)})\|_F$
1	$0.108E+09$	$0.224E+01$
2	$0.476E+02$	$0.498E+00$
3	$0.313E-08$	$0.533E-01$
4	$0.364E-10$	$0.772E-03$
5	$0.682E-14$	$0.151E-06$
6	$0.234E-21$	$0.488E-14$
7	$0.174E-36$	$0.322E-29$

$\delta = 4.04 \cdot 10^{-10}, \theta = 8.46 \cdot 10^{-3}$

Table 4. $B = DR(N)$
N = 150, I1 = 8, I2 = −20 I3 = −20

r	$\|\Omega(B^{(rN)})\|_F$	$\|\Omega(B_S^{(rN)})\|_F$
1	$0.364E+13$	$0.910E+01$
2	$0.877E+10$	$0.236E+01$
3	$0.149E+07$	$0.530E+00$
4	$0.123E-01$	$0.672E-01$
5	$0.347E-18$	$0.161E-02$
6	$0.221E-21$	$0.100E-05$
7	$0.672E-28$	$0.342E-12$
8	$0.673E-41$	$0.475E-25$

$\delta = 8.21 \cdot 10^{-20}, \theta = 7.21 \cdot 10^{-4}$

Table 5. $B = R(N)D$
N = 150, I1 = 2, I2 = −20 I3 = −15

r	$\|\Omega(B^{(rN)})\|_F$	$\|\Omega(B_S^{(rN)})\|_F$
1	$0.580E+06$	$0.288E+03$
2	$0.841E+04$	$0.189E+01$
3	$0.360E+00$	$0.556E+00$
4	$0.150E-09$	$0.153E+00$
5	$0.179E-13$	$0.187E-01$
6	$0.397E-15$	$0.412E-03$
7	$0.439E-19$	$0.420E-07$
8	$0.168E-27$	$0.140E-15$

$\delta = 8.35683 \cdot 10^{-19}, \theta = 5.27 \cdot 10^{-3}$

The absolute and relative gaps have been computed a posterior: $\delta \approx 4.041 \cdot 10^{-10}$, $\theta \approx 8.464 \cdot 10^{-3}$. Since $\theta \gg \delta$, one can expect that the scaled iterates behave more regularly than the ordinary iterates. Table 2 shows that straightforwardly.

Table 6. $B = DR(N)$
$\mathtt{N} = 300, \mathtt{I1} = 6, \mathtt{I2} = -20\ \mathtt{I3} = -15$

r	$\|\Omega B^{(rN)}\|_F$	$\|\Omega B_S^{(rN)}\|_F$
1	$0.456E+12$	$0.110E+02$
2	$0.656E+10$	$0.188E+01$
3	$0.151E+08$	$0.589E+00$
4	$0.302E+04$	$0.139E+00$
5	$0.464E-04$	$0.196E-01$
6	$0.267E-15$	$0.862E-03$
7	$0.253E-18$	$0.853E-06$
8	$0.123E-23$	$0.406E-11$
9	$0.365E-35$	$0.118E-22$

$\delta = 8.75 \cdot 10^{-19}, \theta = 1.63 \cdot 10^{-3}$

We see that due to tiny δ, the quantity $\|\Omega(B^{(rN)})\|_F$ behaves strange: strong reduction in the first cycle and no further quadratic reduction in the subsequent cycles.

On the contrary, due to large θ, $\|\Omega(B_S^{(rN)})\|_F$ exhibits sound quadratic reduction after second cycle. Not only that $\|\Omega(B_S^{(rN)})\|_F$ reduces quadratically, but it gives a lower bound on the number of sure digits in diagonal elements as approximations of singular values. Even better bound for that is the quantity $2\|\Omega(B_S^{(rN)})\|_F^2/\theta$ (see Theorem 1.1(i)). Note that in cases like here, when $\|\Omega(B_S^{(rN)})\|_F < \theta/3$, there are simple and cheap ways to compute the needed lower bound for all θ_i, $1 \le i \le p$ and hence for θ (see [19]).

The clue for strange behavior of $\|\Omega(B^{(rN)})\|_F$ lies in the fact that the classical result (5.1) ensures the quadratic convergence of $\|\Omega(B^{(rN)})\|_F/\delta$ and not of $\|\Omega(B^{(rN)})\|_F$ alone, provided that it is smaller than one.

However, if $\delta = 2\sigma_{s_p}$ and σ_{s_p} is tiny, we just have to blame the result (5.1) or the definition of δ which depends on the minimum singular value. In such a case, the classical result is misleading and useless. On the contrary, the new result solves the problem. Since θ does not depend on σ_{s_p} and since θ can be large even if singular values cluster around zero, the quadratic convergence $\|\Omega(B_S^{(rN)})\|_F$ can be noticed and used even in early stages of the process.

We end the paper by another four examples of matrices. For each diagonalization process we display a table with relevant data.

References

[1] Barlow, J., Demmel, J. W.: Computing accurate eigensystems of scaled diagonally dominant matrices. SIAM J. Num. Anal. *27*, 762–791 (1990).
[2] Charlier, J. P., Van Dooren, P.: On Kogbetliantz's SVD algorithm in the presence of clusters. Linear Algebra Appl. *95*, 135–160 (1987).
[3] Charlier, J. P., Vanbegin, M., Van Dooren, P.: On efficient implementation of Kogbetliantz's algorithm for computing the singular value decomposition. Numer. Math. *52*, 279–300 (1988).
[4] Demmel, J. W., Veselić, K.: Jacobi's method is more accurate then QR. SIAM J. Matrix Anal. Appl. *13*, 1204–1245 (1992).
[5] Drmač, Z.: A posteriori computation of the singular vectors in a preconditioned Jacobi SVD algorithm. IMA J. Numer. Anal. *19*, 191–213 (1999).

[6] Fernando, K. V.: Linear convergence of the row cyclic Jacobi and Kogbetliantz methods. Numer. Math. *56*, 73–94 (1989).
[7] Forsythe, G. E., Henrici, P.: The cyclic Jacobi method for computing the principal values of a complex matrix. Trans. Amer. Math. Soc. *94*, 1–23 (1960).
[8] Hari, V.: On the quadratic convergence of Jacobi algorithms. Radovi Matematički *2*, 127–146 (1986).
[9] Hari, V., Veselić, K.: On Jacobi methods for singular value decompositions. SIAM J. Sci. Stat. Comput. *8*, 741–754 (1987).
[10] Hari, V.: On the quadratic convergence bounds of the serial singular value decomposition Jacobi methods for triangular matrices. SIAM J. Sci. Stat. Comput. *10*, 1076–1096 (1989).
[11] Hari, V.: On sharp quadratic convergence bounds for the serial Jacobi methods. Numer. Math. *60*, 375–406 (1991).
[12] Hari, V., Drmač, Z.: On scaled almost diagonal Hermitian matrix Pairs. SIAM J. Matrix Anal. Appl. *18(4)*, 1000–1012 (1997).
[13] Hari, V.: Parallel SVD methods for matrices in butterfly form. Technical report, University of Zagreb (Croatian language) 1999.
[14] Kogbetliantz, E.: Diagonalization of general complex matrices as a new method for solution of linear equations. Proc. Intern. Congr. Math. Amsterdam *2*, 356–357 (1954).
[15] Kogbetliantz, E.: Solutions of linear equations by diagonalization of coeficient matrices. Quart. Appl. Math. *13*, 123–132 (1955).
[16] Matejaš, J.: Quadratic convergence of scaled iterates by diagonalization methods. University of Zagreb, Ph. D. Thesis (Croatian language) 1996.
[17] Matejaš, J.: Quadratic convergence of scaled matrices in Jacobi method. Numer. Math. *87*, 171–199 (1999).
[18] Matejaš, J.: Convergence of scaled iterates in Jacobi method. Linear Algebra Appl. (forthcoming).
[19] Matejaš, J., Hari, V.: Scaled almost diagonal matrices with multiple singular values. ZAMM *78*, 121–131 (1998).
[20] Matejaš, J., Hari, V.: Scaled iterates by Kogbetliantz method. Proc. of the Conf. on Appl. Math. and Comp., Dubrovnik 13–18. Sept. 1999, pp. 1–20.
[21] Paige, C., Van Dooren, P.: On the quadratic convergence of Kogbetliantz's algorithm for computing the singular value decomposition. Linear Algebra Appl. *77*, 301–313 (1986).
[22] Voevodin, V. V.: Cislennye metody linejnoj algebry. Nauka, Moscow (in Russian) (1966).
[23] Wilkinson, J. H., Reinsch, C.: Handbook of automatic computation II: Linear algebra. Berlin Heidelberg New York: Springer 1971.

Vjeran Hari
Department of Mathematics
University of Zagreb
Bijenička 30
10000 Zagreb
Croatia

Josip Matejaš
Faculty of Economics
University of Zagreb
Kennedyjev trg 6
10000 Zagreb
Croatia

Computing [Suppl] 16, 107–119 (2002)

Computing

On a Method for Computing Inclusions of Solutions of the Basic GPS Equations*

G. Heindl

Abstract

A main step in the computation of the coordinates of a GPS navigator from received satellite signals is to solve a system of four equations

$$p_k = \|x_k - x\|_2 + b, \quad k = 1, \ldots, 4, \tag{E}$$

for the unknowns $x \in \mathbb{R}^3$ and $b \in \mathbb{R}$, where $\|\cdot\|_2$ denotes the standard Euclidean norm on $\mathbb{R}^3$ and $p_k \in \mathbb{R}$ and $x_k \in \mathbb{R}^3$, $k = 1, \ldots, 4$, are given.
The aim of the paper is to present conditions for p_k and x_k, $k = 1, \ldots, 4$, guaranteeing existence and local uniqueness of solutions of (E) and to develop a method for computing inclusions of solutions of (E).

Mathematics Subject Classification: 86A30, 86-08, 65G20.

Keywords: GPS-equations, automatic result verification.

1. Statement of the Problem

The problem to be considered in this paper is

(P) : Given $n \geq 2$,
$x_1, \ldots, x_{n+1} \in \mathbb{R}^n$, and $p_1, \ldots, p_{n+1} \in \mathbb{R}$,
find all pairs $(x, b) \in \mathbb{R}^n \times \mathbb{R}$ such that
(E) $\quad p_k = \|x_k - x\|_2 + b, \quad k = 1, \ldots, n+1,$
where $\|\cdot\|_2$ denotes the standard Euclidean norm on $\mathbb{R}^n$.

Although we are mainly interested in the case $n = 3$, the case $n = 2$ is important too as e.g. [5] shows. But there will be no need to restrict our investigations to these special cases.

Remark. It is easily shown that for any solution (x, b) of (P) the sphere

$$\{y \in \mathbb{R}^n : \|y - x\|_2 = |b|\}$$

*Dedicated to Prof. Dr. J. Herzberger on the occasion of his 60th birthday.

is tangent to the $n+1$ spheres

$$\{y \in \mathbb{R}^n : \|y - x_k\|_2 = |p_k|\}, \quad k = 1, \ldots, n+1.$$

Hence (P) can be considered also as a generalization of the well-known problem of Apollonius to find a circle in a plane that is tangent to three given circles in the plane. This observation is due to H.J. Buhl ([3]).

The mostly used algorithms for solving (P) in the case $n = 3$ are the algorithms of S. Bancroft ([2]) and A. Kleusberg ([4]), presented also in [6] and implemented by K. Borre as MATLAB files `bancroft` and `kleus` (available from web page `http://www.i4.auc.dk/borre/matlab`).

Unfortunately, there are several open questions concerning the range of applicability of the underlying methods. E.g. both authors compute solutions of modified problems which lead not always back to solutions of (P). Therefore, the algorithms of Bancroft and Kleusberg cannot serve directly for deriving inclusions of solutions of (P).

But combining and modifying ideas of both authors and using a result from the paper [1] of J.S. Abel and J.W. Chaffee, it was possible to answer the open questions and to develop formulas on which verifying methods for solving (P) for arbitrary $n \geq 2$ can be based.

In order to exclude exceptional cases in which local uniqueness of solutions of (P) cannot be guaranteed, we will assume however that $x_1, \ldots, x_{n+1}$ are affinely independent.

2. The Reduced Problems ($\mathbf{P}_j$), $j = 1, \ldots, n+1$

Let S denote the set of all solutions of (P), and $\tilde{S}_j$ the set of all $x \in \mathbb{R}^n$ such that

$$p_k - p_j = \|x_k - x\|_2 - \|x_j - x\|_2 \quad \forall k \in \{1, \ldots, n+1\} \setminus \{j\}, \quad j = 1, \ldots, n+1.$$

Obviously the equations used to define $\tilde{S}_j$ are obtained by subtracting the j-th equation of (E) from the remaining equations in order to eliminate b. But in fact $\tilde{S}_j$ does not depend on j. It is easily seen that

$$\begin{aligned}\tilde{S}_j = \tilde{S} :&= \{x \in \mathbb{R}^n : p_k - p_j = \|x_k - x\|_2 - \|x_j - x\|_2 \quad \forall k, j \in \{1, \ldots, n+1\}\}\\ &= \{x \in \mathbb{R}^n : \exists b \in \mathbb{R} : (x, b) \in S\}\end{aligned}$$

and

$$S = \{(x, p_j - \|x_j - x\|_2) : x \in \tilde{S}_j\}, \quad j = 1, \ldots, n+1.$$

Hence (P) can be solved by choosing a $j \in \{1, \ldots, n+1\}$ and solving the problem

(P_j): Find all $x \in \mathbb{R}^n$ such that

$$p_k - p_j = \|x_k - x\|_2 - \|x_j - x\|_2 \quad \forall k \in \{1, \ldots, n+1\} \setminus \{j\}.$$

3. Basic Necessary Conditions for the Existence of Solutions of (P)

If $(x, b) \in S$ then $x \in \tilde{S}$ and consequently

$$|p_k - p_j| = |\,\|x_k - x\|_2 - \|x_j - x\|_2\,| \leq \|x_k - x_j\|_2$$

for all $k, j \in \{1, \ldots, n+1\}$.

Hence, introducing the differences

$$\Delta_{j,k} := \|x_k - x_j\|_2 - |p_k - p_j| \quad (= \Delta_{k,j}) \quad \text{for all } j, k \in \{1, \ldots, n+1\},$$

$$\text{(AC)}: \qquad \Delta_{j,k} \geq 0 \quad \forall (j,k) : k \neq j$$

is a necessary condition for $S \neq \emptyset$.

This was observed first by J.S. Abel and J.W. Chaffee in [1].

Since (AC) is not considered in the algorithms of Bancroft and Kleusberg, these algorithms can deliver 'solutions of (P)' if $S = \emptyset$. This is the case, e.g. for the data $n = 3$,

$$x_1 = \begin{pmatrix} 0 \\ 0 \\ 0 \end{pmatrix}, \quad x_2 = \begin{pmatrix} 1 \\ 0 \\ 0 \end{pmatrix}, \quad x_3 = \begin{pmatrix} 0 \\ 1 \\ 0 \end{pmatrix}, \quad x_4 = \begin{pmatrix} 0 \\ 0 \\ 1 \end{pmatrix},$$

$$p_1 = 4.560477932, (\approx \sqrt{3} + 2\sqrt{2}), \ p_2 = p_3 = p_4 = 1.414213562 \quad (\approx \sqrt{2}).$$

Since (AC) does not hold, S must be empty. However, the Matlab program `bancroft` delivers

$$(x, b) = ((1, 1, 1)^{\mathrm{T}}, 2.8284)$$

as a "solution of (P)" and the Matlab program `kleus` computes

$$(1, 1, 1)^{\mathrm{T}} \text{ and } -6.8989\ldots\mathrm{e}{-01}(1, 1, 1)^{\mathrm{T}}$$

as "solutions of (P_1)".

There is a refinement of (AC) based essentially on

Lemma 1. *Assume that* $\Delta_{k,j} = 0$ *for a pair* (k, j) *with* $k \neq j$. *Then for any* $x \in \mathbb{R}^n$, $p_k - p_j = \|x_k - x\|_2 - \|x_j - x\|_2$ *holds iff*

$$x \in \{x_k + \lambda(x_j - x_k) : \lambda \geq 1\} \quad \textit{in case } p_k - p_j > 0,$$

$$x \in \{x_j + \lambda(x_k - x_j) : \lambda \geq 1\} \quad \textit{in case } p_j - p_k > 0.$$

Proof: It suffices to consider the case $p_k - p_j > 0$. Then $x \in \{x_k + \lambda(x_j - x_k) : \lambda \geq 1\}$ and $\Delta_{k,j} = 0$ obviously imply $p_k - p_j = \|x_k - x\|_2 - \|x_j - x\|_2$.

Conversely, $\Delta_{k,j} = 0$ and $p_k - p_j = \|x_k - x\|_2 - \|x_j - x\|_2$ imply

$$\|(x_k - x_j) + (x_j - x)\|_2 = \|x_k - x_j\|_2 + \|x_j - x\|_2.$$

Squaring the right and the left side of this equation leads to

$$(x_k - x_j)^{\mathrm{T}}(x_j - x) = \|x_k - x_j\|_2 \|x_j - x\|_2,$$

a Cauchy-Schwarz inequality in which equality holds. Since $x_k - x_j \neq 0$, we can conclude that there is a $\mu \geq 0$ such that $x_j - x = \mu(x_k - x_j)$.

Hence $x = x_j - \mu(x_k - x_j) = x_k + \lambda(x_j - x_k)$, where $\lambda = 1 + \mu \geq 0$. □

Corollary 1. *If* $x_j \in \tilde{S}$ *for some* $j \in \{1, \ldots, n+1\}$, *then* $\tilde{S} = \{x_j\}$.

Proof: Assume $x_j \in \tilde{S}$. Then $\Delta_{k,j} = 0$ for all $k \in \{1, \ldots, n+1\}$. Since $n \geq 2$, there is a subset $\{k, \ell\}$ of $\{1, \ldots, n+1\} \setminus \{j\}$.

Consider now any $x \in \tilde{S}$. Lemma 1 shows that x must be on the line ℓ_1 through x_j and x_k and on the line ℓ_2 through x_j and x_ℓ. But the intersection of ℓ_1 and ℓ_2 is $\{x_j\}$. □

Corollary 2. (*Refinement of (AC)*)

If $\tilde{S} \neq \emptyset$ *then either*

$$\tilde{S} = \{x_j\} \textit{ for some } j \in \{1, \ldots, n+1\},$$

or

(AC) is satisfied with $\Delta_{k,j} = 0$ *for at most one pair* (k, j) with $k < j$.

Proof: Assume $\tilde{S} \neq \emptyset$ and $\tilde{S} \neq \{x_j\}$ for all $j \in \{1, \ldots, n+1\}$. Then $x_j \notin \tilde{S}$ for all $j \in \{1, \ldots, n+1\}$ by Corollary 1. Consequently, there is an $x \in \tilde{S} \setminus \{x_1, \ldots, x_{n+1}\}$.

If $\Delta_{k,j} = \Delta_{\ell,m} = 0, k < j, \ell < m$ and w.l.o.g. $k \leq \ell$, then x is on the line ℓ_1 through x_k and x_j and on the line ℓ_2 through x_ℓ and x_m. Since $\ell_1 \neq \ell_2$ would imply that x_k, x_j, x_ℓ, x_m are all different and in the plane spanned by ℓ_1 and ℓ_2, contradicting the affine independence of $x_1, \ldots, x_{n+1}$, we must have $\ell_1 = \ell_2$. But now the affine independence of $x_1, \ldots, x_{n+1}$ implies $x_\ell = x_k$ and $x_m = x_j$ or $x_\ell = x_j$ and $x_m = x_k$. Finally, $k(\leq l) < m$ excludes the second possibility. □

An immediate consequence of Corollary 2 is

Corollary 3. *If $\tilde{S} \neq \emptyset$ and $\tilde{S} \neq \{x_j\}$ for all $j \in \{1, \ldots, n+1\}$, then for all but one $j \in \{1, \ldots, n+1\}$ the strict inequalities*

$$\Delta_{k,j} > 0$$

must hold for all $k \in \{1, \ldots, n+1\} \setminus \{j\}$.

4. Solving (P)

In order to solve (P), we check first whether the necessary conditions for $\tilde{S} \neq \emptyset$ stated in Corollary 2 are satisfied. If not, then $S = \emptyset$. Otherwise we find a j^* such that $S = \{(x_{j^*}, p_{j^*})\}$ or $\Delta_{k,j^*} > 0$ for all $k \in \{1, \ldots, n+1\} \setminus \{j^*\}$.

Assuming that all p_j and x_j are machine representable, then the necessary checks can be carried out rigorously using the precise dot product of PASCAL-XSC.

It remains to consider the case $\Delta_{k,j^*} > 0$ for a $j^* \in \{1, \ldots, n+1\}$ and all $k \in \{1, \ldots, n+1\} \setminus \{j^*\}$. W.l.o.g. we assume $j^* = n+1$ and try to solve (P_{n+1}), i.e. to find all $x \in \mathbb{R}^n$ such that

$$(\mathrm{E}_k) \qquad d_k = \|x_k - x\|_2 - \|x_{n+1} - x\|_2, \quad k = 1, \ldots, n,$$

where $d_k := p_k - p_{n+1}$, and

$$(\mathrm{C}) \qquad |d_k| \leq \|x_k - x_{n+1}\|_2, \quad k = 1, \ldots, n.$$

(C) implies the possibility of a useful transformation of (P_{n+1}) which will be given in

Theorem 1. *Let V denote the nonsingular matrix*

$$(x_1 - x_{n+1}, \ldots, x_n - x_{n+1}) \in \mathbb{R}^{n \times n},$$

a the vector

$$\tfrac{1}{2}(\|x_1 - x_{n+1}\|_2^2 - d_1^2, \ldots, \|x_n - x_{n+1}\|_2^2 - d_n^2)^{\mathrm{T}},$$

and d the vector

$$(d_1, \ldots, d_n)^{\mathrm{T}}.$$

Then, if (C) holds, $x \in \mathbb{R}^n$ *is a solution of* $(P_{n+1})(x \in \tilde{S}_{n+1} = \tilde{S})$ *iff* $x = x_{n+1} + \xi$, where $\xi \in \mathbb{R}^n$ *satisfies the equation*

$$(\tilde{\mathrm{E}}) \qquad V^{\mathrm{T}}\xi = a - \|\xi\|_2 d.$$

Proof: If x satisfies (E_k) then

$$d_k + \|x_{n+1} - x\|_2 = \|x_k - x\|_2,$$

hence

$$(d_k + \|x_{n+1} - x\|_2)^2 = \|(x_k - x_{n+1}) + (x_{n+1} - x)\|_2^2,$$

equivalent to

$$d_k^2 + 2d_k\|x_{n+1} - x\|_2 = \|x_k - x_{n+1}\|_2^2 + 2(x_k - x_{n+1})^{\mathrm{T}}(x_{n+1} - x),$$

equivalent to

$$(x_k - x_{n+1})^{\mathrm{T}}\xi = a_k - \|\xi\|_2 d_k, \quad \text{with } \xi := x - x_{n+1}.$$

This shows that any $x \in \tilde{S}_{n+1}$ can be written as $x_{n+1} + \xi$ where ξ satisfies $(\tilde{\mathrm{E}})$. Conversely, let ξ satisfy $(\tilde{\mathrm{E}})$. Then $x := x_{n+1} + \xi$ satisfies

$$(d_k + \|x_{n+1} - x\|_2)^2 = \|x_k - x\|_2^2 \quad \forall k \in \{1, \ldots, n\},$$

i.e.

$$(d_k + \|x_{n+1} - x\|_2 + \|x_k - x\|_2)(d_k + \|x_{n+1} - x\|_2 - \|x_k - x\|_2) = 0 \quad \forall k \in \{1, \ldots, n\}.$$

If (C) holds then

$$d_k + \|x_{n+1} - x\|_2 + \|x_k - x\|_2 \geq d_k + \|x_{n+1} - x_k\|_2 \geq -|d_k| + \|x_{n+1} - x_k\|_2 > 0 \quad \forall k \in \{1, \ldots, n\}.$$

Hence $x \in \tilde{S}_{n+1}$. □

Obviously (C) holds iff all components of a are positive. We will see however that the analytical expressions for the elements of the set

$$S' := \{\xi \in \mathbb{R}^n : V^{\mathrm{T}}\xi = a - \|\xi\|_2 d\},$$

which will be derived now, depend only on the assumption $a \neq 0$.

Introducing the notations

$$\tilde{a} := (V^{\mathrm{T}})^{-1}a, \qquad \tilde{d} := (V^{\mathrm{T}})^{-1}d,$$

$$R := (\tilde{a}^{\mathrm{T}}\tilde{d})^2 - (1 - \|\tilde{d}\|_2^2)\|\tilde{a}\|_2^2,$$

$$\xi_\ell := \tilde{a} - \frac{\|\tilde{a}\|_2^2}{2\tilde{a}^{\mathrm{T}}\tilde{d}}\tilde{d} \quad \text{if } \|\tilde{d}\|_2 = 1 \quad \text{and} \quad \tilde{a}^{\mathrm{T}}\tilde{d} > 0,$$

$$\lambda_{\underset{(-)}{+}} := -\frac{\tilde{a}^{\mathrm{T}}\tilde{d}}{1 - \|\tilde{d}\|_2^2} \underset{(-)}{+} \frac{\sqrt{R}}{|1 - \|\tilde{d}\|_2^2|}, \qquad \text{and}$$

$$\xi_{\underset{(-)}{+}} := \tilde{a} - \lambda_{\underset{(-)}{+}}\tilde{d} \qquad \text{if } \|\tilde{d}\|_2 \neq 1 \quad \text{and} \quad R \geq 0,$$

we can state

Theorem 2. (1) $S' = \emptyset$ *iff*

$$\|\tilde{d}\|_2 = 1 \textit{ and } \tilde{a}^{\mathrm{T}}\tilde{d} \leq 0$$

or

$$\|\tilde{d}\|_2 \neq 1 \textit{ and } R < 0$$

or

$$\|\tilde{d}\|_2 \neq 1 \textit{ and } R \geq 0 \textit{ and } \lambda_+ \leq 0.$$

(2) $S' = \{\xi_\ell\}$ *iff* $\|\tilde{d}\|_2 = 1$ *and* $\tilde{a}^{\mathrm{T}}\tilde{d} > 0$,
$S' = \{\xi_+\}$ *iff* $\|\tilde{d}\|_2 \neq 1, R \geq 0, \lambda_+ > 0,$ *and* ($\lambda_- \leq 0$ *or* $R = 0$).

(3) $S' = \{\xi_-, \xi_+\}$ (*with* $\xi_- \neq \xi_+$) *iff* $\|\tilde{d}\|_2 \neq 1, R > 0,$ *and* $\lambda_- > 0$.

Proof: If $\xi \in S'$ then $\xi = \tilde{a} - \|\xi\|_2\tilde{d}$. Hence $\|\xi\|_2^2 = \|\tilde{a}\|_2^2 - 2\tilde{a}^{\mathrm{T}}\tilde{d}\|\xi\|_2 + \|\xi\|_2^2\|\tilde{d}\|_2^2$, and since $\tilde{a} \neq 0$, $\|\xi\|_2$ is a positive root of the at most quadratic function

$$Q : \lambda \mapsto (1 - \|\tilde{d}\|_2^2)\lambda^2 + 2\tilde{a}^{\mathrm{T}}\tilde{d}\lambda - \|\tilde{a}\|_2^2, \quad \lambda \in \mathbb{R}.$$

If $\|\tilde{d}\|_2 = 1$, this implies $\tilde{a}^{\mathrm{T}}\tilde{d} > 0$, $\|\xi\|_2 = \frac{\|\tilde{a}\|_2^2}{2\tilde{a}^{\mathrm{T}}\tilde{d}}$, and $\xi = \xi_\ell$.

If $\|\tilde{d}\|_2 \neq 1$ then R must be nonnegative and $\lambda_+ > 0$. We have $\|\xi\|_2 = \lambda_+$ and $\xi = \xi_+$ if $\lambda_- \leq 0$ or $R = 0$. We have $\|\xi\|_2 = \lambda_-$ and $\xi = \xi_-$ or $\|\xi\|_2 = \lambda_+$ and $\xi = \xi_+$ if $R > 0$ and $\lambda_- > 0$.

On the other hand, if λ is a positive root of Q, then $\xi := \tilde{a} - \lambda\tilde{d} \in S'$ since $V^{\mathrm{T}}\xi = a - \lambda d$ and $\|\xi\|_2^2 = \|\tilde{a}\|_2^2 - 2\tilde{a}^{\mathrm{T}}\tilde{d}\lambda + \|\tilde{d}\|_2^2\lambda^2 = \lambda^2$ imply $\lambda = \|\xi\|_2$. □

Theorem 2 characterizes S' in accordance with the approach of S. Bancroft in [2]. An alternative characterization of S' can be derived from the ideas of A. Kleusberg presented in [4]:

As before, we define $\tilde{a}$, $\tilde{d}$ by $\tilde{a} := (V^{\mathrm{T}})^{-1}a$, $\tilde{d} := (V^{\mathrm{T}})^{-1}d$. In addition, we put

$$e := \frac{\tilde{a}^{\mathrm{T}}\tilde{d}}{\|\tilde{a}\|_2^2}\tilde{a} - \tilde{d} \text{ and in case } \|e\|_2 \leq 1 :$$

$$N_{\underset{(-)}{+}} := \tilde{a}^{\mathrm{T}}\tilde{d} \underset{(-)}{+} \sqrt{1 - \|e\|_2^2}\|\tilde{a}\|_2,$$

$$\xi_+ := \frac{\|\tilde{a}\|_2^2}{N_+}\left(e + \frac{\sqrt{1 - \|e\|_2^2}}{\|\tilde{a}\|_2}\tilde{a}\right) \quad \text{if } N_+ > 0,$$

$$\xi_- := \frac{\|\tilde{a}\|_2^2}{N_-}\left(e - \frac{\sqrt{1 - \|e\|_2^2}}{\|\tilde{a}\|_2}\tilde{a}\right) \quad \text{if } N_- > 0.$$

Then it can be verified by elementary calculations:

Theorem 3. (1)

$$S' = \emptyset \textit{ iff}$$
$$\|e\|_2 > 1$$
$$\textit{or}$$
$$\|e\|_2 \leq 1 \textit{ and } N_+ \leq 0.$$

(2)

$$S' = \{\xi_+\} \textit{ iff}$$
$$\|e\|_2 < 1, N_+ > 0, \textit{and } N_- \leq 0$$
$$\textit{or}$$
$$\|e\|_2 = 1 \textit{ and } (N_- =)\ N_+ > 0.$$

(3)

$$S' = \{\xi_-, \xi_+\} (\textit{with } \xi_- \neq \xi_+) \textit{ iff}$$
$$\|e\|_2 < 1 \textit{ and } N_- > 0.$$

Remark. The data $x_i \in \mathbb{R}^3$, $p_i \in \mathbb{R}$, $i = 1, 2, 3, 4$, derived from GPS observations usually guarantee the existence of at least one solution of (P).

If there are two solutions, one must be singled out as the true one. This however can be done only by using additional information about the possible constellations of the receiver and the satellites, or the possible range of b.

Theorem 2 was used for writing a PASCAL-XSC program HERZB_60 for including the solutions of (P) in the case $n = 3$. In order to avoid input/output conversion errors, a PASCAL-XSC version based on a decimal floating point system was chosen.

Let us see the results computed for some constructed test examples and two real life examples in comparison with the results obtained by the Matlab programs bancroft and kleus.

5. Numerical Examples

Example 1.

$$x_1 = \begin{pmatrix} 7 \\ 3 \\ -1 \end{pmatrix}, \quad x_2 = \begin{pmatrix} 10 \\ 3 \\ -1 \end{pmatrix}, \quad x_3 = \begin{pmatrix} 7 \\ 7 \\ -1 \end{pmatrix}, \quad x_4 = \begin{pmatrix} 7 \\ 3 \\ 2 \end{pmatrix},$$
$$p_1 = 5, \qquad p_2 = 6, \qquad p_3 = 1, \qquad p_4 = 6.$$

HERZB_60 showed that there is a unique solution (x, b) of (P), namely $(x_3, 1)$. x_3 was found when checking whether x_j is a solution of (P_j), $j = 1, \ldots, 4$. bancroft computed a complex result

$$(x' + iy', b' + ic'),$$

where $x' \approx x_3$, $b' \approx 1$ and $y', c' \approx 10^{-7}$.

kleus failed.

Example 2.

$$x_1 = \begin{pmatrix} 0 \\ 0 \\ 0 \end{pmatrix}, \quad x_2 = \begin{pmatrix} 1 \\ 0 \\ 0 \end{pmatrix}, \quad x_3 = \begin{pmatrix} 0 \\ 1 \\ 0 \end{pmatrix}, \quad x_4 = \begin{pmatrix} 0 \\ 0 \\ 1 \end{pmatrix},$$
$$p_1 = 1.1, \qquad p_2 = 2.1, \qquad p_3 = p_4 = 1.514213562 (\approx \sqrt{2} + 0.1).$$

HERZB_60 showed that there is a unique solution (x, b) of (P), where

$$x \in \begin{pmatrix} -1.00000000127_4^3 \\ 0.0 \\ 0.0 \end{pmatrix}, \qquad b \in 0.09999999872_6^7$$

(x is on line through x_1 and x_2).

`bancroft` computed (the good result)

$$\left(\begin{pmatrix}-1.0000\\0.0000\\0.0000\end{pmatrix}, 0.1000\right).$$

The result for x obtained by `kleus` was

$$\begin{pmatrix}\mathtt{NaN}\\\mathtt{NaN}\\\mathtt{NaN}\end{pmatrix}.$$

Example 3. Same as Example 2, but $\begin{pmatrix}x_1\\p_1\end{pmatrix}$ and $\begin{pmatrix}x_3\\p_3\end{pmatrix}$ interchanged.

`HERZB_60` and `bancroft` obtained the same result as in Example 2. But `kleus` computed now two possible x,

$$\begin{pmatrix}-1.000\ldots\mathtt{e}+00\\-1.825\ldots\mathtt{e}-08\\-1.825\ldots\mathtt{e}-08\end{pmatrix} \quad\text{and}\quad \begin{pmatrix}-9.999\ldots\mathtt{e}-01\\1.825\ldots\mathtt{e}-08\\1.825\ldots\mathtt{e}-08\end{pmatrix}.$$

This shows that `kleus` depends strongly on the numbering of the x_j.

Example 4.

$$x_1 = \begin{pmatrix}1\\1\\1\end{pmatrix}, \quad x_2 = \begin{pmatrix}1\\0\\0\end{pmatrix}, \quad x_3 = \begin{pmatrix}0\\1\\0\end{pmatrix}, \quad x_4 = \begin{pmatrix}0\\0\\1\end{pmatrix},$$
$$p_1 = 0.3, \qquad p_2 = p_3 = p_4 = 0.1.$$

`HERZB_60` showed that there is a unique solution (x, b) of (P), where

$$x \in \begin{pmatrix}4.073500107751^{7}_{6}\mathrm{E}-01\\ \text{"}\\ \text{"}\end{pmatrix}, \quad b \in -7.26499892483^{1}_{4}\mathrm{E}-01.$$

`bancroft` computed $\left(\begin{pmatrix}\mathtt{NaN}\\\mathtt{NaN}\\\mathtt{NaN}\end{pmatrix}, \mathtt{NaN}\right).$

`kleus` derived for x:

$$\begin{pmatrix}5.8234\ldots\mathtt{e}-01\\ \text{"}\\ \text{"}\end{pmatrix} \quad\text{and}\quad \begin{pmatrix}4.073500107516751\mathtt{e}-01\\ \text{"}\\ \text{"}\end{pmatrix}.$$

Whereas the second vector is a very good approximation of the true x, the first one approximates a solution of an auxiliary problem constructed for (P_1) but not the solution of the original (P_1).

Example 5. (from [1])

$$x_1 = \begin{pmatrix} 0 \\ 0 \\ 1 \end{pmatrix}, \quad x_2 = \begin{pmatrix} 0.866 \\ 0 \\ 0.5 \end{pmatrix}, \quad x_3 = \begin{pmatrix} -0.433 \\ 0.75 \\ 0.5 \end{pmatrix}, \quad x_4 = \begin{pmatrix} -0.433 \\ -0.75 \\ 0.5 \end{pmatrix},$$
$$p_1 = 0.4, \qquad p_2 = 1.1, \qquad p_3 = 0.8, \qquad p_4 = 1.2.$$

HERZB_60 showed that there are exactly two solutions (x_1, b_1), (x_2, b_2) of (P), where

$$x_1 \in \begin{pmatrix} -5.83253125^{24}_{31}\mathrm{E}{-02} \\ 2.41994882^{61}_{58}\mathrm{E}{-01} \\ 8.19497692^{39}_{28}\mathrm{E}{-01} \end{pmatrix}, \quad b_1 \in 9.25191902^{9}_{4}\mathrm{E}{-02},$$

$$x_2 \in \begin{pmatrix} -1.63854321 08^{1}_{8}\mathrm{E}{-01} \\ 6.0754736 82^{5}_{2}\mathrm{E}{-01} \\ 2.555871999^{17}_{07}\mathrm{E}{-00} \end{pmatrix}, \quad b_2 \in -1.278302630^{83}_{94}\mathrm{E}{+00}.$$

`bancroft` and `kleus` worked well on this example.

Example 6. (example used by K. Borre for testing `bancroft`)

$$\underbrace{\begin{pmatrix} -11716227.778 \\ -10118754.628 \\ 21741083.973 \end{pmatrix}}_{x_1}, \underbrace{\begin{pmatrix} -12082643.974 \\ -20428242.179 \\ 11741374.154 \end{pmatrix}}_{x_2}, \underbrace{\begin{pmatrix} 14373286.650 \\ -10448439.349 \\ 19596404.858 \end{pmatrix}}_{x_3}, \underbrace{\begin{pmatrix} 10278432.244 \\ -21116508.618 \\ -12689101.970 \end{pmatrix}}_{x_4},$$
$$p_1 = 22163882.092, \quad p_2 = 21492579.823, \quad p_3 = 21492492.771, \quad p_4 = 25284588.982.$$

HERZB_60 showed that there is a unique solution (x, b) of (P), where

$$x \in \begin{pmatrix} 5.950250053^{8}_{6}\mathrm{E}{+05} \\ -4.85650122^{07}_{11}\mathrm{E}{+06} \\ 4.07832998 1^{6}_{2}\mathrm{E} + 06 \end{pmatrix}, \quad b \in 1.4733^{10}_{06}\mathrm{E}{+02}.$$

`bancroft` delivered

$$\begin{pmatrix} 5.949993\ldots\mathrm{e}{+05} \\ -4.856504\ldots\mathrm{e}{+06} \\ 4.07833\ldots\mathrm{e}{+06} \end{pmatrix} \quad \text{for } x \text{ and } 1.47859\ldots\mathrm{e}{+02} \text{ for } b.$$

`kleus` computed a very accurate approximation of x and

$$\begin{pmatrix} -1.709\ldots\mathrm{e}{+}06 \\ 1.051\ldots\mathrm{e}{+}07 \\ -7.696\ldots\mathrm{e}{+}06 \end{pmatrix}$$

for a second (non-existing) solution of (P_1).

Example 7. (from [4])

$$\underbrace{\begin{pmatrix} 14832308.66 \\ -20466715.89 \\ -7428634.75 \end{pmatrix}}_{x_1}, \underbrace{\begin{pmatrix} -15799854.05 \\ -13301129.17 \\ 17133838.24 \end{pmatrix}}_{x_2}, \underbrace{\begin{pmatrix} 1984818.91 \\ -11867672.96 \\ 23716920.13 \end{pmatrix}}_{x_3}, \underbrace{\begin{pmatrix} -12480273.19 \\ -23382560.53 \\ 3278472.68 \end{pmatrix}}_{x_4},$$

$$p_1 = 24310664.064, \quad p_2 = 22914500.784, \quad p_3 = 20628709.405, \quad p_4 = 23422277.972.$$

HERZB_60 showed that there is a unique solution (x, b) of (P), where

$$x \in \begin{pmatrix} 1.11159904^{6000}_{5993}\mathrm{E}{+}06 \\ -4.34825863^{08}_{10}\mathrm{E}{+}06 \\ 4.527351820^{3}_{2}\mathrm{E}{+}06 \end{pmatrix}, \quad b \in [4.7\mathrm{E}{-}04, 6.3\mathrm{E}{-}04].$$

The Matlab programs `bancroft` and `kleus` computed in addition to an approximation of the unique solution a second wrong solution. However, Kleusberg himself has published in [4] only one (very precise) approximation for x, namely

$$\begin{pmatrix} 1111590.46 \\ -4348258.63 \\ 4527351.82 \end{pmatrix}.$$

6. Concluding Remarks

For real-life applications, HERZB_60 should be extended to interval input and such that it can utilize also the information received from more than four satellites. Although it seems not to be a simple task to construct such an extension without introducing non-acceptable overestimations, the plan for an update HERZB_65 exists.

Acknowledgements

I am indebted to my assistant A. Rogat for preparing the TEX version of the paper and to the referee for several valuable suggestions and corrections of misprints.

References

[1] Abel, J. S., Chaffee, J. W.: Existence and uniqueness of GPS solutions. IEEE Trans. Aerospace and Electronic Systems 27(6), Nov. 1991.

[2] Bancroft, S.: An algebraic solution of the GPS equations. IEEE Trans. on Aerospace and Electronic Systems 21(7), Jan. 1985.

[3] Buhl, H. J.: private communication 1999.

[4] Kleusberg, A.: Die direkte Lösung des räumlichen Hyperbelschnitts. Zeitschrift für Vermessungswesen *4* (1994).

[5] Schmidt, R. O.: A new approach to geometry of range difference location. IEEE Trans. Aerospace and Electronic Systems 8(6), Nov. 1972.

[6] Strang, G., Borre, K.: Linear algebra, geodesy, and GPS. Wellesley: Cambridge Press 1997 (containing descriptions of the algorithms used in the MATLAB programs `bancroft` and `kleus` available from web page `http://www.i4.auc.dk/borre/matlab`).

G. Heindl
Bergische Universität Wuppertal
Department of Mathematics
Gaußstraße 20
D-42097 Wuppertal
Germany

Computing [Suppl] 16, 121–133 (2002)

Computing

Construction of Bounds for the Positive Root of a General Class of Polynomials with Applications

J. Herzberger and **A. Langer**

Abstract

We consider classes or polynomials arising in numerical analysis when determining the order of convergence and in classical mathematics of finance when calculating the effective rate of interest. Those polynomials always have a unique positive root according to Descartes' rule of signs. In the presence of parameters, for example, we want to give a priori bounds for this root with an accuracy satisfying the demands of the applications. Starting with a review of the cases of polynomials treated in the literature and the bounds given there, we propose a kind of pertubation method using the monotonicity principle first established in Herzberger [5] using some slightly improved bounds for these cases. Thus we are able to construct a priori bounds for classes of polynomials with more general coefficients compared with those treated in earlier papers. Some numerical examples demonstrate how good our method works out in practice.

Mathematics Subject Classification: 26C05, 40A05, 65H05, 90A09.

Keywords: Polynomial roots, mathematics in finance, order of convergence of infinite sequences.

1. Introduction

One frequently posed problem in numerical analysis is the solution of a nonlinear equation which in its ideal form means to approximate a simple and isolated zero x^* of an equation

$$f(x) = 0$$

by a convergent sequence of iterates $\{x_k\}$ up to a certain accuracy.

Many popular iteration methods dealing with this task, for example, the secant method, Newton's method or Müller's method, belong to a more general class of iteration methods of solving nonlinear equations. They are based on the very important principle of interpolation or, more precisely, on approximation by interpolation. This common principle is successfully used in several fields of numerical analysis, for example, in numerical quadrature for deriving the Newton-Côtes-formulas, in numerical differentiation and in describing the Richardson extrapolation process in both fields mentioned. Applying this principle, one approximates the nonlinear function f in question by an "easier to handle" function, here by proper polynomials, and calculates the zero of this substituting polynomial as an approximation for the zero x^* of the original f. In practice, this approximation is done by using Hermite interpolation of f at several recent iterates. Secant method and Newton's method use linear interpolation, which in both cases leads to a simple and easily explicit solvable linear equation producing a comparatively simple-structured iteration formula. For Müller's method it still gives an explicit solvable quadratic equation.

According to their construction as sketched here, those iteration methods are referred to as so-called interpolation iteration methods, short (IIM). An extensive study of those methods can already be found in the monograph of Traub [17] or in a more condensed form in the textbook [7].

The (IIM) are defined by an integer vector of n parameters

$$m = (a_0, a_2, \ldots, a_{n-1}), \quad a_i \geq 0, \quad a_i \in \mathbb{N} \quad \text{and} \quad 0 \leq i \leq n-1. \tag{1}$$

It can be shown (see Traub [17]) that the error recursion for the sequence of errors $\{e_k\}$ of the iterates $\{x_k\}$ of an (IIM) has the form

$$e_k \leq c(k) \prod_{j=0}^{n-1} e_{k-j-1}^{a_j}, \quad k \geq n \quad \text{where} \quad e_k = |x^* - x_k| > 0. \tag{2}$$

By means of a simple transformation it can be shown that for $\sum_{i=0}^{n-1} a_i > 1$ under some weak conditions such an error sequence fulfilling the above recursion always is locally convergent to 0 (see Traub [17], or Herzberger [8]). The calculation of the Q-order used in Traub's monograph [17] or of the R-order used in the book [8] is finally shown to lead to the determination of the – according to Descartes' rule of signs – uniquely existing single simple positive root τ of a polynomial

$$p(t) = t^n - \sum_{i=0}^{n-1} a_i t^{n-i-1}, \quad a_i \text{ as in (1)}. \tag{3}$$

For the Q- or R-order of convergence of iterative numerical processes, see, for example, Herzberger [7], and the references therein; the same is true for the derivation of the polynomial equation (3) which is basically derived from the recursion (2) by taking the logarithm.

The parameter $a_i - 1$ finally determines the maximum order of the derivatives of the function f which are used in the iteration formula at a certain interpolation point. Since, on the other hand, all these points are shifted backwards throughout the iteration process, it is useful from the point of making efficient use of information only to consider iteration methods with "equal information" which means we set

$$a_i = a, \quad 0 \leq i \leq n-1, \quad a \geq 1, \quad a \in \mathbb{N}. \tag{4}$$

For this case, the polynomial p in (3) has the more simpler form

$$p(t) = t^n - a \sum_{i=1}^{n} t^{n-i}, \quad a \geq 1. \tag{5}$$

The determination of the zero τ of such an equation is a comparatively simple task due to the shape of the graph of p in the neighbourhood of τ if the parameters a and n are known by value. But if Eq. (3) or (4) contains an unknown parameter which, for example, gives rise to a further optimization process, then the question is how to derive explicit, easy-to-compute and tight a priori bounds for the zero τ depending on this parameter. Beginning with Ostrowski [13] and followed by Traub [17], there exist many such bounds in the literature, sometimes even for more generalized polynomials. Those bounds already allow estimating the asymptotic behavior of the value of τ and thus the asymptotic behavior of the speed of convergence of an (IIM). With regard to the accuracy of the generated numerical values, some of those bounds are not yet of satisfactory quality.

It turned out that polynomial equations like (5) play an important role in classical mathematics of finance, obviously a completely different area in applied mathematics. The determining equation for a simple ordinary annuity use the following notations:

S = original debt,

A = amount of the annuity, i.e., of the regular payments,

n = number of regular payments (usually the term),

$x = 1 + i$, with interest rate i in decimals.

The regular payments with amount A pay back the original debt including all interest due.

If we denote $a = \frac{A}{S}$, then the defining equation for an annuity can be shown to be of the same form as Eq. (5). This equation (5) then contains three parameters a, i (as $t = 1 + i$) and n. If a and n are given, then we again have the problem of calculating the positive root τ of Eq. (5) which gives by $\tau - 1$ the effective rate of the annuity. A more detailed discussion can be found in textbooks in classical finance, for example, in S.P. Shao and L.P. Shao [15], and Herzberger [10].

In [9], a special problem for estimating the effective rate of a "changed" annuity was solved. This change was done by reducing the regular payments by a factor of $\frac{1}{k}$ and by enlarging the number of payments by the factor k at the same time. The parameters a, i and n of the original annuity are assumed to be given. The corresponding parameters of the changed annuity are $\tilde{n} = nk$ and $\tilde{a} = \frac{a}{k}$ and the problem now is to give bounds for the effective rate j of the changed annuity in terms of the known parameters a, i and n of the original annuity as well as the factor k.

In the original paper [9] this result is interpreted into the language of the estimation of the order of convergence as described before and then used to give a procedure for calculating improved bounds for certain types of (IIM). Starting with a possibly fictive iteration method of order $na - 1$, the parameter vector of which is one dimensional and denoted by $(na - 1)$, one tries to give bounds for an (IIM) with the parameter vector of n components given by $(a, \dots, a)$. This exactly corresponds to the situation described for the annuity problem above, where $i = na - 1$ is the interest rate of the original annuity with only one payment and the changed annuity has n payments of amount a (corresponding to the parameter vector $(a, \dots, a)$). The derived bounds for the effective rate of the changed annuity can now be used as bounds for the root τ (or order of convergence) for the (IIM) defined by the parameter vector $(a, \dots, a)$.

2. Bounds for Polynomial Roots

The classical bounds for the unique positive root of the polynomial in (5) can already be found in [17] and are given in their improved version by the formulas

$$a + 1 - \frac{a}{(a+1)^n}\left(1 + \frac{1}{n}\right)^n < \tau < a + 1 - \frac{a}{(a+1)^n}. \tag{6}$$

They can be derived by an improvement process from the trivial a priori bounds

$$\frac{n}{n+1}(a+1) < \tau < a + 1,$$

which already indicate that for $n \to \infty$ the value of τ approaches $a + 1$. Different kinds of bounds for Eq. (10) were derived by means of Linear Algebra type conclusions by M.S. Petković and Lj. Petković in [14]. Now we want to apply the reinterpreted result from the annuity problem mentioned in Sect. 1. The result put into the setting of the calculation of the order of convergence of iterative processes reads as follows:

Given an (IIM) with parameter vector $m = (a, \dots, a)$ (n components, $n \geq 2$). The order of convergence of this method then is $\sigma = 1 + j^*$ where j^* is the unique positive root of the polynomial equation (5) *and $j^* > \frac{i^*}{n}$ with $i^* = na - 1$. Furthermore, j^* fulfills the inequalities*

$$\begin{aligned} &\frac{i^*}{n} + \frac{\frac{1}{n}}{1 - \frac{1}{\left(1+\frac{i^*}{n}\right)^n\left(1-\frac{na}{i^*}\right)+\frac{na}{i^*}}} < j^* \quad \textit{and} \\ &\left(\frac{i^*}{n}\right) - \frac{\left(1+\frac{i^*}{n}\right)^n\left(1-\frac{na}{i^*}\right)+\frac{na}{i^*}}{\left(1+\frac{i^*}{n}\right)^{n-1}\left(n - a\frac{(n-1)\frac{i^*}{n}-1}{\left(\frac{i^*}{n}\right)^2}\right) - \frac{a}{\left(\frac{i^*}{n}\right)^2}} > j^*. \end{aligned} \tag{7}$$

This result was already used in [9].

Let us now consider a numerical example for the formulas (7).

Example.

(a) $m = (1,1,1)$ (Müller's method)gives the bounds

$$1.816326531 < \sigma = 1.839286755 < 1.87037037$$

(b) $m = (4,4,4,4)$

$$4.992619075 < \sigma = 4.993566957 < 5.033524259$$

(c) $m = (2,2,2,2,2,2)$

$$2.99644945 < \sigma = 2.99724133 < 3.054572409$$

The last three numerical examples indicate that the upper bounds calculated in such a way are sometimes worse than the trivial upper bound $a+1$ given above. This can be proved in general.

Lemma 1. *For the upper bounds in* (7) *we have for sufficiently large values of n that*

$$\lim_{n\longrightarrow\infty} \tau = a+1$$

and

$$\tau > a+1, \quad \text{for } n \geq n_0.$$

The elementary but lengthy proof can be found in [12].

The consequence of this lemma is that the upper bound in (7) in most cases can be improved by substituting it by the trivial bound $a+1$.

3. More General Classes of Polynomials

In the calculation of the order of convergence of iterative processes polynomials with more general coefficients sometimes occur. Those polynomial equations are of the form

$$p(t) = t^n - \sum_{i=1}^{n} a_i t^{n-i} = 0, \quad a_i \geq 0, \;\; 1 \leq i \leq n, \;\; \sum_{i=1}^{n} a_i > 1. \tag{8}$$

First results about such polynomials can be found in Tornheim [16] for its unique positive root $\tau > 1$. In Herzberger [4] are the bounds

$$1 < \max\left\{ \max_{1\le k\le n}\left(\sum_{j=1}^{n} a_j\right)^{\frac{1}{k}}, \max_{0\le k\le n-1}\left(\frac{n-k}{n-k-1}\right)\left(\min_{1\le i\le n-k} a_i + 1\right)\right\}$$
$$\le \tau \le \min\left\{\left(\max_{1\le k\le n} a_k + 1\right), 2\max_{1\le k\le n} a_k^{\frac{1}{k}}\right\} \tag{9}$$

derived for the polynomial in (8). They contain for

$$a_1 \ge a_2 \ge a_3 \ge \cdots \ge a_2 \quad \text{with} \quad a_1 = \max_{1\le i\le k} a_i.$$

Tornheim's result $a_1 \le \tau = a_1 + 1$ as well as for $a_k = a > 0, 1 \le k \le n$ the trivial bounds $\frac{n}{n+1}(a+1) \le \tau \le a+1$ of Traub [17].

Remark. We observe that for the polynomial equation

$$p(t) = t^n - a\sum_{i=1}^{n-1} t^{n-i} - (a+1) = 0$$

the unique positive root is given by $\tau = a + 1$.

Another generalization of Eq. (5) was treated in [5]. This class of polynomial equation is given by

$$p(t) = t^n - at^{n-1} - b\sum_{i=1}^{n-1} h^i t^{n-1-i} = 0, \quad a > 0,\ b > 0,\ h > 0. \tag{10}$$

The bounds for its unique positive root τ are derived as

$$\frac{1}{2}(n+1)^{-1}\left[n(a+h) + \sqrt{g^2 n_2^2 + 4h(a-b)}\right] < \tau < \frac{1}{2}(a+h+|g|), \tag{11}$$

where $g^2 = (a+h)^2 - 4h(a-b)$.

The generalization of the polynomial equations in (5) by

$$p(t) = t^n - a\sum_{i=0}^{n-1} h^i t^{n-1-i} = 0, \quad h > 0 \text{ and } a > 0, \tag{12}$$

which not only arises for certain types of iterative methods (see [4]) but is even more important in the treatment of annuities in finance when the regular payments are time-dependent by the formula

$$A_j = h^{j-1}A, \quad 1 \le j \le n.$$

Those annuities have geometrically growing payments.

In [1] it was shown that by a simple variable transformation

$$u = \frac{t}{h} \quad \text{or} \quad t = hu.$$

Equation (12) can be obtained out of Eq. (5). Likewise, the bounds for the positive root τ of Eq. (12) can be transformed by the same variable transformation out of the bounds (6) for the positive root of Eq. (5) and they are given by

$$(a+h) - \frac{ah^n}{(a+h)^n}\left(1+\frac{1}{n}\right)^n < \tau < (a+h) - \frac{ah^n}{(a+h)^n} \quad \text{if } n > \frac{h}{a}. \qquad (13)$$

This result was first obtained by Kjurkchiev in [11] applying an analysis as done by Traub [17] directly from Eq. (12).

This list of generalizations of Eq. (5) shows that all these new equations can be derived from (5) by changing only one parameter or introducing one additional parameter also with geometric behavior. The only exception is Tornheim's result treating a more general situation but getting rather crude bounds.

In the next section, we are trying to give bounds for polynomial equations which differ from Eq. (5) far more than the examples given above.

4. Bounds for New Classes of Polynomials

As a tool for getting bounds for more general polynomials, we are going to use the so-called monotonicity principle which was explicitly used in connection with bounding positive roots in [6]. The resulting bounds (9) were derived in the same way and we just want to formulate this basic tool:

Let p and q be two polynomials on an interval (a,b), $a \geq 0$, both polynomials are assumed to have only one positive root each, x_p and x_q, respectively, the asymptotic behavior for large values of x, i.e., for $x \to +\infty$ is $p(x)$, $q(x) \to +\infty$, then we have the property

$$p(x) \geq q(x) \ \text{ for } x \in (a,b) \Longleftrightarrow x_q \geq x_p.$$

This very simple to verify principle is only useful in application if one of both polynomials p or q can be chosen in such a simple form that its positive unique zero can easily be calculated in an explicit way, for example, by taking a root. Then, according to this principle, this obtained value is either an upper or a lower bound for the positive root of the second polynomial under investigation.

In application to polynomials in finance we have the fact that the positive root t in question is an accumulation factor $t = 1 + i$, where i is the interest rate as a decimal fraction and usually of the order of magnitude of around 0.08. This means that, for example, substituting higher powers t^m by lower ones or conversely, usually does not introduce a too big error due to the fact that the

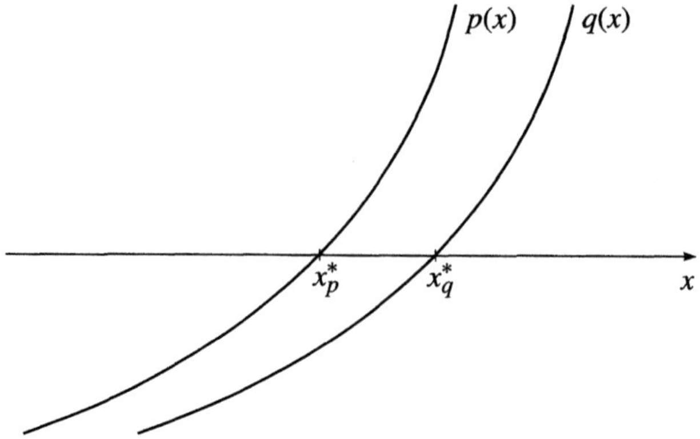

Fig. 1. The monotonicity principle

positive zero is well-conditioned. This argument is not valid for these polynomials in numerical analysis, where the positive root and at the same time order of convergence is usually not at all around the value of 1.

Now, let us return to our problem of bounding the positive root of a polynomial equation of the general form (8). For getting proper bounds we want to use the bounds already achieved for the special situations as in Eq. (5). Let us denote by p the polynomial in (5)

$$p_a(t) = t^n - a\sum_{i=1}^{n-1} t^{n-1-i}, \quad a_i > 0, \;\; 1 \le i \le n$$

which can also be written as

$$p_a(t) = t^n - aq(t), \quad \text{where} \quad q(t) = \sum_{i=0}^{n-1} t^i, \;\; n \ge 2.$$

If we want to change the coefficients of the polynomials p_a, then the change in the coefficients of p_a can be expressed by subtracting a certain polynomial r from p_a with

$$r(t) = \sum_{i=0}^{m} b_i t^i, \quad m < n, \;\; b_i \ge 0, \;\; 0 \le i \le m.$$

The polynomial pertubed this way is called $\hat{p}$ and is given by

$$\hat{p}(t) = p_a(t) - r(t).$$

On the other hand, we denote another polynomial of the form of Eq. (5) by

$$p_{a+c}(t) = p_a(t) - cq(t).$$

For this rather special polynomial we can apply the formulas for the bounds as given in (6) and then we try to use the polynomial p_{a+c} itself in monotonicity principle for estimating the positive root of $\hat{p}$.

In order to get the polynomial p_{a+c} related to the pertubed polynomial $\hat{p}$, we fix the parameter c such that the two polynomials p_{a+c} and $\hat{p}$ have an intersection in the interval $q \in [2, \infty)$. This is done by equating

$$p_a(t) - r(t) = p_a(t) - cq(t) \quad \text{at} \quad t = d.$$

Thus, we get

$$c = \frac{r(d)}{q(d)}.$$

It is clear from the shape of the graphs of both polynomials with the intersection point at $t = d$ that there is no other intersection point in the interval $[2, +\infty)$. Analytically, this can be proved by showing that $\frac{r(t)}{q(t)}$ does not change its sign in this interval. The lengthy but technical proof can be found in [12]. As a result we get the following lemma.

Lemma 2. *For $d > 2$ the inequality $\hat{p}(2) < p_{a+c}(2)$ is valid whereas for sufficient large values of t the inequality $p_{a+c}(t) < \hat{p}(t)$ holds true.*

The proof uses for both properties the monotonicity of the expression $\frac{r(t)}{q(t)}$ as mentioned above and the fact that $d > 2$ (see [12]).

After these preparations we are now able to prove our main result for the application.

Theorem 3. *Let $\hat{t}$ be the unique positive root of the polynomial $\hat{p}$ and t_{a+c} the unique positive root of the polynomial p_{a+c} with $c = \frac{r(d)}{q(d)}$. The following properties are valid:*

(i) If $d > 2$ is an upper bound for $\hat{t}$, then t_{a+c} is a lower bound for $\hat{t}$.

(ii) If $d \geq 2$ is a lower bound for $\hat{t}$, then t_{a+c} is an upper bound for $\hat{t}$.

Proof: In both cases we have the equality $\hat{p}(d) = p_{a+c}(d)$ for $t = d$.

Ad (i): In case of $t \in [2, d)$ an application of the first part of Lemma 2 together with the uniqueness of the intersection point of both polynomials $\hat{p}$ and p_{a+c} gives $\hat{p}(t) \leq p_{a+c}(t)$ and the monotonicity principle leads to the asserted inequality $t_{a+c} \leq \hat{t}$.

Ad (ii): For $t \in [2, \infty)$ we have, regarding the second part of Lemma 2 and again by virtue of the uniqueness of the intersection point of the two polynomials p_{a+c} and $\hat{p}$, the inequality $\hat{p}(t) \geq p_{a+c}(t)$ from which we conclude the asserted inequality $\hat{t} \leq t_{a+c}$ using the monotonicity principle. □

Analogously, we can prove similar results for pertubed polynomials $\hat{p}$ generated by

$$\hat{p}(t) = p_a(t) + r(t).$$

The consequences are that the constant c changes its sign and that inequalities in Lemma 2 are reversed. In the same way in the second part of the assertions (i) and (ii) of Theorem 3 lower and upper bounds change their roles.

According to the result of Theorem 3 and the last comment, the positive root t_{a+c} of the polynomial p_{a+c} can be taken as a lower or an upper bound – depending on the value of d – for the positive root $\hat{t}$ of the pertubed polynomial $\hat{p}$ in question. Since the polynomial p_{a+c} has the form of the polynomials in (5), we can use, for example, the bounds (6) for t_{a+c}. Then those bounds (for the bound t_{a+c}) can be taken as bounds for $\hat{t}$.

We summarize the achieved results in the following table. Denote $\underline{d}$ or $\bar{d}$, respectively, a lower or an upper bound for the positive root $\hat{t}$ of the polynomial $\hat{p}$. Furthermore, let ℓ_{a+c} or u_{a+c}, resp., lower or upper bounds for the zero $\hat{t}_{a+c}$ of the polynomial p_{a+c}. Then the following table shows the possible inequalities for getting bounds for $\hat{t}$.

	$d = \underline{d}$	$d = \bar{d}$	
$\hat{p} = p_a - r$	$\hat{t} \leq u_{a+c}$	$\ell_{a+c} \leq \hat{t}$	with $c = \frac{r(d)}{q(d)}$
$\hat{p} = p_a + r$	$\ell_{a+c} \leq \hat{t}$	$\hat{t} \leq u_{a+c}$	with $c = -\frac{r(d)}{q(d)}$

5. Calculation of Bounds

After the derivation of the resulting bounds for the positive root $\hat{t}$ of the pertubed polynomial $\hat{p}$ we want to give some numerical examples.

The last table in Sect. 4 uses for this purpose lower and upper bounds for the positive root of the polynomial p_{a+c}. Such bounds can be found, for example, in (6). We are using here improved bounds for the polynomial p_{a+c} given by

$$1 + a + c - \frac{a + c - \frac{1}{n}}{\left(1 + a + c - \frac{1}{n}\right)^n - 1} < \hat{t} < 1 + a + c - \frac{a + c}{(1 + a + c)^n - 1}. \tag{14}$$

The proof of this compared with (6) slightly improved bounds can be found in [12].

It can be shown that the polynomial p_a is convex in the interval $[1 + a - \frac{1}{n}, \infty)$ and then one explicit Newton-step starting at the $1 + a$ leads to the upper bound in (14).

Examples. Using the bounds in (14) we give the following numerical examples:

(a) $p_a(t) = t^4 - 4t^3 - 4t^2 - 4t - 4$, $p(t) = t^3 + t^2 + t + 1$, $r(t) = 1$; then we get $\hat{p}(t) = t^4 - 4t^3 - 4t^2 - 4t - 3$; setting $c = \frac{-r(d)}{q(d)}$ with $\underline{d} = 4$ and $\bar{d} = 5$, we get the bounds

$$4.980804 < \hat{t} = 4.9870878 < 4.9871568.$$

(b) $p_a(t) = t^4 - 3t^3 - 3t^2 - 3t - 3$, $p(t) = t^3 + t^2 + t + 1$, $r(t) = t^2 + 2t + 1$; $\hat{p}(t) = t^4 - 3t^3 - 4t^2 - 5t - 4$; setting $c = \frac{r(d)}{q(d)}$ with $\underline{d} = 4$ and $\bar{d} = 5$ we get the bounds

$$4.2188515 < \hat{t} = 4.2644932 < 4.2844009.$$

By a proper choice of the parameter a in the initial polynomial p_a we can achieve that the positive zero t_{a+c} of the polynomial p_{a+c} necessary for determining the bounds for $\tilde{t}$ according to the table given at the end of Sect. 4 can explicitly be calculated.

For this we consider the polynomial

$$p_a(t) = t^n - aq(t) - 1$$

and the pertubed polynomial

$$\hat{p}(t) = p_a(t) - r(t)$$

and lastly the polynomial

$$p_{a+c}(t) = p_a(t) - cq(t) = t^n - (a + c) \sum_{i=0}^{n-1} t^i - 1.$$

The positive root of the last polynomial p_{a+c} is obviously given by $t_{a+c} = a + c + 1$.

Again with $c = \frac{r(d)}{q(d)}$ we get the following situation:

Lemma 4. *Let d and D be, respectively, a lower or an upper bound for the positive root $\hat{t}$ of the polynomial $\hat{p}$. With $c = \frac{r(z)}{q(z)}$, where $z \in \{d, D\}$, we get the bounds*

$$z = d \Longrightarrow \hat{t} \le t_{a+c} = a + c + 1 = a + 1 + \frac{r(d)}{q(d)},$$

$$z = D \Longrightarrow t_{a+c} = a + c + 1 = a + 1 + \frac{r(D)}{q(D)} \le \hat{t}.$$

We want to demonstrate the method by a numerical example.

Example. $p_a(t) = t^4 - t^3 - t^2 - t - 2$, $r(t) = t^2 + t$, $q(t) = t^3 + t^2 + t + 1$; then $\hat{p}(t) = t^4 - t^3 - 2t^2 - 2t - 2$ and with $d = 2$ and $D = 3$ we get the bounds

$$2.3 \leq \hat{t} = 2.35930 \leq 2.4.$$

6. Conclusions

After a review of the most important classes of polynomials the unique positive root of which is of interest in the determination of the order of convergence of iterative numerical processes and in classical mathematics of finance as well and bounds for it in the literature, we described a method to get bounds for more general polynomials. The polynomials treated in the literature have coefficients which are built in a systematic manner. In Sect. 4, we could treat polynomials with a more general pattern of the – still positive – coefficients and derive bounds for their unique positive root which are of similar accuracy as those in the literature. For the application of our method proposed here we derived and used slightly improved bounds for a classical case from the literature.

The essence of our approach is a sophisticated use of the so-called monotonicity principle for achieving easy-to-compute bounds for positive roots of polynomials as suggested in [6]. In this sense our approach can be considered to be an important step towards the treatment of polynomials with nearly arbitrary choice of the positive coefficients.

References

[1] Abels, A., Herzberger, J.: A unified approach for bounding the positive root of certain classes of polynomials with applications. Computing Suppl. 15, 1–7 (2001).

[2] Haase, G., Herzberger, J.: Schranken für die positive Wurzel von Polynomen in der Finanzmathematik. ZAMM *78*, 931–932 (1998).

[3] Haase, G., Herzberger, J.: Über die Näherungsformeln für Polynomwurzeln in der Finanzmathematik. ZAMM *79*, 861–862 (1999).

[4] Herzberger, J.: On the *R*-order of some recurrences with applications to inclusion methods. Computing *36*, 175–180 (1986).

[5] Herzberger, J.: Bounds for the *R*-order of certain iterative numerical processes. BIT *26*, 259–262 (1986).

[6] Herzberger, J.: Untere Schranken für die positiven Wurzeln von Polynomen bei der Berechnung der *R*-Ordnung von Iterationsverfahren. ZAMM *75*, 639–640 (1995).

[7] Herzberger, J.: Berechnung der Konvergenzordnung von Folgen bei iterativen numerischen Prozessen. In: Wissenschaftliches Rechnen – eine Einführung in das Scientific Computing (J. Herzberger, ed.). Berlin: Akademie Verlag 1995, pp. 279–315.

[8] Herzberger, J.: Einführung in das wissenschaftliche Rechnen. Bonn: Addison-Wesley 1997.

[9] Herzberger, J.: Bounds for the positive root of a certain class of polynomials with applications. BIT *39(2)*: 366–372 (1999).

[10] Herzberger, J.: Einführung in die Finanzmathematik. München: R. Oldenbourg 1999.

[11] Kjurkchiev, N.: Note on the estimation of the order of convergence of some iterative processes. BIT *32*, 525–528 (1992).

[12] Langer, A.: Über explizite Schranken für positive Polynomwurzeln mit Anwendungen. Diplom-thesis, Universität Oldenburg 2002 (not available).

[13] Ostrowski, A.: Solution of equations and systems of equations. New York: Academic Press 1960.
[14] Petković, M. S., Petković, L. D.: On the bounds of the *R*-order of some iterative methods. ZAMM *69*, 197–198 (1989).
[15] Shao, S. P., Shao, L. P.: Mathematics for management and finance, 6th ed. Cincinatti: South-Western Publishing 1990.
[16] Tornheim, L.: Convergence of multipoint iterative methods. J. ACM *11*, 210–220 (1964).
[17] Traub, J. F.: Iterative methods for the solution of equations. Englewood-Cliffs, N.J.: Prentice-Hall 1964.
[18] Vandergraft, J. S.: Introduction to numerical computations. New York: Academic Press 1978.

J. Herzberger
Fachbereich Mathematik
Universität Oldenburg
D-26111 Oldenburg
Germany
e-mail: meyer@mathematik.uni-oldenburg.de

A. Langer
Steenkenweg 26
D-26135 Oldenburg
Germany

Computing [Suppl] 16, 135–142 (2002)

Computing

Rounding Near Zero*

Ulrich W. Kulisch

Abstract

This paper deals with arithmetic on a discrete subset S of the real numbers $\mathbb{R}$ and with floating-point arithmetic in particular. We assume that arithmetic on S is defined by semimorphism. Then for any element $a \in S$ the element $-a \in S$ is an additive inverse of a, i.e. $a \oplus (-a) = 0$. The first part of the paper describes a necessary and sufficient condition under which $-a$ is the unique additive inverse of a in S. In the second part this result is generalized. We consider algebraic structures M which carry a certain metric, and their semimorphic images on a discrete subset N of M. Again, a necessary and sufficient condition is given under which elements of N have an unique additive inverse. This result can be applied to complex floating-point numbers, real and complex floating-point intervals, real and complex floating-point matrices, and real and complex floating-point interval matrices.

Mathematics Subject Classification: 94B40, 94B99.

Keywords: Theoretical foundation of computer arithmetic, interval arithmetic, complex arithmetic, matrix and vector arithmetic, rounding.

1. The One-dimensional Case

Let $\mathbb{R}$ denote the set of real numbers and S a discrete subset of $\mathbb{R}$ which is symmetric with respect to zero, i.e. $0 \in S$ and for all $a \in S$ also $-a \in S$. A semimorphism defines arithmetic operations $\oplus$ and $\otimes$ on S by the following rules:

(RG) $\quad a \circledcirc b := \bigcirc(a \circ b) \quad$ for all $a, b \in S$ and $\circ \in \{+, *\}$.

In (RG) $\bigcirc$ is a rounding $\bigcirc : \mathbb{R} \to S$ with the properties

(R1)	$\bigcirc(a) = a$	for all $a \in S$	(projection).
(R2)	$a \leq b \Rightarrow \bigcirc(a) \leq \bigcirc(b)$	for $a, b \in \mathbb{R}$	(monotonicity).
(R3)	$\bigcirc(-a) = -\bigcirc(a)$	for all $a \in \mathbb{R}$	(antisymmetry).

That is, the rounding is a monotone and antisymmetric projection of $\mathbb{R}$ onto S.

A semimorphism leads to the best possible arithmetic in S in the sense that between the computed result of an operation and the correct result there is never another element of S. The computed result is rounding dependent. In case of

* This article was prepared while the author was staying at the Electronical Laboratory, Agency of Industrial Science and Technology, MITI, at Tsukuba, Ibaraki 305-8568, Japan in 1999.

rounding to nearest it is accurate to 1/2 unit in the last place (ulp). In all other cases it is accurate to 1 ulp.

Typical monotone and antisymmetric roundings are the rounding to the nearest element in S, the rounding towards zero, and the rounding away from zero.

Since $0 \in S$, (R1) and (RG) yield immediately that for every $a \in S$ the element $-a$ is an additive inverse of a in S:

$$a \oplus (-a) = \bigcirc(a + (-a)) = \bigcirc(0) = 0 \quad \text{for all } a \in S.$$

However, in a normalized floating-point system $-a$ is not in general the only additive inverse of a in S. We briefly recall the definition of normalized floating-point numbers:

A normalized floating-point number is a real number of the form:

$$x = \circ m \cdot b^e.$$

Here $\circ \in \{+, -\}$ is the sign of the number, m is the mantissa, b is the base of the number system in use and e is the exponent. b is an integer greater than unity. The exponent is an integer between two fixed integer bounds e_1, e_2, and in general $e_1 \leq 0 \leq e_2$. The mantissa is of the form

$$m = \sum_{i=1}^{r} d_i \cdot b^{-i}.$$

The d_i are the digits of the mantissa. They have the property $d_i \in \{0, 1, \ldots, b-1\}$ for all $i = 1(1)r$ and $d_1 \neq 0$. Without the condition $d_1 \neq 0$, floating-point numbers are said to be unnormalized. The set of normalized floating-point numbers does not contain zero. So zero is adjoined to S. For a unique representation of zero it is often assumed that $m = 0.00 \cdots 0$ and $e = 0$. A floating-point system depends on the constants b, r, e_1, and e_2. We denote it by $S = S(b, r, e_1, e_2)$.

The floating-point numbers are not equally spaced between successive powers of b and their negatives. This spacing changes at every power of b. In particular, there are relatively large gaps around zero which contain no further floating-point number. Figure 1 shows a simple floating-point system $S = S(2, 3, -1, 2)$ consisting of 33 elements.

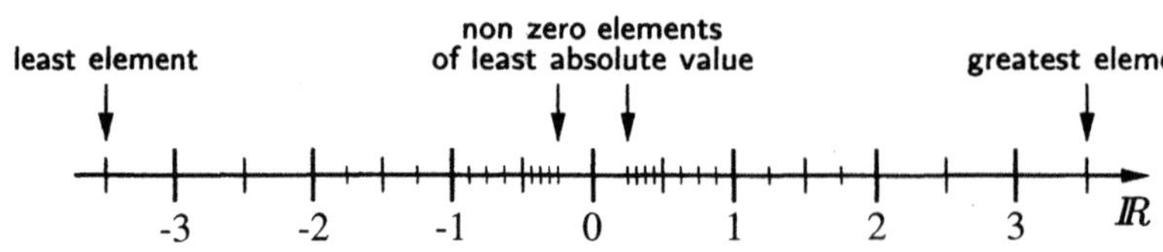

Fig. 1. The characteristic spacing of a floating-point system

If, for instance, the rounding towards zero is chosen, the entire interval $(-1/4, 1/4)$ is mapped onto zero. So whenever the real sum of two numbers of S falls into this interval (e.g. $1/4 - 3/8$) their sum in S is zero, $a \oplus b = 0$, and the two elements form a pair of additive inverses.

The following theorem characterizes a discrete subset S of $\mathbb{R}$ by a necessary and sufficient condition under which the element $-a$ is the unique additive inverse of a in S:

Theorem 1. *If S is a symmetric, discrete subset of $\mathbb{R}$ with $0 \in S$, $\bigcirc : \mathbb{R} \to S$ a semimorphism, and $\varepsilon > 0$ the least distance between distinct elements of S, then for all $a \in S$ the element $b = -a$ is the unique additive inverse of a if and only if*

$$\bigcirc^{-1}(0) \subseteq (-\varepsilon, \varepsilon). \tag{1}$$

Here $\bigcirc^{-1}(0)$ denotes the inverse image of 0 and $(-\varepsilon, \varepsilon)$ is the open interval between $-\varepsilon$ and ε.

Proof: At first we show that (1) is sufficient: We assume that $b \neq -a$ is an additive inverse of a in S. Then $a \oplus b = 0$ and by (RG) $a + b \in \bigcirc^{-1}(0)$. This means by (1)

$$-\varepsilon < a + b < \varepsilon. \tag{2}$$

Since $-b \neq a$ we have by definition of $\varepsilon : |a - (-b)| = |a + b| \geq \varepsilon$. This contradicts (2), so the assumption is false. Under the condition (1) there is no additive inverse of a other than $-a$. In other words (1) is sufficient.

Now we show that (1) is necessary also: Since $0 \in S$ by (R1) $\bigcirc(0) = 0$ and $0 \in \bigcirc^{-1}(0)$. Since $\bigcirc$ is monotone, $\bigcirc^{-1}(0)$ is convex. Since $\bigcirc$ is antisymmetric, $\bigcirc^{-1}(0)$ is symmetric with respect to zero. We assume now that (1) is not true, then $\bigcirc^{-1}(0) \supset (-\varepsilon, \varepsilon)$. We take two elements $a, b \in S$ with distance ε. Then $a \neq b$ and $|a - b| = \varepsilon$, i.e. $a + (-b) \in \bigcirc^{-1}(0)$ or $a \oplus (-b) = 0$. This means that $-b \neq -a$ is inverse to a. Thus (1) is necessary because otherwise there would be more than one additive inverse to a. □

(1) holds automatically if the number ε itself is an element of S. Then, because of (R1), $\bigcirc(\varepsilon) = \varepsilon$ and, because of the monotonicity of the rounding (R2) $\bigcirc^{-1}(0) \subseteq (-\varepsilon, \varepsilon)$. In other words: if the least distance ε of two elements of a discrete subset S of $\mathbb{R}$ is an element of S, then for all $a \in S$ the element $-a \in S$ is the unique additive inverse of a in S. (This holds under the assumption that the mapping $\bigcirc : \mathbb{R} \to S$ is a semimorphism).

Such subsets of the real numbers do indeed occur. For instance, the integers or the so called fixed-point numbers are subsets of $\mathbb{R}$ with this property. Sometimes the normalized floating-point numbers are extended into a set with this property. Such a set is obtained if in the case $e = e_1$ unnormalized mantissas are permitted. Then ε itself becomes an element of S and for all $a \in S$ the element $-a$ is the unique additive inverse of a. This is the case, for instance, if IEEE arithmetic with denormalized numbers is implemented.

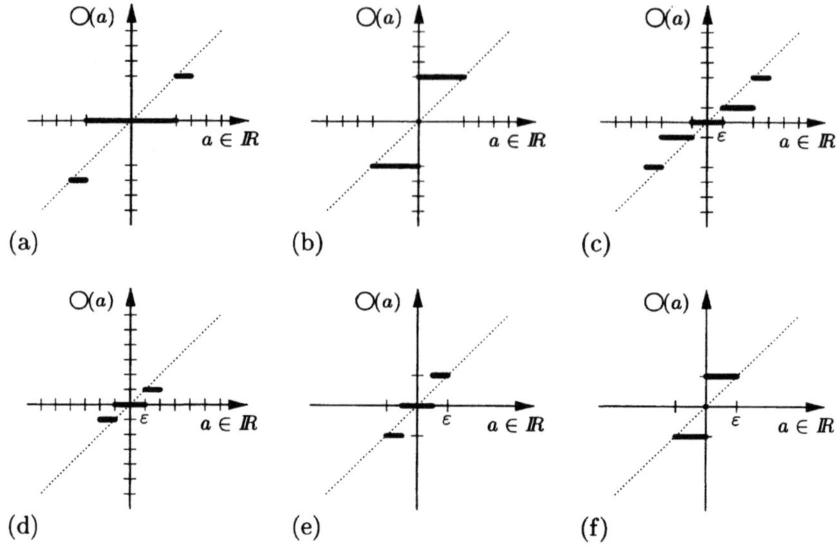

Fig. 2. The behavior of frequently used roundings near zero

Figure 2 illustrates the behavior of typical roundings in the neighborhood of zero. (R1) means that for floating-point numbers the rounding function coincides with the identity mapping:

(a) shows the conventional behavior of the rounding in case of normalized floating-point numbers. In this case (1) does not hold and we have no uniqueness of the additive inverse.

(b) shows the rounding away form zero in case of normalized floating-point numbers. In this case (1) holds and we have unique additive inverses.

(c) here and in the following cases ε is an element of S and we have unique additive inverses.

(d) shows the rounding toward zero near zero in the case where denormalized numbers are permitted for $e = e_1$. In the IEEE arithmetic standard this situation is called gradual underflow.

(e) shows the rounding to nearest in the neighborhood of zero. The roundings (d) and (e) are provided by the IEEE arithmetic standard.

(f) shows the rounding away from zero in the neighborhood of zero with denormalized numbers permitted for $e = e_1$. This rounding has all required properties. It is $\bigcirc(a) = 0$ if and only if $a = 0$, a property which can be very important for making a clear distinction between a number that is zero and a number that is not, however small it might actually be. This rounding is not provided by the IEEE arithmetic standard.

2. Rounding in Product Spaces

In addition to the real numbers, other spaces frequently occur in numerical mathematics. Such spaces are the complex numbers, the real and complex intervals, the real and complex matrices, and the real and complex interval matrices. In their computer representable subsets subtraction is no longer the inversion of addition nor is division the inversion of multiplication. Nevertheless subtraction is not an independent operation. If arithmetic in the computer representable subspaces is defined by semimorphism, subtraction can be defined by addition and multiplication with the negated multiplicative unit. For clarity we briefly show the definition here.

Let M denote any one of the sets listed above and N its computer representable subset. A semimorphism defines arithmetic operations $\oplus$ nd $\circledast$ in N by

(RG) $a \circledcirc b := \bigcirc(a \circ b)$ for all $a, b \in N$ and $\circ \in \{+, *\}$.

Here $\bigcirc : M \to N$ is a mapping with the following properties

(R1) $\bigcirc(a) = a$ for all $a \in N$ (rounding).
(R2) $a \leq b \Rightarrow \bigcirc(a) \leq \bigcirc(b)$ for $a, b \in M$ (monotonicity).
(R3) $\bigcirc(-a) = -\bigcirc(a)$ for all $a \in M$ (antisymmetry).

In case of the interval spaces, the order relation $\leq$ means set inclusion $\subseteq$. In this case the rounding is required to have the additional property

(R4) $a \leq \bigcirc(a)$ for all $a \in M$ (upwardly directed).

With this definition it is shown in [3], [4] for all spaces mentioned above that the multiplicative unit e has a unique additive inverse $\ominus e$ in N. With this quantity the minus operator (negation) and subtraction are defined by

$$\ominus b := (\ominus e) \odot b \quad \text{and} \quad a \ominus b := a \oplus (\ominus b). \tag{3}$$

This preserves all rules of the minus operator in the computer representable subspaces, [3], [4].

The proof that $\ominus e$ is unique is intricate and not easy in all the individual cases [3], [4]. A generalization of the conditions given in Sect. 1 for the existence and uniqueness of additive inverses for the product spaces listed above could simplify the situation considerably. This is now done.

We assume that the basic set M is mapped into a discrete subset N by semimorphism, where N is symmetric, i.e. $0 \in N$ and for all $a \in N$ also $-a \in N$. It follows from (RG) and (R1) that an element $a \in N$ which has an additive inverse $-a$ in M has the same additive inverse in N:

$$a \oplus (-a) := \bigcirc(a + (-a)) = \bigcirc(0) = 0.$$

Now we assume additionally that M is a metric space with a distance function $d : M \times M \to \mathbb{R}$ which has the property:

$$d(a+c, b+c) = d(a,b) \quad \text{for all } a,b,c \in M \quad \text{(translation invariant).} \qquad (4)$$

With d, the property of N being discrete can now be expressed by

$$d(a,b) \geq \varepsilon > 0 \quad \text{for all } a,b \in N \text{ with } a \neq b, \qquad (5)$$

where $\varepsilon > 0$ is the least distance of distinct elements of N. With these concepts the following theorem holds:

Theorem 2. *For all elements a of N which have an unique additive inverse $-a$ in M, $-a$ is also the unique additive inverse of a in N if and only if*

$$d(x,0) < \varepsilon \qquad \textit{for all } x \in M \textit{ with } \bigcirc(x) = 0. \qquad (6)$$

Proof: At first we show that (6) is sufficient: We assume that $b \neq -a$ is an additive inverse of a in N. Then we obtain by (RG) and (6):

$$a \oplus b = 0 \quad \Rightarrow \quad \bigcirc(a+b) = 0 \quad \Rightarrow \quad d(a+b,0) < \varepsilon. \qquad (7)$$

On the other hand we get by the definition of ε and by (4):

$$d(b,-a) \geq \varepsilon \quad \Rightarrow \quad d(a+b, a+(-a)) = d(a+b,0) \geq \varepsilon.$$

This contradicts (7), so the assumption that there is an additive inverse b of a other than $-a$ is false. In other words (6) is sufficient.

Now we show that (6) is necessary also: By (R1) we obtain $0 \in \bigcirc^{-1}(0)$. As a consequence of (R2) $\bigcirc^{-1}(0)$ is convex and by (R3) $\bigcirc^{-1}(0)$ is symmetric, since for all $a \in M$

$$a \in \bigcirc^{-1}(0) \Rightarrow \bigcirc(a) = 0 \Rightarrow \bigcirc(-a) = -\bigcirc(a) = 0 \Rightarrow -a \in \bigcirc^{-1}(0).$$

Should (6) not hold there would be elements $x \in M$ with

$$d(x,0) \geq \varepsilon > 0 \quad \text{and} \quad \bigcirc(x) = 0.$$

ε is the least distance of distinct elements of N. Now we choose two different elements $a, b \in N$ with distance ε, for instance $a = x$ and $b = 0$. Then $d(a,b) = \varepsilon$ and $\bigcirc(a-b) = 0$, i.e. $a - b \in \bigcirc^{-1}(0)$ or $a \oplus (-b) = 0$. This means that $-b$ is inverse to a and $-b \neq -a$. In other words: if (6) does not hold there are elements in N which have more than one additive inverse. This shows that (6) is necessary which completes the proof. □

To fully establish the theorem of Sect. 2 we still have to demonstrate that in all the basic spaces under consideration a metric does indeed exist which is translation invariant, see (4). We just mention the appropriate metric and leave the demonstration of (4) to the reader:

(i) If M is the set of real numbers $\mathbb{R}$, then $d(a,b) = |a-b|$.

(ii) If M is the set of complex numbers $\mathbb{C}$, the distance of two complex numbers $a = a_1 + ia_2$ and $b = b_1 + ib_2$ is defined by $d(a,b) := |a_1 - b_1| + |a_2 - b_2|$.

(iii) If M is the set of real intervals $I\mathbb{R}$ the distance of two intervals $a = [a_1, a_2]$ and $b = [b_1, b_2]$ is defined by $d(a,b) := \max(|a_1 - b_1|, |a_2 - b_2|)$.

(iv) If M is the set of complex intervals $I\mathbb{C}$ the distance of two complex intervals $a = a_1 + ia_2$ and $b = b_1 + ib_2$ with real intervals a_1, a_2, b_1 and b_2 is defined by $d(a,b) := d(a_1, b_1) + d(a_2, b_2)$.

(v) In case of two matrices $a = (a_{ij})$ and $b = (b_{ij})$ with components a_{ij}, b_{ij} of $\mathbb{R}$, $\mathbb{C}$, $I\mathbb{R}$, or $I\mathbb{C}$, the distance is defined in each case as the maximum of the distances of corresponding matrix components: $d(a,b) := \max(d(a_{ij}, b_{ij}))$.

In all the basic structures under consideration, which are the sets $\mathbb{R}$, $\mathbb{C}$, $I\mathbb{R}$, $I\mathbb{C}$, and the matrices with components of these sets, the multiplicative unit e has an unique additive inverse $-e$. Under the condition (1) and (6) respectively, therefore, $-e$ is also the unique additive inverse in any discrete computer representable subset. This allows the definition of the minus operator and subtraction in the computer representable subsets as shown in (3) with all its consequences (see [3], [4]).

A closer look at interval spaces is particularly interesting. Again we consider the sets $\mathbb{R}$ and $\mathbb{C}$ as well as the matrices $M\mathbb{R}$ and $M\mathbb{C}$ with components of $\mathbb{R}$ and $\mathbb{C}$. All these sets are ordered with respect to the order relation $\leq$. If M denotes any one of these sets the concept of an interval is defined by

$$A = [a_1, a_2] := \{a \in M \mid a_1, a_2 \in M, a_1 \leq a \leq a_2\}.$$

In the set IM of all such intervals arithmetic operations can be defined, see [1], [3], [4]. For all $M \in \{\mathbb{R}, \mathbb{C}, M\mathbb{R}, M\mathbb{C}\}$ the elements of IM in general are not computer representable and the arithmetic operations in IM are not computer executable. Therefore, subsets $N \subseteq M$ of computer representable elements have to be chosen. An interval in IN is defined by

$$A = [a_1, a_2] := \{a \in M \mid a_1, a_2 \in N, a_1 \leq a \leq a_2\}.$$

Arithmetic operations in IN are defined by semimorphism, i.e. by (RG) with the monotone and antisymmetric rounding $\Diamond : IM \to IN$ which is upwardly directed. $\Diamond$ is uniquely defined by these properties (R1), (R2), (R3), and (R4). This process leads to computer executable operations in the interval spaces IN of computer representable subsets of $\mathbb{R}, \mathbb{C}, M\mathbb{R}$, and $M\mathbb{C}$.

In this treatise the inverse image of zero with respect to a rounding plays a key role. Since the rounding $\diamondsuit : IM \to IN$ is upwardly directed with respect to set inclusion as an order relation the inverse image of zero $\diamondsuit^{-1}(0)$ can only be zero itself. Thus the necessary and sufficient criterion (6) for the existence of unique additive inverses evidently holds for IN. Among others this establishes the fact with all its consequences that the unit interval $[e, e]$ has a unique additive inverse $[-e, -e]$ in IN for all discrete subsets N of $M \in \{\mathbb{R}, \mathbb{C}, M\mathbb{R}, M\mathbb{C}\}$.

Acknowledgement

I wish to thank the Electrotechnical Laboratory (ETL) at Tsukuba, Japan and the Agency of Industrial Science and Technology, MITI, for their support and for the opportunity to prepare this article in a pleasant scientific atmosphere. Especially I owe thanks to Satoshi Sekiguchi for being a wonderful host.

References

[1] Alefeld, G., Herzberger, J.: An introduction to interval computations. New York: Academic Press 1983.
[2] Kaucher, E.: Über metrische und algebraische Eigenschaften einiger beim numerischen Rechnen auftretender Räume. Dissertation, Universität Karlsruhe 1973.
[3] Kulisch, U.: Grundlagen des numerischen Rechnens – Mathematische Begründung der Rechnerarithmetik. Reihe Informatik, Band 19, Bibliographisches Institut, Mannheim Wien Zürich 1976.
[4] Kulisch, U., Miranker, W. L.: Computer arithmetic in theory and practice. New York: Academic Press 1981.
[5] Yohe, J. M.: Roundings in floating-point arithmetic. IEEE Trans. Computers C-22(6), June 1973, pp. 577–586.
[6] American National Standards Institute/Institute of Electrical and Electronic Engineers: A Standard for Binary Floating-point Arithmetic. ANSI/IEEE Std. 754-1985, New York 1985 (reprinted in SIGPLAN 22, 2 pp. 9–25, 1987). Also taken over as IRC Standard 559: 1989.
[7] American National Standards Institute/Institute of Electrical and Electronic Engineers: A Standard for Radix-independent Floating-point Arithmetic. ANSI/IEEE Std. 854-1987, New York 1987.

Ulrich W. Kulisch
Universität Karlsruhe (TH)
D-76128 Karlsruhe
Germany

Computing [Suppl] 16, 143–149 (2002)

Computing

A Note on the Convergence of the SOR-like Weierstrass Method*

N. Kyurkchiev

Abstract

Convergence properties of the SOR-like Weierstrass method for the simultaneous approximation of polynomial roots are considered. Then the choice of the acceleration parameter is discussed.

Mathematics Subject Classification: 65H05, 30C15, 65B99.

Keywords: Zeros of polynomials, simultaneous methods, SOR-methods.

1. Introduction

Let

$$f(x) = x^n + a_{n-1}x^{n-1} + \cdots + a_1 x + a_0$$

be the polynomial with the simple complex zeros x_i, $i = 1, \ldots, n$ and let $x_i^k, i = 1, \ldots, n$ be distinct reasonably close approximations of these zeros. The method, which, in short, will always be referred to as the (WGSS)-method, (Weierstrass method in the Gauss-Seidel style) can be defined by the sequences

$$x_i^{k+1} = x_i^k - \frac{f(x_i^k)}{\prod\limits_{j<i}(x_i^k - x_j^{k+1})\prod\limits_{j>i}(x_i^k - x_j^k)}, \tag{1}$$
$$i = 1, 2, \ldots, n;\ \ k = 0, 1, 2, \ldots,$$

where the x_i^{k+1} denote the new approximations.

It is known [1] that the R-order of convergence is at least $\tau_n + 1 > 2$, where τ_n is the unique positive root of the equation $\tau^n - \tau - 1 = 0$.

In this paper we are dealing with the successive overrelaxation method (SOR-WGSS method)

---

*Dedicated to Prof. Dr. J. Herzberger on the occasion of his 60th birthday.

$$x_i^{k+1} = x_i^k - h_k \frac{f(x_i^k)}{\prod\limits_{j<i}(x_i^k - x_j^{k+1})\prod\limits_{j>i}(x_i^k - x_j^k)}, \tag{2}$$
$$i = 1, 2, \dots, n; \quad k = 0, 1, 2, \dots,$$

where h_k is an acceleration parameter.

In [3] it is shown that the SOR-WGSS method converges for $0 < h_k < 2$.

If h_k is close to 2 then the convergence is quite slow. The optimal value of h_k is not known.

It should be noted that the SOR-WGSS method has a form of a prediction-correction method.

Consider

$$W_i(x_i^k) = \frac{f(x_i^k)}{\prod\limits_{j\neq i}^{n}(x_i^k - x_j^k)}, \quad \max_{1\leq i\leq n}|W_i(x_i^k)| = W,$$
$$P(x_i^k) = \prod_{j=1}^{i-1}\frac{x_i^k - x_j^k}{x_i^k - x_j^{k+1}}, \quad \min_{j\neq i}|x_i^k - x_j^k| = d,$$
$$\max_{1\leq i\leq n}|P(x_i^k)| = P.$$

From (2) it follows

$$x_i^{k+1} = x_i^k - h_k W_i(x_i^k)P(x_i^k).$$

In practice, conditions for the convergence of the method (2) are given in the form of an inequality like

$$h_k WP < \frac{\omega d}{2(n-1)}, \tag{3}$$

where $\omega > 0$ is a constant. The inequality (3) will be used in our analysis of the SOR-WGSS method. We assume that $h_k \in (1, 2)$ and $\omega < \frac{h_k}{2}$. Later on, we will find the upper bound for ω.

2. Main Results

Before establishing the convergence theorem for the method (2), we give some necessary estimates using the previous notations.

Lemma 1. *If the inequality* (3) *holds, then for $i \in I_n$, we have*

$$(i) \quad |x_i^{k+1} - x_j^k| > d\tfrac{2n-2-\omega}{2(n-1)};$$
$$(ii) \quad |x_i^{k+1} - x_j^{k+1}| > d\tfrac{n-1-\omega}{n-1};$$
$$(iii) \quad h_k|P(x_i^k)|\sum_{j\neq i}^{n}\tfrac{|W_j(x_j^k)|}{|x_i^{k+1} - x_j^k|} < \tfrac{\omega}{2-\omega};$$

$$(iv)\quad |P_1(x_i^k)| = \prod_{j\neq i}^{n}\left|\frac{x_i^{k+1}-x_j^k}{x_i^{k+1}-x_j^{k+1}}\right| < \left(1+\frac{\frac{\omega}{2}}{n-1-\omega}\right)^{n-1} < e^{\frac{\omega}{2}};$$

$$(v)\quad |P(x_i^k)| < e^{\frac{\omega}{2}};$$

$$(vi)\quad |P(x_i^k)| > \frac{1}{e^{2\omega}}.$$

Proof: Of *(i)*:

$$|x_i^{k+1}-x_j^k| \geq |x_i^k - x_j^k| - |x_i^{k+1}-x_i^k| > d - \frac{\omega d}{2(n-1)} = d\frac{2n-2-\omega}{2(n-1)}.$$

Of *(ii)*:

$$|x_i^{k+1}-x_j^{k+1}| \geq |x_i^k - x_j^k| - |x_i^{k+1}-x_i^k| - |x_j^{k+1}-x_j^k| > d - 2\frac{\omega d}{2(n-1)} = d\frac{n-1-\omega}{n-1}.$$

Of *(iii)*:

$$h_k|P(x_i^k)|\sum_{j\neq i}^{n}\frac{|W_j(x_j^k)|}{|x_i^{k+1}-x_j^k|} \leq \frac{(n-1)Wh_kP2(n-1)}{d(2n-2-\omega)} = \omega\frac{n-1}{2n-2-\omega} \leq \frac{\omega}{2-\omega}.$$

The last inequality $\frac{n-1}{2n-2-\omega} \leq \frac{1}{2-\omega}$ holds for each $n \geq 2$.

Of *(iv)*:

$$|P_1(x_i^k)| = \prod_{j\neq i}^{n}\left|\frac{x_i^{k+1}-x_j^k}{x_i^{k+1}-x_j^{k+1}}\right| \leq \prod_{j\neq i}^{n}\left(1+\frac{|x_j^{k+1}-x_j^k|}{|x_i^{k+1}-x_j^{k+1}|}\right) < \left(1+\frac{\frac{\omega d}{2(n-1)}}{\frac{d(n-1-\omega)}{n-1}}\right)^{n-1}$$

$$= \left(1+\frac{\frac{\omega}{2}}{n-1-\omega}\right)^{n-1} < e^{\frac{\omega}{2}}.$$

Of *(v)*:

$$|P(x_i^k)| = \prod_{j=1}^{i-1}\left|\frac{x_i^k - x_j^k}{x_i^k - x_j^{k+1}}\right| < \left(1+\frac{\frac{\omega d}{2(n-1)}}{\frac{d(2n-2-\omega)}{2(n-1)}}\right)^{n-1} = \left(1+\frac{\frac{\omega}{2}}{n-1-\frac{\omega}{2}}\right)^{n-1} < e^{\frac{\omega}{2}}.$$

Of *(vi)*:

$$|P(x_i^k)| \geq \prod_{j=1}^{i-1}\left(1-\frac{|x_j^{k+1}-x_j^k|}{|x_i^k - x_j^{k+1}|}\right) > \prod_{j=1}^{i-1}\left(1-\frac{\omega}{2n-2-\omega}\right)$$

$$> \prod_{j=1}^{i-1}\left(1-\frac{\omega}{n}\right) > \frac{1}{\prod_{j=1}^{i-1}\left(1+\frac{2\omega}{n}\right)} > \frac{1}{(1+\frac{2\omega}{n})^n} > \frac{1}{e^{2\omega}}.$$

Using the estimates given in Lemma 1 we can prove the following lemma:

Lemma 2. *Suppose the inequality* (3) *holds, then*

$$|W_i(x_i^{k+1})| < \left(h_k e^{\frac{\omega}{2}} - 1 + \frac{\omega}{2-\omega}\right) e^{\frac{\omega}{2}} |W_i(x_i^k)|. \tag{4}$$

Proof: Evidently, we have

$$f(x) = \prod_{j=1}^{n}(x - x_j^k)\left(1 + \frac{W_i(x_i^k)}{x - x_i^k} + \sum_{j \neq i}^{n} \frac{W_j(x_j^k)}{x - x_j^k}\right). \tag{5}$$

From (2), we obtain

$$\frac{W_i(x_i^k)}{x_i^{k+1} - x_i^k} = -\frac{1}{h_k P(x_i^k)}.$$

According to this, the relation (5) for $x = x_i^{k+1}$ becomes

$$f(x_i^{k+1}) = (x_i^{k+1} - x_i^k)\prod_{j \neq i}^{n}(x_i^{k+1} - x_j^k)\left(1 - \frac{1}{h_k P(x_i^k)} + \sum_{j \neq i}^{n} \frac{W_j(x_j^k)}{x_i^{k+1} - x_j^k}\right).$$

After dividing by $\prod_{j \neq i}^{n}(x_i^{k+1} - x_j^{k+1})$, we obtain

$$W_i(x_i^{k+1}) = \frac{f(x_i^{k+1})}{\prod_{j \neq i}^{n}(x_i^{k+1} - x_j^{k+1})} = (x_i^{k+1} - x_i^k) P_1(x_i^k) B, \tag{6}$$

where

$$B = 1 - \frac{1}{h_k P(x_i^k)} + \sum_{j \neq i}^{n} \frac{W_j(x_j^k)}{x_i^{k+1} - x_j^k}.$$

Starting from (6) and using estimates of Lemma 1 we can find

$$h_k |P(x_i^k)||B| \leq h_k P - 1 + h_k P \sum_{j \neq i}^{n} \frac{W_j(x_j^k)}{x_i^{k+1} - x_j^k} \leq h_k e^{\frac{\omega}{2}} - 1 + \frac{\omega}{2-\omega},$$

$$\begin{aligned} |W_i(x_i^{k+1})| &\leq |x_i^{k+1} - x_i^k||P_1||B| \leq h_k |W(x_i^k)||P(x_i^k)||P_1(x_i^k)||B| \\ &< \left(h_k e^{\frac{\omega}{2}} - 1 + \frac{\omega}{2-\omega}\right) e^{\frac{\omega}{2}} |W_i(x_i^k)|. \end{aligned}$$

We consider the contraction factor Q for the SOR-WGSS method appearing in the relation

$$|x_i^{k+2} - x_i^{k+1}| < Q|x_i^{k+1} - x_i^k|.$$

Let us assume that the ratio $\frac{h_{k+1}}{h_k}$ is very close to 1. Then we have the following theorem.

Theorem 1. *For the contraction factor Q the following formula is valid*

$$Q = q(\omega, h_k)\alpha,$$

where

$$q(\omega, h_k) = \left(h_k e^{\frac{\omega}{2}} - 1 + \frac{\omega}{2-\omega}\right)e^{3\omega},$$

$$\alpha = \frac{h_{k+1}}{h_k}.$$

Proof: Let $c_i^t = x_i^t - x_i^{t+1}$, $t = k, k+1$. Then

$$|c_i^k| = h_k|W_i(x_i^k)||P(x_i^k)| \le h_k e^{\frac{\omega}{2}}|W_i(x_i^k)|.$$

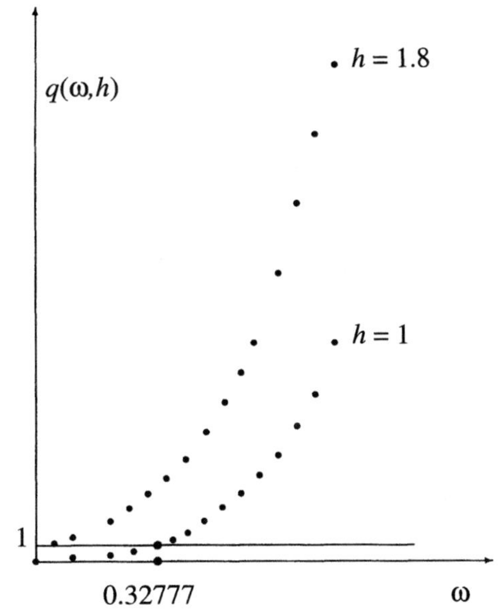

Fig. 1. Contraction of the method (2)

By induction, we get

$$\begin{aligned}
|c_i^{k+1}| &\le h_{k+1}e^{\frac{\omega}{2}}|W_i(x_i^{k+1})| \\
&\le h_{k+1}e^{\frac{\omega}{2}}\left(h_k e^{\frac{\omega}{2}} - 1 + \frac{\omega}{2-\omega}\right)e^{\frac{\omega}{2}}|W_i(x_i^k)|\frac{h_k}{h_k}\frac{|P(x_i^k)|}{|P(x_i^k)|} \\
&\le \frac{h_{k+1}}{h_k}e^{\omega}\left(h_k e^{\frac{\omega}{2}} - 1 + \frac{\omega}{2-\omega}\right)e^{2\omega}|c_i^k|,
\end{aligned}$$

$$|x_i^{k+2} - x_i^{k+1}| < \alpha q(\omega, h_k)|x_i^{k+1} - x_i^k| = Q|x_i^{k+1} - x_i^k|.$$

This proves the theorem.

According to the relation in Fig. 1 above, we represent the quantity $q(\omega, h)$ as a function of ω, taking h as a parameter. The upper bound, corresponding to $q = 1$, is denoted by ω_h. From the equation

$$q(\omega, 1) = \left(e^{\frac{\omega}{2}} - 1 + \frac{\omega}{2-\omega}\right)e^{3\omega} = 1,$$

we find the upper bound for $\omega : \omega < \omega_h \approx 0.32777$.

Other results have been found by Petković, Herceg and Ilić [7], Petković and Kyurkchiev [8], and Kyurkchiev [4].

Different aspects concerning these results can also be found in [9], [2], [8] and [5].

References

[1] Alefeld, G., Herzberger, J.: On the convergence speed of some algorithms for the simultaneous approximation of polynomial roots. SIAM J. Numer. Anal. *11*, 237–243 (1974).

[2] Ishihara, K., Muroya, Y., Yamamoto, T.: On nonlinear SOR-like methods, II – Convergence of the SOR-Newton method for mildly nonlinear equations. Japan J. Indust. Appl. Math. *14*, 99–110 (1997).

[3] Kanno, S., Kyurkchiev, N., Yamamoto, T.: On some methods for the simultaneous determination of polynomial zeros. Japan J. Indust. Appl. Math. *13(2)*, 267–288 (1996).

[4] Kyurkchiev, N.: Initial approximations and root finding methods. Berlin: Wiley VCH *104*, 1–180 (1998).

[5] Kyurkchiev, N., Hristov, V.: Distribution of the critical points of Weierstrass procedure. Mathematics and education in mathematics. Proc. of 29th Spring Conf. of Bulg. Math., Lovetch, April 3–6, 2000.

[6] Kyurkchiev, N., Petković, M.: On the behavior of approximations of the SOR Weierstrass method. Comput. Math. Appl. *32*, 117–121 (1996).

[7] Petković, M., Herceg, D., Ilić, S.: Point estimation theory and its applications. Mala knjiga Novi Sad (1997).

[8] Petković, M., Kyurkchiev, N.: A note on the convergence of the Weierstrass SOR method for polynomial roots. JCAM *80*, 163–168 (1997).
[9] Yamamoto, T.: On nonlinear SOR-like metods, I – Applications to simultaneous methods for polynomial zeros. Japan J. Indust. Appl. Math. *14*, 87–97 (1997).

Prof. Dr. N. Kyurkchiev
Institute of Mathematics
Bulgarian Academy of Sciences
"Acad. G. Bonchev Str.", bl. 8
BG - 1113 Sofia
Bulgaria

Computing [Suppl] 16, 151–163 (2002)

Computing

Boundary Regularity Aspects in Solving Contact Problems

D. Langemann

Abstract

The nonlinear contact problem consists in finding the unknown contact zone. Predictor/corrector iterations over the partitions of the boundary approximate the contact zone. Each step contains the solution of a mixed problem of linear elasticity, whose behaviour is well understood. In particular square-root singularities occur in the collision points between regions with boundary conditions of different types, except for the correct partition including the searched contact region.

This property is used to construct a numerical algorithm providing the solution of a quasi-stationary contact problem with dry friction. A relation between macroscopical input parameters of the contact problem is stated. It is checked with known theoretical results and applied – as an example – for discussing the influence of centrifugal forces to the deformation behaviour within the contact zone.

AMS Classification: 65C20, 73T05, 35B65.

Keywords: Contact, dry friction, regularity, modeling, numerics, implementation.

1. Introduction

Within contact mechanics in general, the investigation of rolling contact is of special theoretical and practical interest. The contact zone and its possible partition is unknown. The deformation problem is formulated as free boundary value problem with mixed boundary conditions. Of course, rolling contact is influenced by the actual outer forces like the load, the drag force, the driving moment and so on. They result in a certain transport and rotational velocity and a certain slip between the rolling bodies. These quantities are called macroscopical inputs.

A critical part in the investigation of contact problems is the friction law. Dry friction is often described by Coulombs law, cf. [5]. It distinguishes between sticking and sliding of two particles in contact. So the contact zone is divided into stick zones and slip zones. Of course, they are unknown too. A first simplification of the free boundary value problem is the concentration on quasi-stationary rolling contact. That means that all observed quantities are constant with respect to a steadily moving co-ordinate system.

A widely used method of the computational handling of rolling contact problems is given in [4]. The deformation problem inside both bodies in contact is shifted to the contacting surface and formulated there in terms of Fredholm's integral equations. A predictor/corrector scheme is used to decide whether a certain patch belongs to the contact zone and – in case it does – to the stick zone.

Since this procedure is quite expensive, approximate descriptions are found, e.g. in [3]. Numerous situations of rolling are investigated and the results are used to save a macroscopical friction law in a simplified relation or in a table look up, which can be easily used in multi-body system dynamics. This ansatz is essential for the modeling of multi-body systems because the computation of contact mechanics is much too expensive to be done while the simulation of system dynamics.

A different idea is followed in this paper. Suppose a given partition of the smooth surface $\partial\Omega$ of the body in the domain Ω into a zone with contact conditions and the rest with vanishing Neumann conditions. The zone with contact condition is parted again into zones with stick and slip conditions. The solution of the respective mixed problem of linear elasticity is unique modulo a rigid displacement, cf. [2] and [7]. Square-root singularities occur in the collision points between regions with boundary conditions of different types, except for the correct partition including the searched contact region. This is proved for the Signorini contact problem in [1] where is shown that its solution belongs even to $H^2(\Omega)$ which implies the disappearance of these singularities. This property is applied to construct a numerical algorithm for the solution of quasi-stationary rolling contact problems here.

In Sect. 2, the boundary value problem for the quasi-stationary solution is derived from the time-dependent deformation problem. We get a free boundary value problem with additional restrictions and unknown macroscopical inputs. The additional restrictions produce a relation between the position of the contact zone and its partition into stick and slip zone on the one hand and the acting forces and momenta and the present velocities in the quasi-stationary solution on the other hand.

Section 3 demonstrates this for a simplified mixed boundary value problem for the Laplacian. The additional restrictions are used to select at most one admissible partition of $\partial\Omega$ providing a solution without the named singularities.

As a result we get an algorithm for the numerical handling of rolling contact problems which is applied for the solution of the quasi-stationary rolling contact problem in Sect. 5 and Sect. 6. It is based on the solution of usual boundary value problems with standard methods. So, existing implementations can be used.

The results of the algorithm has been checked with existing ones, and as a selected new one, the influence of the centrifugal forces to the deformation behavior within the contact zone is presented in Sect. 5.

2. Modeling

Let us regard a reference configuration Ω with a smooth boundary whose particles X are mapped to the positions $x(t)$. In the present formulation Ω is a disk with radius R. The displacement is given by $u(X,t) = x(t) - X$. A time-dependent deformation problem is formulated as the boundary value problem

$$\begin{aligned} \varrho\ddot{u}(X,t) &= \nabla_X \cdot T_X(X,t) + f(X,t) \quad \text{in } X \in \Omega, \\ p(X,t) &= T_X(X,t) \cdot n_X(X) \quad \text{on } X \in \partial\Omega, \end{aligned} \tag{1}$$

where ϱ is the mass density, $T_X = T_X(X,t;u(X,t))$ is a stress tensor not yet specified, f is the inner force density, p is the traction at the boundary and n is the outer normal to the reference configuration.

Quasi-stationary rolling is defined by the existence of a moving reference configuration Ω' whose virtual particles Y are mapped to $y = x(t)$ independently of time. This implies that the rolling body rotates with a constant rotational speed ω and that the moving reference configuration is translated with the constant transport velocity b in the direction e_1, s. Fig. 1.

The existence of nontrivial quasi-stationary rolling implies smoothness and rotational symmetry of the reference configuration Ω and thus of the moving reference configuration Ω', too.

Using the transformation of co-ordinates defined by

$$x = X + u(X,t) = Y + bte_1 + v(Y,t) \quad \text{with } Y = Q(\omega t)X \tag{2}$$

and the rotational matrix $Q(\omega t)$, we get the boundary value problem (1) in co-ordinates of the moving reference configuration. Setting all time-derivatives to zero, it is

$$\begin{aligned} \varrho\omega^2\left[\left(\frac{\partial^2 v}{\partial Y^2} \cdot DY\right) \cdot DY - \left(I + \frac{\partial v}{\partial Y}\right)Y\right] &= \nabla_Y \cdot T_Y(Y) + f(Y) \quad \text{in } Y \in \Omega', \\ p(Y) &= T_Y(Y) \cdot n_Y(Y) \quad \text{on } Y \in \partial\Omega', \end{aligned} \tag{3}$$

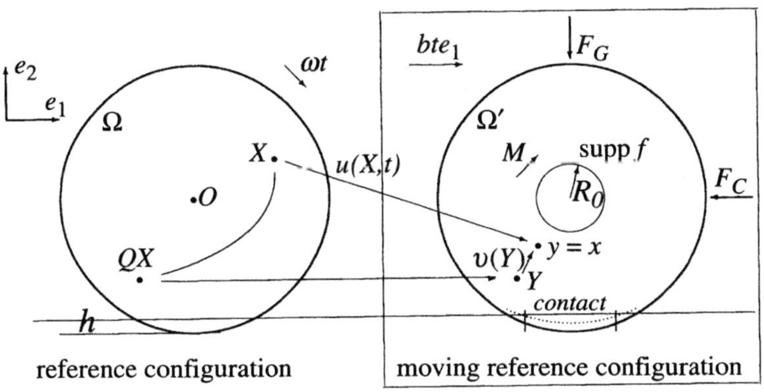

Fig. 1. Model set-up, quasi-stationary rolling

where D is given by $\omega DY = \frac{d}{dt}(Q(\omega t)X)$. By the way, the co-ordinates Y are called arbitrary Eulerian-Lagrangian co-ordinates.

The slip is the relative velocity of two particles in contact to each other. Formulated in the original co-ordinates it is $s(X) = \dot{x}(t)$ if $X \in \partial\Omega$ and X in contact to the support which is not deformable in e_1-direction. If a completely deformable support is considered, its motion has be taken into account too. After the transformation of co-ordinates (2), it holds

$$s(Y) = -\omega Y_2 + b + \omega(-v_{1,1}Y_2 + v_{1,2}Y_1). \tag{4}$$

If Y is a particle in contact, then $Y \in \partial\Omega'$ and $R \approx -Y_2$ is a rather good approximation. In engineering applications $s_0 = \omega R + b$ is called the *rigid-body slip* of the rolling system.

Now, let be $\mathscr{C} \subset \partial\Omega'$ the part of the boundary of the moving reference configuration staying in contact with the support. Then $p(Y) = 0$ holds for all $Y \in \partial\Omega' \backslash \mathscr{C}$. Inside $\mathscr{C}$, it holds $Y_2 + v_2 = -R + h$ expressing that the rolling body lies on a rigid ground.

The contact zone is divided into *stick zones* $\mathscr{A}$ and *slip zones* $\mathscr{B}$. For simplicity, we assume to have only one simply connected stick zone situated in the contact zone. We will see, that this restriction holds and furthermore that – in case of a driven wheel – the stick zone starts at the incoming border of the contact zone. Nevertheless, we state the problem including a hypothetical slip zone $\mathscr{B}_2$ in front of the stick zone and a slip zone $\mathscr{B}_1$ in the back of the stick zone.

Let μ_f be the frictional coefficient in Coulomb's law, s. [5]. We get one of the possible partitions of $\mathscr{C}$ with

$$\begin{array}{llll} \text{in } \mathscr{B}_1 & p_1 = \mu_f p_2 & \text{and} & s(Y) < 0, \\ \text{in } \mathscr{A} & s(Y) = 0 & \text{and} & p_1 \in [-\mu_f p_2, \mu_f p_2], \\ \text{in } \mathscr{B}_2 & p_1 = -\mu_f p_2 & \text{and} & s(Y) > 0. \end{array} \tag{5}$$

Hence we find mixed boundary condition on a-priori unknown boundary segments. Supplementary to the boundary conditions, each part of the contact zone has an additional restriction to the boundary values. These are a unilateral constraints of the slip and thus of the tangential derivative in the slip zone and an inclusion of the tangential traction in the stick zone.

The standard algorithms like in [4] use the additional restrictions for searching suitable $\mathscr{A}$ and $\mathscr{B}_1$. Expensive predictor/corrector methods for integral equations of first Fredholm type are used. In general these methods do not find a steady situation where the predictor stays unchanged, but a periodical situation of several possible predictors.

To explain the present approach, a simplified problem is stated in the following. The idea of the numerical algorithm is explained. We refer to [1] for analytical details and proofs which are done there for stationary contact.

3. A Simplified Analogous Problem

3.1. Problem Statement

In this section we formulate a simplified problem which has analogous properties and thus it is suitable to demonstrate typical effects and methods with less technical effort. We use the usual notations from the theory of partial differential equations and come back to the deformation problem (3) later.

Again, we regard a circle domain Ω of radius 1 and a one-dimensional u related vaguely to one component of the displacement. The polar co-ordinates are (r, φ). Within a stick zone $\mathscr{A}$, we have Dirichlet boundary conditions and within a slip zone $\mathscr{B}$, we find Neumann boundary conditions. That is

$$\begin{aligned} -\Delta u &= f(x) && \text{in} \quad x \in \Omega, \\ \frac{\partial u}{\partial n} &= u_n = 0 && \text{on} \quad x \in \partial\Omega \backslash (\mathscr{A} \cup \mathscr{B}), \\ u_n &= p = -q && \text{on} \quad \mathscr{B}, \\ s(x) &= s_0 - u_\varphi = 0 && \text{on} \quad \mathscr{A}. \end{aligned} \tag{6}$$

Here q plays the role of a given normal pressure, u_φ is the tangential derivative in $\mathscr{A}$, comp. Eq. (4). The additional restrictions to Eq. (6) are

$$s(x) > 0 \quad \text{on } x \in \mathscr{B} \quad \text{and} \quad u_n = p \in [-q, q] \text{ on } x \in \mathscr{A}. \tag{7}$$

We remark that the solution u of Eq. (6) is invariant to an additional constant and hence it is equivalent to

$$\begin{aligned} -\Delta u &= f(x) && \text{in} \quad x \in \Omega, \\ u_n &= 0 && \text{on} \quad x \in \partial\Omega \backslash (\mathscr{A} \cup \mathscr{B}), \\ u_n &= p = -q && \text{on} \quad \mathscr{B}, \\ u &= s_0 \varphi && \text{on} \quad \mathscr{A}. \end{aligned} \tag{8}$$

We denote $F = \int_\Omega f(x) dx$ and the support $\Omega_0 = \operatorname{supp} f$ may have a positive distance to the boundary $\partial\Omega$.

We know from theory that there is always a solution $u \in W^{1,2}$ for each s_0 and each $f(x)$ and every choice of $\mathscr{A}$ and $\mathscr{B}$, cf. [9]. Of course, this solution may not fulfill the additional restrictions (7) in general. In particular, the normal derivative u_n has in general a square-root singularity in the collision point between $\mathscr{A}$ and $\mathscr{B}$. So, $u_n \in [-q, q]$ from Eq. (7) can be fulfilled only if the named singularity vanishes.

3.2. Regularization

To have a closer look to the square-root singularity, we introduce regularized co-ordinates (ξ_1, ξ_2) in the corner point x_0 between $\mathscr{A}$ and $\mathscr{B}$. Without loss of generality, we set $x_0 = (1, 0)$ in polar coordinates and $\varepsilon\xi_1 = \varphi$ and $\varepsilon\xi_2 = 1 - r$ with a small parameter ε. With

$$\frac{\partial u}{\partial \varphi} = \frac{1}{\varepsilon}\frac{\partial u}{\partial \xi_1} \quad \text{and} \quad \frac{\partial u}{\partial r} = -\frac{1}{\varepsilon}\frac{\partial u}{\partial \xi_2}$$

we get

$$\Delta u = \frac{1}{\varepsilon^2}\left(\frac{\partial^2 u}{\partial \xi_2^2} + \frac{1}{(1-\varepsilon\xi_2)^2}\frac{\partial^2 u}{\partial \xi_1^2} - \frac{\varepsilon}{1-\varepsilon\xi_2}\frac{\partial u}{\partial \xi_2}\right).$$

Close to x_0 follows

$$\Delta u \sim \frac{1}{\varepsilon^2}\left(\frac{\partial^2 u}{\partial \xi_2^2} + \frac{\partial^2 u}{\partial \xi_1^2}\right) \quad \text{for } \varepsilon \to 0. \tag{9}$$

In the π-angle domain round x_0, the boundary conditions are

$$u(\xi_1, \xi_2) = \varepsilon s_0 \xi_1 \text{ on } (\xi_1, \xi_2) \in \mathscr{A} \quad \text{and} \quad -u_{,2}(\xi_1, \xi_2) = -\varepsilon q(\varepsilon\xi_1) \quad \text{on } x \in \mathscr{B}. \tag{10}$$

In [6], the behavior of u in angle domains is investigated intensively. The singularity depends on the differential operator, here the Laplacian, and the angle of aperture, here π. It does not depend on the boundary values as long as they fulfill the necessary regularity assumptions.

Since q is related to the bounded normal traction p_2 in Eq. (3), it can be assumed to be bounded itself. The typical behavior of the normal force component, s. [3] and [4], provides even more, namely $q \in H^{\frac{1}{2}-\varepsilon}(\partial\Omega)$. Close to x_0 the function q is in fact bounded but non-vanishing and has a bounded derivative. So, the results from [6] can be used.

Having introduced new local polar co-ordinates (z, ψ) with the origin x_0 belonging to the regularized co-ordinates via $\xi_1 = z\cos\psi$ and $\xi_2 = z\sin\psi$, we get

$$u(z, \psi) \sim \alpha_{\frac{1}{2}} z^{\frac{1}{2}} \sin\frac{1}{2}\psi \quad \text{for } z \to 0 \tag{11}$$

and only the leading term in the expansion determines the behavior of u in the corner, i.e. the singularity. Hence

$$u\left(z, \frac{\pi}{2}\right) \sim \frac{\sqrt{2}}{2}\alpha_{\frac{1}{2}} z^{\frac{1}{2}} \quad \text{for } z \to 0.$$

Thus, the normal derivative $u_n(x_0) = -\varepsilon^{-1}u_{,2}(x_0)$ is given for each fixed ε by

$$\frac{\partial u}{\partial \xi_2}(x_0) = \frac{\partial u}{\partial z}\left(z, \frac{\pi}{2}\right) \sim \frac{\sqrt{2}}{4}\alpha_{\frac{1}{2}} z^{-\frac{1}{2}} \quad \text{for } z \to 0. \tag{12}$$

Finally, the normal derivative $p = u_n$ in x_0 at the boundary is singular if and only if the first coefficient $\alpha_{\frac{1}{2}}$ vanishes. Having u close to x_0 these coefficients determining the singularity can be computed via setting z sufficiently small but fixed and using

$$\alpha_{\frac{1}{2}} = \frac{2}{\pi}\lim_{z\to 0}\frac{1}{\sqrt{z}}\int_0^{\pi} u(z,\psi)\sin\frac{\psi}{2}\,d\psi. \tag{13}$$

All following terms in the expansion are vanishing faster for $z \to 0$. Let us remark that the leading coefficient $\alpha_{\frac{1}{2}}$ is called *stress intensity factor* of u in x_0.

The smoothness condition of the boundaries of both bodies in contact is essential, because a singular boundary with a corner or an edge would in general yield a non-vanishing stress intensity factor in these critical points, s. [7]. The application of the presented formalism to such situations is a question of theoretical interest.

3.3. Uniqueness

This section deals with the question whether the stick zone and the slip zone determine completely the rigid-body slip s_0 and the integrated acting force F. In particular it will be considered if two solutions u_1 and u_2 exist fulfilling (8) and the additional conditions (7) with the same stick zone $\mathscr{A}$ and slip zone $\mathscr{B}$ but with different rigid-body slips s_1 and s_2 or different F_1 and F_2.

Case 1: Assume two solutions u_1 and u_2 with $s_1 = s_2$ but $F_1 > F_2$. The difference $u = u_1 - u_2$ obeys

$$-\Delta u(x) = f(x) \quad \text{in} \quad x \in \Omega \quad \text{and} \quad \int_\Omega f(x)\,dx > 0,$$

$$u_n = 0 \quad \text{on} \quad x \in \partial\Omega\backslash\mathscr{A},$$

$$u = 0 \quad \text{on} \quad x \in \mathscr{A}.$$

The minimum principle yields that the minimal u in $\Omega\backslash\Omega_0$ lies on the boundary $\partial\Omega$, in particular on $\mathscr{A}$. Eq. (13) provides with $u(z,\psi) \geq 0$ and $u(z,\psi) \not\equiv 0$ a non-vanishing $\alpha_{\frac{1}{2}}$ and hence a singular $p = u_n$. One of the solutions has singular normal derivatives, contradiction.

Case 2: Assume two solutions u_1 and u_2 with $s_1 > s_2$ but $F_1 = F_2$. The difference $u = u_1 - u_2$ obeys

$$\begin{aligned} -\Delta u(x) &= 0 && \text{in} \quad x \in \Omega \\ u_n &= 0 && \text{on} \quad x \in \partial\Omega \backslash \mathscr{A}, \\ u &= (s_1 - s_2)\varphi && \text{on} \quad x \in \mathscr{A}. \end{aligned}$$

The minimum and the maximum lie on the boundary again. In particular, the minimum of u is in x_0 and $u(x_0) = 0$. By Eq. (13), we get a non-vanishing stress intensity factor $\alpha_{\frac{1}{2}}$. One of the solutions has singular normal derivatives, contradiction.

Case 3: Assume two solutions u_1 and u_2 with $s_1 > s_2$ but $F_1 \neq F_2$. The difference $u = u_1 - u_2$ obeys

$$\begin{aligned} -\Delta u(x) &= f(x) && \text{in} \quad x \in \Omega \\ u_n &= 0 && \text{on} \quad x \in \partial\Omega \backslash \mathscr{A}, \\ u &= (s_1 - s_2)\varphi && \text{on} \quad x \in \mathscr{A}. \end{aligned}$$

If $F_1 - F_2 > 0$, the minimum of u is in x_0 and the argumentation is exactly the same as in case 2. Otherwise, the maximum of u is in the other corner of $\mathscr{A}$ called x_0' and the argumentation follows again case 2.

For given $\mathscr{A}$ and $\mathscr{B}$, there exists exactly one couple s_0 and F so that the boundary value problem (8) has a solution u fulfilling the additional restrictions (7). It holds $u \in W^{1,\infty}$ then. Remark that it is a necessary condition and not a sufficient one.

The existence of a solution $u \in W^{1,2}$ for every s_0 and F without any respect to the additional restrictions was upgraded to the existence of $u \in W^{1,\infty}$ for a single couple s_0 and F. It may be conjectured that always $u \in C^1(\bar{\Omega})$ holds for a solution u of Eq. (8) under the restrictions (7).

4. Numerical Implementation and Results

The results from the previous sections can be easily used in an algorithm to find suitable s_0 and F for given zones $\mathscr{A}$ and $\mathscr{B}$ or vice versa to find suitable zones for given s_0 and F. Here, we demonstrate the algorithm in the case of given stick and slip zones and searched s_0 and F. The corners of the stick zone $\mathscr{A}$ are called x_0 and x_0'.

Finite element or finite difference methods provide a $W^{1,2}$-solution for all s_0 and F. Eq. (13) provides the leading coefficients $\alpha_{\frac{1}{2}}(x_0)$ and $\alpha_{\frac{1}{2}}(x_0')$. These are the stress intensity factors of u in x_0 and x_0'. Now, a solver for nonlinear equations finds the solutions s_0 and F for which $(\alpha_{\frac{1}{2}}(x_0), \alpha_{\frac{1}{2}}(x_0')) = (0, 0)$ holds. This is a 2×2-system which is solved in a straightforward manner. The stress intensity factors depend continuously and monotonously on s_0 and F in a neighborhood of the $W^{1,\infty}$-solution.

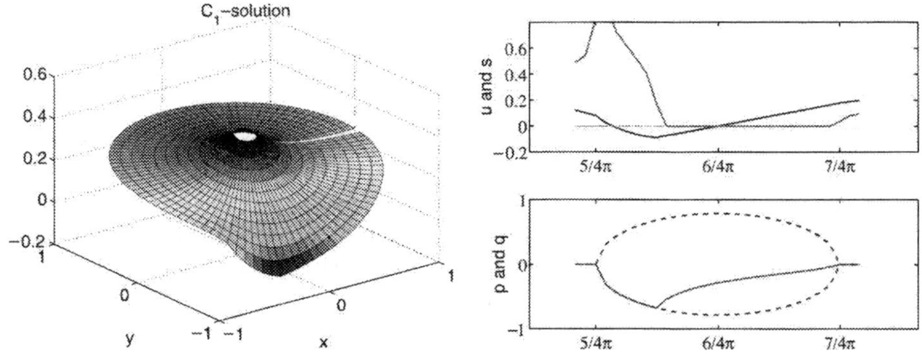

Fig. 2. Typical solution u of the simplified boundary value problem (8) and the behavior at the boundary near the contact zone

A typical solution is presented in Fig. 2. The algorithm works quite exactly, e.g. a discretization with 4600 knots provides a defect in the range of 10^{-11} just by using standard procedures. The couple s_0 and F as a solution of the system of non-linear equations is found without any problems. The number of steps needed by Newton's method is in the range of 10, and s_0 and F are found with machine accuracy.

Comparing this to the conflicting convergence behaviour of predictor/corrector methods named in Sect. 2, this is an advantage of the presented algorithm. Further, Fredholm's integral equations of the first type have an uncomfortable error behaviour if the discretization is refined. This is avoided here, because the presented algorithm is based on the sequential solution of boundary value problems. Grid adaptation and refinement close to critical points or edges is well studied.

Theoretical results for driven wheels are confirmed. Compare the fact that the stick zone starts at the incoming border of a contact zone, first discussed in [8]. For instance, if the algorithm is applied to Eq. (6) including a non-empty slip zone $\mathscr{B}_2$ in front of the stick zone, a $W^{1,\infty}$-solution is found. But then $s(x) > 0$ and $p(x) > 0$ hold together in $\mathscr{B}_2$. So, the additional restrictions are injured.

This example shows that $u \in W^{1,\infty}$ is a necessary but non-sufficient condition. On the one hand, this means that a solution must be checked whether a suitable partition of the contact zone into stick and slip zones was been found. On the other hand, this may serve to find and evaluate theoretical results about the existence of certain zones.

Furthermore, the algorithm has been used to find a relation between s_0 and F. Such a relation is analogous to a macroscopical friction law. It is similar to rolling friction with saturation, see [5]. The rigid-body slip was nearly proportional to the length of the slip zone. So, initial guesses for s_0 and F needed in Newton's method are easily to be found, even if an analytical proof of these facts is difficult. Of course, the results of the simplified boundary value problem cannot be applied directly to the original problem (3) and thus they are presented merely in a short form.

5. The Two-dimensional Deformation Problem Including Centrifugal Forces

5.1. Centrifugal Forces

The algorithm presented in the previous section shall be applied for solving the time-independent boundary value problem (3). It is written in the form

$$\begin{aligned} \varrho\omega^2(Wv - Y) &= Lv + f \quad \text{in} \quad Y \in \Omega', \\ p &= Bv \qquad\quad \text{on} \quad Y \in \partial\Omega'. \end{aligned} \tag{14}$$

In this abbreviated notation, L denotes the Lamé operator, W is the second order operator on the left-hand side of Eq. (3), f are the inner forces consisting in the load F_G, the drag force F_C and the driving momentum M. The boundary operator B is the natural boundary operator to L – and not to the combined operator $L - \varrho\omega^2 W$, what is the reason for some difficulties described in the following.

The investigations require an elliptic operator $L - \varrho\omega^2 W$. Its type changes for higher rotational speed ω. In this case, basically different considerations are needed.

The horizontal equilibrium of forces yields

$$F_C - \varrho\omega^2 \int\limits_{\Omega'} Wv\, dY + \int\limits_{\partial\Omega'} p_1\, dY = 0 \tag{15}$$

and the equilibrium of momenta provides

$$M - \varrho\omega^2 \int\limits_{\Omega'} Wv \times Y\, dY + \int\limits_{\partial\Omega'} p \times Y\, dY = 0. \tag{16}$$

The conditions (15) and (16) are necessary for the existence of a solution of the boundary value problem (14), resp. (3), and hence for the existence of a quasi-stationary solution of the time-dependent boundary value problem (1).

We assume that the drag force depends in a known manner on the transport speed b and M is given. Then the corrector terms

$$k_C = \int\limits_{\Omega'} Wv\, dY \quad \text{and} \quad k_M = \int\limits_{\Omega'} Wv \times Y\, dY \tag{17}$$

complete Eqs. (15) and (16) to an equation for determining the rotational speed $\omega < 0$ and thus the rigid-body slip s_0. The named difficulties arise form the fact that both corrector terms k_C and k_M are depending on the displacement v.

To handle this, an iteration $v^{(i)} \to v^{(i+1)}$ converging to the solution $\lim_{i\to\infty} v^{(i)} = v$ is stated:

Step 1: Determine the corrector terms k_M and k_C from the displacement $v^{(i)}$ via Eq. (17).

Step 2: Calculate the rotational velocity ω and the transport speed b form $v^{(i)}$ and the corrector terms via Eqs. (15) and (16).

Step 3: Compute $v^{(i+1)}$ from Eq. (14). Since the boundary conditions do not obey the equilibrium conditions for $v^{(i+1)}$ but for $v^{(i)}$, its discretization become contradictory. A generalized solution is used instead, which minimizes the error in a suitable norm on $\bar{\Omega}'$.

The iteration is aborted if the change of the corrector terms k_C and k_M is sufficiently small. Then the equilibrium conditions in step 3 are fulfilled with respect to the chosen accuracy.

5.2. Combined Calculation of Contact and Stick Zone

The algorithm which was used to determine the stick zone in Sect. 4 for a given contact zone, can be upgraded to compute the contact zone $\mathscr{C} = \mathscr{A} \cup \mathscr{B}$ and the stick zone $\mathscr{A}$ simultaneously.

We get a 4×4 nonlinear system for the two stress intensity factors in tangential direction in the corners of the stick zone and for the two stress intensity factors in normal direction in the corners of the contact zone.

In the case of a given contact zone $\mathscr{C}$ and a given stick zone $\mathscr{A}$, the driving momentum M, the load F_G and the rigid-body slip s_0 are searched. Then, the lacking quantities like b and ω are computed by Eqs. (15) and (16).

The system looses its symmetry, and thus not every contact zone is feasible. We introduce a displacement of the contact zone as a whole to complete the 4×4-

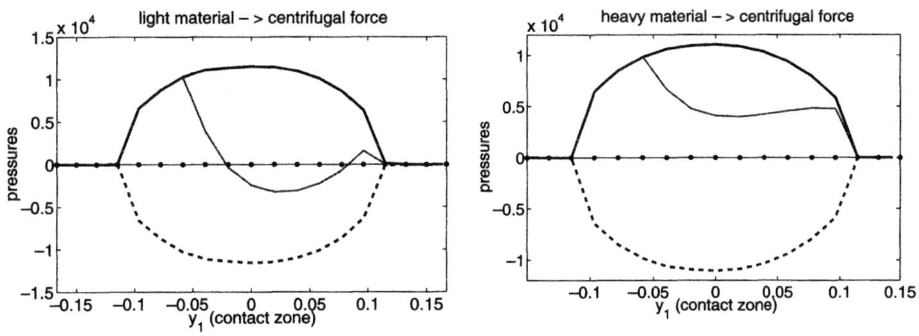

Fig. 3. Solution of Eq. (3), saturation of the tangential force in the contact zone in the cases of light and heavy material

system. The results are presented in Fig. 3. The influence of the sticking is the less the heavier the material is.

The algorithm can be applied vice versa for given M, F_G and a known relation between the transport speed and the drag force, e.g. $-c_w b^2 = F_C$. In this case the positions of the contact zone and the stick zone are computed.

It is analogously found that the additional restrictions can only be fulfilled if $\mathscr{B}_2 = \emptyset$, cf. (5). That means that the stick zone of a driven wheel starts at the incoming border according to the theory, [8].

6. Application to Three-dimensional Problems

In a three-dimensional problem of rolling contact, the zones $\mathscr{A}$, $\mathscr{B}$ and $\mathscr{C}$ are not anymore intervals on a one-dimensional boundary but patches on a two-dimensional surface.

The idea to make the respective stress intensity factors vanishing stays the same. Special care has to be taken to the normal component of the outer traction at the boundary of the contact zone $\mathscr{C}$. At the boundary of the stick zone $\mathscr{A}$, the critical component is the tangential one. It is no longer possible to solve a finite system of non-linear equations. But the stress intensity factors vanish on whole the named boundaries.

A way to come out is given by the fact that the stress intensity factors depend in a monotonous manner on the shape of the boundary, comp. the problem of a punch on a half space in [7]. A strong analytical base like given in [1] for the two-dimensional case seems to be still vacant.

The boundary $\partial\mathscr{A} \subset \mathscr{C}$ of the stick zone may be described by $Y_{\mathscr{A}}(a)$ with the parameter a. The respective stress intensity factor in a point at this boundary is $\alpha_{\frac{1}{2}}(a) = \alpha_{\frac{1}{2}}(Y_{\mathscr{A}}(a))$. Now, we state the boundary evolution problem

$$\frac{\partial}{\partial\tau} Y_{\mathscr{A}}(a,\tau) = k \cdot \alpha_{\frac{1}{2}}(a,\tau), \tag{18}$$

where τ is a virtual time. Eq. (18) is discretized spatially by the line method. With a suitable choice of the constant factor k, standard methods for stiff ordinary differential equations provide a solution $Y_{\mathscr{A}}(a,\tau)$ of the discretized Eq. (18) with the property

$$\lim_{\tau\to\infty} Y_{\mathscr{A}}(a,\tau) = Y_{\mathscr{A}}(a). \tag{19}$$

The final steady state $Y_{\mathscr{A}}(a)$ describes the boundary $\partial\mathscr{A}$ of the stick zone with vanishing stress intensity factors. In case of an unknown contact zone itself, the same procedure is applied for the evolution of its boundary $\partial\mathscr{C}$.

7. Conclusion

Contact problems are characterized by high regularity requirements of the solution. The fact was used to construct an algorithm to handle contact problems with dry friction in two and in three dimensions.

It consists in solving a system of nonlinear equations between the macroscopical values like the rigid-body slip, the drag force, the transport and the rotational speed on the one side, and the position and length of the contact zone and the stick zone on the other side. In the three-dimensional case, the shape of the contact zone and of the stick zone is determined via an auxiliary boundary evolution problem.

The basic idea, fundamentally discussed in [1], consists in considering the stress intensity factors and in letting them vanish. So each step in the solution of the named non-linear system requires a solution of the boundary value problem what can be done by standard methods like finite elements or even finite differences.

The advantages are first the possibility to adapt the accuracy close to the corner points and second the convergence of the method and hence a simple abortion criterion. Additional conditions to the boundary values of partial differential equations enforce higher regularity on the boundary and so, they serve in finding the boundary partition and in finding a relation between macroscopical values, which might describe a wider class of problems.

References

[1] Eck, C., Nazarov, S. A., Wendland, W. L.: Asymptotic analysis for a mixed boundary-value contact problem: Arch. Rational Mech. Anal. *156*, 275–316 (2001).
[2] Gurtin, M. E.: The linear theory of elasticity, encyclopedia of physics, VIa/2: Mechanics of solids. Berlin: Springer 1972
[3] Johnson, K. L.: Contact mechanics. Cambridge: University Press 1985.
[4] Kalker, J. J.: Three-dimensional elastic bodies in rolling contact. Solid Mech. Appl. *2*. Dordrecht: Kluwer Academic Publisher 1990.
[5] Kragelski, I. V.: Friction and wear. Washington, DC: Butterworth 1965.
[6] Mazja, V. G., Nazarov, S. A., Plamenevskij, B. A.: Asymptotic theory of elliptic boundary value problems in singularly pertubated domains. Basel: Birkhäuser 2000.
[7] Nowacki, W.: Theory of elasticity. Warsaw 1978.
[8] Poritsky, H.: Stresses and deflections of cylindrical bodies in contact with application to contact of gears and of locomotive wheels. J. App. Mech. *72*, 191–201 (1950).
[9] Renardy, M., Rogers, R. C.: An introduction to partial differential equations. New York: Springer 1992.

Dirk Langemann
University of Rostock
Department of Mathematics
D-18051 Rostock
Germany
e-mail: lgm@alf.math.uni-rostock.de

Computing [Suppl] 16, 165–170 (2002)

Computing

Convex-decomposable Operators and Inclusive Algorithms

Y.-x. Ling, H.-l. Cao, and **H-f. Sheng**

Abstract

If an operator has the nature $Fy - Fx \leq (D_1(x,y) + D_2(x,y))(y-x)$ for $x \leq y$, where D_1 and D_2 are isotone and antitone respectively, we denote it a convex-decomposable operator. We will show that it is a rather large class of operators. For these operator equations, some monotone inclusive iterations are given. Under some usual assumptions, the iteration converges to the minimum and maximum solution.

In this paper, we discuss some algorithm for solving a class of nonlinear systems $Fx = 0,\ x \in \Omega$, where $\Omega = \{x \in R^n \mid a \leq x \leq b\}$ is a finite ordered interval.

Mathematics Subject Classification: 47H17, 47H99.

Keywords: Nonlinear operator equations, convex decomposable operators.

1. Definition of a Convex-decomposable Operator

In [1] [2], we have introduced the definition of a convex-decomposable operator. Now we will study it further more.

Let F be an operator, $F : \Omega \subseteq R^n \longrightarrow R^n$, if there exists $D(x,y) \in L(R^n)$ such that

$$Fy - Fx \leq D(x,y)(y-x), \qquad \forall x \leq y, \qquad x,y \in \Omega, \tag{1}$$

then $D(x,y)$ is called a unilateral generalized divided difference of F.

The operator $F : \Omega \subseteq R^n \longrightarrow R^n$ is called convex-decomposable, if there exist $D_i : \Omega \times \Omega \longrightarrow L(R^n)$, $i = 1,2$, such that the unilateral generalized divided differentience $D(x,y)$ of F can be splitted into

$$D(x,y) = D_1(x,y) + D_2(x,y), \tag{2}$$

where

$$D_1(x,y) \leq D_1(\bar{x},\bar{y}) D_2(x,y) \geq D_2(\bar{x},\bar{y})\} \quad \forall x \leq \bar{x} \leq \bar{y}, \quad x \leq y \leq \bar{y}, \quad x,y,\bar{x},\bar{y} \in \Omega.$$

In [1], the author has proved that the convex-decomposable operator $F : \Omega \subseteq R^n \longrightarrow R^n$ is a large class of operators, for example, it includes:

(1) Ordered Lipschitz operators: if there exists a nonnegative linear operator $P \in L(R^n)$ such that

$$Fy - Fx \leq P(y - x) \quad \forall x \leq y \quad x, y \in \Omega \subseteq R^n.$$

(2) G-differentiable order-convex operators whose G-derivatives monotone-decomposable, i.e. the G-derivative $F'(x)$ is the sum of an isotone and an antitone operator.

In [2], we have proved:

(3) If the operator $F : \Omega \subseteq R^n \longrightarrow R^n$ is twice differentiable and its second derivative F'' is bounded in Ω, then F is a convex-decomposable operator.

(4) Under the same conditions we can get a more general formulation: where $\forall x \leq y, \quad x, y \in \Omega$ and $\bar{D}_1, \bar{D}_2 : \Omega \times \Omega \longrightarrow L(R^n)$ are isotone and $D_1, D_2 : \Omega \times \Omega \longrightarrow L(R^n)$ are antitone.

(5) Suppose that the operator $F : \Omega \subseteq R^n \longrightarrow R^n$ is continuously differentiable and $F'(x)$ is Lipschitz-continuous, then there exist $D_1, D_2 : \Omega \times \Omega \longrightarrow L(R^n)$ such that

$$(D_1(x) + D_2(y))(y - x) \leq Fy - Fx \leq (D_1(y) + D_2(x))(y - x),$$

for any $x \leq y$, $x, y \in \Omega$ and $D_1, D_2 : \Omega \times \Omega \longrightarrow L(R^n)$ are isotone and antitone, respectively.

2. Monotone Inclusive Algorithm

In [2] we have given a Steffenson-like inclusive method for finding the solution of a convex-decomposable operator equation. First we give a simple algorithm to solve an operator equation in $C^2(\Omega)$:

Suppose that $F : [x^0, y^0] \subset R^n \rightarrow R^n$ satisfies the following inequality:

$$Fy - Fx \leq (D_1(y) + D_2(x))(y - x), \qquad \forall x \leq y, x, \quad y \in [x^0, y^0], \tag{3}$$

where $D_1, D_2 : [x^0, y^0] \subset R^n \rightarrow L(R^n)$ are isotone and antitone, respectively. If

(i) $D_1(y^0) + D_2(x^0)$ has a nonnegative nonsingular left-subinverse,

(ii) $Fx^0 \leq 0 \leq Fy^0$,

then

(1) the equation $Fx = 0$ has a minimum solution x^* and a maximum solution y^* in $[x^0, y^0]$,

(2) the iteration

$$\begin{aligned} &(D_1(y^k) + D_2(x^k))(x^{k+1} - x^k) + Fx^k \\ &\quad = 0(D_1(y^k) + D_2(x^{k+1}))(y^{k+1} - y^k) + Fy^k = 0\} \\ &k = 1, 2, \ldots \end{aligned} \tag{4}$$

is well-defined, and the interval sequence $[x^k, y^k]$ provides monotone enclosures for the solutions of $Fx = 0$:

$$[x^*, y^*] \subset [x^{k+1}, y^{k+1}] \subset [x^k, y^k], \quad k = 1, 2, \cdots [x^*, y^*] = [\lim x^k, \lim y^k] = \bigcup_{k=0}^{\infty} [x^k, y^k]. \tag{5}$$

For a more complex situation, we consider a class of Steffenson-like iterations

$$[D_1(\tilde{u}_1^k, \tilde{v}_1^k) + D_2(\tilde{u}_2^k, \tilde{v}_2^k)](x^{k+1} - x^k) + Fx^k = 0, \tag{6}$$

$$[D_1(\bar{u}_1^k, \bar{v}_1^k) + D_2(\bar{u}_2^k, \bar{v}_2^k)](y^{k+1} - y^k) + Fy^k = 0, \tag{7}$$

where

$$\begin{array}{ll} \tilde{u}_1^k \in [x^k, y^0], & \tilde{u}_2^k \in [x^{p_k}, x^k] \\ \tilde{v}_1^k \in [x^0, y^{-1}], & \tilde{v}_2^k \in [x^k, x^k - \tilde{B}_k Fx^k] \\ \bar{u}_1^k \in [y^k - \bar{B}_k Fy^k, y^k], & \bar{u}_2^k \in [x^{-1}, y^*] \\ \bar{v}_1^k \in [y^k, y^{q_k}], & \bar{v}_2^k [x^0, y^k] \end{array}$$

and $\tilde{B}_k$ is a nonnegative nonsingular left-subinverse of $D_1(\tilde{u}_1^k, \tilde{v}_1^k) + D_2(x^{p_k}, x^k)$, $0 \leq p_k \leq k-1$, $\bar{B}_k$ is a nonnegative nonsingular left-subinverse of $D_1(y^k, y^{q_k}) + D_2(\bar{u}_2^k, \bar{v}_2^k)$, $0 \leq q_k \leq k-1$.

For these Steffenson-like iterations, we can prove:

If $F : \Omega \subset R^n \to R^n$ is a continuous convex-decomposable operator and $[x^0, y^0] \subset [x^{-1}, y^{-1}] \subset \Omega$, such that $D_1(u_1, v_1) + D_2(u_2, v_2)$ has nonnegative non-singular left-subinverse B for $(u_i, v_i) \in [x^{-1}, y^{-1}]$, $i = 1, 2$, and $Fx^0 \leq 0 \leq Fy^0$, then

(i) the equation $Fx = 0$ has a minimum solution x^* and a maximum solution y^* in $[x^0, y^0]$,

(ii) the iterations (2.3), (2.4) are well defined, and the interval sequence $[x^k, y^k]$ provides monotone enclusures for the solution of the equation $Fx = 0$.

Proof: Suppose that, B_0 is the nonnegative nonsingular left-subinverse of $D_1(y^0, y^{-1}) + D_2(x^{-1}, x^0)$, we define an operator H as

$$Hz = z - B_0 Fz, \quad z \in [x^0, y^0].$$

Since F is a continuous convex-decomposable operator, and for every $z_1 \leq z_2$, $z_1, z_2 \in [x^0, y^0]$, we get

$$\begin{aligned} Hz_1 - Hz_2 &= z_1 - z_2 - B_0(Fz_1 - Fz_2) \\ &\geq z_1 - z_2 - B_0[D_1(z_1, z_2) + D_2(z_1, z_2)](z_1 - z_2) \\ &\geq z_1 - z_2 - B_0[D_1(y^0, y^{-1}) + D_2(x^{-1}, x^0)](z_1 - z_2) \geq 0, \end{aligned}$$

thus, the operator H is an isotone operator, and by the condition

$$\begin{aligned} Hx^1 &= x^0 - B_0Fx^0 \geq x^0 \\ Hy^1 &= y^0 - B_0Fy^0 \leq y^0, \end{aligned}$$

the Lemma of Kantorovich guarantees the existence of a minimum solution x^* and a maximum solution y^* of the operator H in $[x^0, y^0]$. Since B_0 is nonnegative, it is evident that $Fx^* = Fy^* = 0$, x^* and y^* are also the minimum solution and maximum solution of $Fx = 0$ in $[x^0, y^0]$, respectively.

Secondly, we will use induction to prove that the iteration is well posed and that the iteration sequence $\{x^k\}$ and $\{y^k\}$ has the property

$$x^0 \leq x^1 \leq \cdots \leq x^k \leq x^* \leq y^* \leq y^k \leq \cdots \leq y^1 \leq y^0, \tag{8}$$

$$Fx^k \leq 0 \leq Fy^k, \quad k = 1, 2, \ldots \tag{9}$$

This is trivial for $k = 0$. Let us assume that (2.6), (2.7) hold for some $k \geq 0$. We then consider the operator $\tilde{H}_k$ defined by

$$\tilde{H}_k z = z - \tilde{B}_k[(D_1(\tilde{u}_1^k, \tilde{v}_1^k) + D_2(\tilde{u}_2^k, \tilde{v}_2^k))z + Fx^k] \tag{10}$$

since

$$D_1(\tilde{u}_1^k, \tilde{v}_1^k) + D_2(\tilde{u}_2^k, \tilde{v}_2^k) \leq D_1(\tilde{u}_1^k, \tilde{v}_1^k) + D_2(x^{p_k}, x^k), \tag{11}$$

we have

$$I - \tilde{B}_k(D_1(\tilde{u}_1^k, \tilde{v}_1^k) + D_2(\tilde{u}_2^k, \tilde{v}_2^k)) \geq I - \tilde{B}_k(D_1(\tilde{u}_1^k, \tilde{v}_1^k) + D_2(x^{p_k}, x^k)) \geq 0, \tag{12}$$

proving that $\tilde{H}_k$ is isotone. Setting

$$\tilde{z}_k = x^k - \tilde{B}_k Fx^k \tag{13}$$

(and $\tilde{z}_k \geq x^k$), we have

$$\begin{aligned} 0 \leq Fx^* - Fx^k &\leq [D_1(x^k, x^*) + D_2(x^k, x^*)](x^* - x^k) \\ &\leq [D_1(\tilde{u}_1^k, \tilde{v}_1^k) + D_2(x^{p_k}, x^k)](x^* - x^k) \end{aligned}$$

and then we get

$$\begin{aligned} -\tilde{B}_k Fx^k &= \tilde{B}_k(Fx^* - Fx^k) \\ &\leq \tilde{B}_k[D_1(\tilde{u}_1^k, \tilde{v}_1^k) + D_2(x^{p_k}, x^k)](x^* - x^k) \leq (x^* - x^k), \end{aligned}$$

i.e.

$$\tilde{z}_k = x^k - \tilde{B}_k F x^k \le x^*. \tag{14}$$

This ensures $\tilde{z}_k \in [x^k, x^*]$. In the sequel, we show that $\tilde{H}_k$ has a fixed point in $[\tilde{z}_k - x^k, x^* - x^k]$. In fact, by virtue of (2.8), (2.10), (2.11) and (2.12) we can conclude

$$\begin{aligned}\tilde{H}_k(\tilde{z}_k - x^k) &= (\tilde{z}_k - x^k) - \tilde{B}_k[(D_1(\tilde{u}_1^k, \tilde{v}_1^k) + D_2(\tilde{u}_2^k, \tilde{v}_2^k))(\tilde{z}_k - x^k) + Fx^k] \\ &\ge -\tilde{B}_k F x^k = \tilde{z}_k - x^k\end{aligned}$$

and furthermore

$$\begin{aligned}\tilde{H}_k(x^* - x^k) &= (x^* - x^k) - \tilde{B}_k[(D_1(\tilde{u}_1^k, \tilde{v}_1^k) + D_2(\tilde{u}_2^k, \tilde{v}_2^k))(x^* - x^k) + Fx^k] \\ &\le -\tilde{B}_k F x^k \le x^* - x^k.\end{aligned}$$

By the Kontorovich lemma, $\tilde{H}_k$ has a minimum fixed point, which we denote by $w^k \in [\tilde{z}_k - x^k, x^* - x^k]$. If we set $x^{k+1} = x^k + w^k$, we know that

$$x^k \le \tilde{z}_k \le x^{k+1} \le x^*,$$

and thus x^{k+1} is the solution of Eq. (2.4).

Similarly, we can show that Eq. (2.5) is solvable, and

$$y^{k+1} \in [y^*, \bar{z}^k] \subset [y^*, y^k],$$

where $\bar{z}^k$ is given by

$$\bar{z}^k = y^k - \bar{B}_k F y^k.$$

These conclusion are verifying (2.6).

Since D_1 and D_2 are isotone and antitone respectively, we have

$$D_1(\tilde{u}_1^k, \tilde{v}_1^k) + D_2(\tilde{u}_2^k, \tilde{v}_2^k) \ge D_1(x^k, x^{k+1}) + D_2(x^k, x^{k+1}) = D(x^k, x^{k+1}),$$

and

$$D_1(\bar{u}_1^k, \bar{v}_1^k) + D_2(\bar{u}_2^k, \bar{v}_2^k) \ge D_1(y^{k+1}, y^k) + D_2(y^{k+1}, y^k) = D(y^{k+1}, y^k),$$

from that we conclude

$$\begin{aligned}Fx^{k+1} &\le Fx^k + D(x^k, x^{k+1})(x^{k+1} - x^k) \\ &\le Fx^k + [D_1(\tilde{u}_1^k, \tilde{v}_1^k) + D_2(\tilde{u}_2^k, \tilde{v}_2^k)](x^{k+1} - x^k) = 0,\end{aligned}$$

$$Fy^{k+1} \geq Fy^k + D(y^{k+1}, y^k)(y^{k+1} - y^k)$$
$$\geq Fy^k + [D_1(\bar{u}_1^k, \bar{v}_1^k) + D_2(\bar{u}_2^k, \bar{v}_2^k)](y^{k+1} - y^k) = 0.$$

It follows that we have verified (2.7).

At last, by use of the regularity of the natural partially ordered R^n, there exist $\tilde{x}^*$ and $\bar{y}^*$ such that

$$\lim_{k \longrightarrow \infty} x^k = \tilde{x}^*, \qquad \lim_{k \longrightarrow \infty} y^k = \bar{y}^*,$$

or

$$[x^*, y^*] \subset [\tilde{x}^*, \bar{y}^*] \subset \bigcap_{k=0}^{\infty} [x^k, y^k].$$

From (2.6) (2.7), we know $F\tilde{x}^* = F\bar{y}^* = 0$, but x^* and y^* are the minimum and maximum solution of $Fx = 0$ in $[x^0, y^0]$, such that $\tilde{x}^* = x^*$, $\bar{y}^* = y^*$.This completes the proof of the theorem. □

References

[1] Ling, Y., Xu, Z.: A monotone iteration of Steffenson-type for nonlinear equations. (in Chinese with English abstract) Xi'an Jiaotone Daxue Xuebao *26*, 59–68 (1992).

[2] Ling, Y.: The convex-decomposable operator equation and its monotonic inclusive iteration. Comput. *57*, 345–356 (1996).

[3] Ortega, J. M., Rhenboldt, W. C.: Iterative solution of nonlinear equations in several variables. New York: Academic Press 1970.

[4] Potra, F. A.: Monotone iterative methods for nonlinear equations. Numer. Funct. Anal. Optim. *9*, 809–843 (1987).

Yong-xiang Ling
Xi'an Jiaotong
University Xi'an
China 710049

Hua-lin Cao
Han-fang Sheng
Qindao
China 266041

Computing [Suppl] 16, 171–184 (2002)

Computing

Fast Inclusion and Residual Iteration for Solutions of Matrix Equations

T. Ogita, S. Oishi and **Y. Ushiro**

Abstract

This paper is concerned with the problem of verifying an accuracy of numerical solutions of matrix equations. In this paper, a fast algorithm of obtaining very sharp verified error bound is proposed. The new verification method is based on combination of the rounding mode controlled double precision computation and the high precision computation. The method can also be used to improve numerical solutions by the residual iteration. Numerical results have also been presented for illustrating that a very sharp verified error bound of numerical solutions can be obtained with a computational cost comparable to that for calculating an approximate solution. Moreover, it is shown using an example that a few residual iterations give extremely improved solutions with little computational cost.

AMS Subject Classifications: 65G05, 65F99.

Keywords: Numerical verification, matrix equation, rounding mode controlled computation, high precision computation, residual iteration.

1. Introduction

In this paper, we are concerned with an accuracy of a numerical solution of the following matrix equation

$$Ax = b, \tag{1}$$

where A is an $n \times n$ real matrix and b is an n-vector. Recently, utilizing rounding mode controlled instructions in IEEE standard 754 of floating point numbers, fast methods [4], [6] have been developed for calculating a rigorous error bound between an exact solution and a numerical solution of Eq. (1). These methods are called rounding mode controlled verification methods [5]. Although the rounding mode controlled verification methods work good for a wide class of linear problems, sometimes numerical results show that obtained error bounds of the numerical solutions are overestimated.

The aim of this paper is to propose a fast algorithm of obtaining a very sharp verified error bound of a numerical solution of Eq. (1). This new verification method is based on combination of the rounding mode controlled double precision computation and the high precision computation. The method can also be used to improve a numerical solution itself by the residual iteration. Numerical results have also been presented for illustrating that a very sharp verified error bound of a numerical solution of Eq. (1) can be obtained with a computational cost comparable to that for calculating an approximate solution of Eq. (1).

Moreover, it is shown using an example that a few residual iterations give extremely improved solutions with little computational cost.

2. Floating Point Arithmetic

In this section, based on Ref. [6], we shall briefly review that for a given $n \times n$ matrix A, an n-vector $\boldsymbol{b}$ and an approximate solution $\tilde{\boldsymbol{x}}$ of $A\boldsymbol{x} = \boldsymbol{b}$ whose elements are double precision floating point numbers of IEEE 754, an inclusion of a residual $A\tilde{\boldsymbol{x}} - \boldsymbol{b}$ can be calculated rigorously using the rounding mode controlled computation. Then, we propose to use the high precision computation in order to enclose the residual $A\tilde{\boldsymbol{x}} - \boldsymbol{b}$ very sharply with guaranteed accuracy.

2.1. Rounding Mode Controlled Computation

Let us start with discussions on floating point numbers. Let $\mathbb{R}$ be the set of real numbers. Let $\mathbb{F}$ be a set of double precision floating point numbers of IEEE 754 floating point standard. It is well known that $\mathbb{F}$ is symmetric, i.e., $x \in \mathbb{F} \Rightarrow -x \in \mathbb{F}$, such that $|x|$ is exact for $x \in \mathbb{F}$. Let $c \in \mathbb{R}$. Four types of rounding modes are defined in IEEE 754. Among them, the following two rounding modes are useful in rounding mode controlled verification:

(1) *rounding downward:* Round $c \in \mathbb{R}$ to the largest floating point number $f \in \mathbb{F}$ satisfying $f \le c$. We represent it by $\bigtriangledown : \mathbb{R} \to \mathbb{F}$.

(2) *rounding upward:* Round $c \in \mathbb{R}$ to the smallest floating point number f satisfying $f \ge c$. We represent it by $\bigtriangleup : \mathbb{R} \to \mathbb{F}$.

The floating point arithmetic abiding by IEEE 754 is defined by the following formula: For $\cdot \in \{+, -, \times, /\}$ and $\bigcirc \in \{\bigtriangleup, \bigtriangledown\}$

$$x \odot y = \bigcirc(x \cdot y), (\text{for any } x, y \in \mathbb{F}). \tag{2}$$

Here, in the left-hand side of Eq. (2), $\odot$ means the floating point arithmetic and in the right-hand side of this equation, $x \cdot y$ represents the ordinary real arithmetic.

We can now present an algorithm of calculating an inclusion of a residual $A\tilde{\boldsymbol{x}} - \boldsymbol{b}$, Algorithm 1. In this paper, we shall express algorithms MATLAB-like.

Algorithm 1. *Rigorous inclusion of $A\tilde{\boldsymbol{x}} - \boldsymbol{b}$*

function $[\underline{\boldsymbol{r}}, \overline{\boldsymbol{r}}] =$ **dincr**$(A, \boldsymbol{b}, \tilde{\boldsymbol{x}})$
setround(*down*); % *rounding downward mode*
$\underline{\boldsymbol{r}} = A * \tilde{\boldsymbol{x}} - \boldsymbol{b}$; % *lower bound of* $A\tilde{\boldsymbol{x}} - \boldsymbol{b}$
setround(*up*); % *rounding upward mode*
$\overline{\boldsymbol{r}} = A * \tilde{\boldsymbol{x}} - \boldsymbol{b}$; % *upper bound of* $A\tilde{\boldsymbol{x}} - \boldsymbol{b}$

Here, the instruction **setround**(*down*) means to adopt the rounding downward mode and the instruction **setround**(*up*) means to adopt the rounding upward mode. We assume that once the rounding mode is changed, it remains unchanged until the next instruction **setround** appears. This assumption is ensured on a wide class of computer systems abiding by the IEEE 754. Under this assumption, the vectors $\underline{\boldsymbol{r}}$ and $\overline{\boldsymbol{r}}$ become the lower bound and the upper bound of $A\tilde{\boldsymbol{x}} - \boldsymbol{b}$, respectively. Thus, the following inequality holds:

$$\underline{\boldsymbol{r}} \leq A\tilde{\boldsymbol{x}} - \boldsymbol{b} \leq \overline{\boldsymbol{r}}.$$

2.2. High Precision Computation

In order to obtain a very sharp inclusion of the residual $A\tilde{\boldsymbol{x}} - \boldsymbol{b}$, we shall now propose a method of enclosing the residual using high precision computations. One of easy and fast methods is to use the quadruple precision floating point arithmetic provided that this precision is supported by a computer on which one execute computations. An alternative is to use the multiple precision arithmetic implemented via software.

In this paper, for the sake of simplicity we assume that the quadruple precision floating point arithmetic is available. Since nowadays the quadruple precision floating point systems can be implemented easily on many computing systems, this assumption does not restrict ourselves severely. To make the discussion concrete we now discuss how to obtain a guaranteed numerical results with the quadruple precision computations taking the case of the computer HITACHI SR8000 as an example. This machine is one of a popular supercomputer with parallel architecture.

Remark 1. *In IEEE 754, the quadruple precision is defined with 128 format width in bits (A sign width is 1, an exponent width is 15 and an actual mantissa width is 112). However, at present even if a name is given as quadruple precision on computers, it does not often abide by the format of IEEE 754. For example, the quadruple precision of HITACHI SR8000 has only 104 actual mantissa width in bits.*

The quadruple precision floating point number of SR8000 consists of two double precision floating point numbers. This quadruple precision number has 104 actual mantissa width in bits, i.e., it has just twice digit for double precision. Therefore, the rounding errors do not occur because the exact result of a multiplication of two double precision numbers is substituted to the quadruple precision number. However, the addition of floating point number causes rounding errors. Therefore, we try to develop algorithms of calculating bounds of the addition of quadruple precision floating point numbers.

Let $\mathbb{Q}$ be a set of quadruple precision floating point numbers of HITACHI SR8000. Then, the following inequality holds: For any $a, b \in \mathbb{Q}$ and $\epsilon = 2^{-104}$

$$|a + b - quad(a + b)| \leq 2\epsilon \cdot \max\{|a|, |b|\}, \tag{3}$$

where $quad(\cdot)$ means the quadruple precision floating point arithmetic. Based on Eq. (3), the following algorithms give rigorous bounds of an addition $a + b$ such that $\underline{s} \leq a + b \leq \overline{s}$. Here, the instruction **abs**(a) is to calculate $|a|$.

Algorithm 2. *Upper bound of $a + b$ using high precision computation*

$$
\begin{aligned}
&\textbf{function } \bar{s} = \textbf{addup}(a, b) \\
&\epsilon_Q = 2^{-103}; \\
&s = a + b; \\
&c = \epsilon_Q * \textbf{max}(\textbf{abs}(a), \textbf{abs}(b)); \\
&\bar{s} = s + c;
\end{aligned}
$$

Algorithm 3. *Lower bound of $a + b$ using high precision computation*

$$
\begin{aligned}
&\textbf{function } \underline{s} = \textbf{adddown}(a, b) \\
&\epsilon_Q = 2^{-103}; \\
&s = a + b; \\
&c = \epsilon_Q * \textbf{max}(\textbf{abs}(a), \textbf{abs}(b)); \\
&\underline{s} = s - c.
\end{aligned}
$$

We here present a remark. When we implement the algorithms **addup** and **adddown**, we should be careful of the existence of underflow. For example, we can make a computer inform us of the appearance of underflow using a compiler option.

Based on the above mentioned discussions, using quadruple precision computation, we now present Algorithm 4 to get tight enclosure of the residual $A\tilde{\boldsymbol{x}} - \boldsymbol{b}$ such that $\underline{\boldsymbol{r}} \leq A\tilde{\boldsymbol{x}} - \boldsymbol{b} \leq \bar{\boldsymbol{r}}$ in case where each elements of A, $\boldsymbol{b}$, $\tilde{\boldsymbol{x}}$, $\underline{\boldsymbol{r}}$ and $\bar{\boldsymbol{r}}$ are the double precision.

Algorithm 4. *Inclusion of $A\tilde{\boldsymbol{x}} - \boldsymbol{b}$ using high precision computation*

$$
\begin{array}{ll}
\textbf{function } [\underline{\mathbf{r}}, \bar{\mathbf{r}}] = \textbf{qincr}(A, \boldsymbol{b}, \tilde{\boldsymbol{x}}) & \\
[m, n] = \textbf{size}(A); & \%\ \textit{matrix size of } A \\
\epsilon_D = 2^{-52}; & \\
\textbf{for } \ i = 1 : m, & \\
\quad \underline{qs} = \overline{qs} = -b_i; & \%\ \underline{qs},\ \overline{qs} :\ \textit{quadruple precision} \\
\quad \textbf{for } \ j = 1 : n, & \\
\quad\quad qc = \textbf{qprod}(A_{ij}, \tilde{x}_j); & \%\ qc :\ \textit{quadruple precision} \\
\quad\quad \overline{qs} = \textbf{addup}(\overline{qs}, qc); & \\
\quad\quad \underline{qs} = \textbf{adddown}(\underline{qs}, qc); & \\
\textbf{end} & \\
\bar{r}_i = \overline{qs} + \epsilon_D * \textbf{abs}(\overline{qs}); & \%\ \textit{round } \overline{qs} \textit{ into double precision} \\
\underline{r}_i = \underline{qs} - \epsilon_D * \textbf{abs}(\underline{qs}); & \%\ \textit{round } \underline{qs} \textit{ into double precision} \\
\textbf{end} &
\end{array}
$$

Here, for two double precision numbers a and b, the instruction $\mathbf{qprod}(a, b)$ returns the quadruple precision number expressing $a \times b$. If Algorithm 4 terminates, then the following inequality holds:

$$\underline{\boldsymbol{r}} \leq A\tilde{\boldsymbol{x}} - \boldsymbol{b} \leq \overline{\boldsymbol{r}}.$$

Here, a difference between $\underline{\boldsymbol{r}}$ and $\overline{\boldsymbol{r}}$ may be very small compared with the result obtained by Algorithm 1.

Remark 2. *If the exact scalar product with long accumulator [2,3] invented by U. Kulisch and W. Miranker can be used, one can obtain more sharp inclusion of residual* $A\tilde{\boldsymbol{x}} - \boldsymbol{b}$. *See [2],[3] for detail.*

Generally, a computing speed for high precision computation is considerably slower than that for double precision computation. However, we can assume that the computational cost of executing quadruple precision floating point arithmetic is constant multiple, say this constant c, of that for executing double precision floating point arithmetic. In fact, this is the case for HITACHI SR8000. Figure 1 shows speed ratio of quadruple precision computation to double precision computation on SR8000 for calculating $A\tilde{\boldsymbol{x}} - \boldsymbol{b}$. In this case, it is seen that $c < 25$ regardless of the size of the problem, n. Thus, if c is less than n, then considering that the computational cost for calculating an inclusion of a residual vector is $O(n^2)$ quadruple precision floating point operations, the computational cost of enclosing $A\tilde{\boldsymbol{x}} - \boldsymbol{b}$ in quadruple precision becomes less than that for computing an LU factorization of A in double precision.

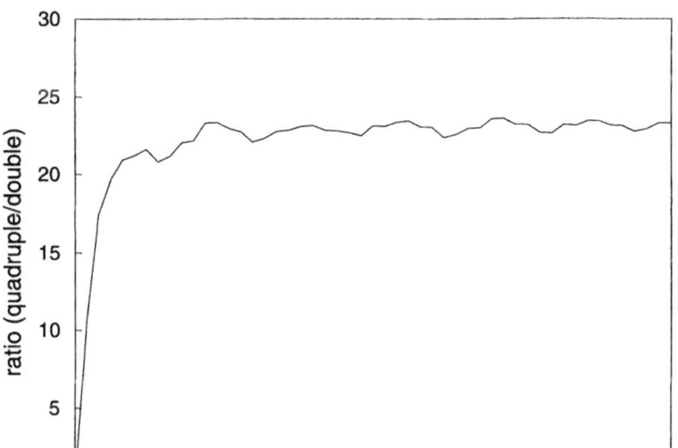

Fig. 1. Speed ratio of quadruple precision computation to double precision computation on HITACHI SR8000

3. Verification Theory

In the following, for a real n-vector $\boldsymbol{x} = (x_1, x_2, \cdots, x_n)^T$, we denote its maximum norm by

$$\|\boldsymbol{x}\|_\infty = \max_{1 \le i \le n} |x_i|. \tag{4}$$

Moreover, for an $m \times n$ real matrix A with i-j element a_{ij}, we denote its maximum norm by

$$\|A\|_\infty = \max_{1 \le i \le m} \sum_{j=1}^{n} |a_{ij}|. \tag{5}$$

Let A be an $n \times n$ real matrix and $\boldsymbol{b}$ an n-vector. Let $\boldsymbol{x}^*$ and $\tilde{\boldsymbol{x}}$ be an exact solution and an approximate solution of $A\boldsymbol{x} = \boldsymbol{b}$, respectively. In this section, we shall show that for a given $n \times n$ real matrix A, a rigorous upper bound of $\|\tilde{\boldsymbol{x}} - \boldsymbol{x}^*\|_\infty$ can be calculated very sharply using the rounding mode controlled computations and the high precision computations.

Let $\boldsymbol{r} = A\tilde{\boldsymbol{x}} - \boldsymbol{b}$ be a residual vector. Then, the following inequality holds:

$$\|\tilde{\boldsymbol{x}} - \boldsymbol{x}^*\|_\infty = \|A^{-1}\boldsymbol{r}\|_\infty \le \|A^{-1}\|_\infty \|\boldsymbol{r}\|_\infty. \tag{6}$$

In order to verify solutions of $A\boldsymbol{x} = \boldsymbol{b}$, it is usual to use bounds of the norm of inverse matrix A^{-1}. To obtain such estimation, there are several methods. Among them, we shall consider the following two methods as examples.

In the first place, we consider the case where A is a general dense matrix. The following is a well-known theorem for calculating an upper bound of $\|A^{-1}\|_\infty$.

Theorem 1. *Let A be an $n \times n$ real matrix, $\boldsymbol{b}$ a real n-vector and R an approximate inverse matrix of A. Let also $\boldsymbol{x}^*$ and $\tilde{\boldsymbol{x}}$ be an exact solution and an approximate solution of $A\boldsymbol{x} = \boldsymbol{b}$, respectively. If $\|RA - I\|_\infty < 1$ is satisfied, then A^{-1} exists and the following inequality holds:*

$$\|A^{-1}\|_\infty \le \frac{\|R\|_\infty}{1 - \|RA - I\|_\infty}, \tag{7}$$

and

$$\|\tilde{\boldsymbol{x}} - \boldsymbol{x}^*\|_\infty \le \frac{\|R(A\tilde{\boldsymbol{x}} - \boldsymbol{b})\|_\infty}{1 - \|RA - I\|_\infty} \tag{8}$$

$$\le \frac{\|R\|_\infty \|A\tilde{\boldsymbol{x}} - \boldsymbol{b}\|_\infty}{1 - \|RA - I\|_\infty}, \tag{9}$$

where I is an $n \times n$ identity matrix. □

Based on this theorem, one of the authors (S. Oishi) and S. Rump have proposed a fast verification method [6] for an upper bound of $\|A^{-1}\|_\infty$ using the LU factorization of A.

When a sparse matrix A is treated, it is important to keep the sparsity of A. Thus, we should not calculate A^{-1} explicitly nor an LU factorization of A. As an example of treatment of a sparse matrix, we consider a special case where a matrix A is sparse and monotone ($A^{-1} \geq O$). The following theorem is useful for calculating an upper bound of $\|A^{-1}\|_\infty$ if A is a monotone matrix.

Theorem 2. (Ogita et al. [4]). *Let A be an $n \times n$ real monotone matrix and $\boldsymbol{e}$ an n-vector where $\boldsymbol{e} = (1, 1, \ldots, 1)^T$. Let also $\tilde{\boldsymbol{y}}$ be an approximate solution of $A\boldsymbol{y} = \boldsymbol{e}$. If $\|A\tilde{\boldsymbol{y}} - \boldsymbol{e}\|_\infty < 1$ is satisfied, then*

$$\|A^{-1}\|_\infty \leq \frac{\|\tilde{\boldsymbol{y}}\|_\infty}{1 - \|A\tilde{\boldsymbol{y}} - \boldsymbol{e}\|_\infty}. \tag{10}$$

□

Utilizing such theorems, we can calculate an upper bound of $\|A^{-1}\|_\infty$ and an error bound of a numerical solution of $A\boldsymbol{x} = \boldsymbol{b}$. Namely, if an upper bound α of $\|A^{-1}\|_\infty$ and a computed solution $\tilde{\boldsymbol{x}}$ of $A\boldsymbol{x} = \boldsymbol{b}$ are given, Algorithm 5 gives an upper bound of $\|\tilde{\boldsymbol{x}} - \boldsymbol{x}^*\|_\infty$.

Algorithm 5. *Rigorous upper bound of $\|\tilde{\boldsymbol{x}} - \boldsymbol{x}^*\|_\infty$*

> **function** $e_{\text{abs}} =$ **errabs1**$(A, \boldsymbol{b}, \tilde{\boldsymbol{x}}, \alpha)$
> $[\underline{\boldsymbol{r}}, \overline{\boldsymbol{r}}] =$ **dincr**$(A, \boldsymbol{b}, \tilde{\boldsymbol{x}})$; % *inclusion of* $A\tilde{\boldsymbol{x}} - \boldsymbol{b}$
> **setround**(up);
> $\widehat{\boldsymbol{r}} =$ **max**(**abs**$(\underline{\boldsymbol{r}})$, **abs**$(\overline{\boldsymbol{r}})$);
> $\overline{r_{\text{norm}}} =$ **norm**$(\widehat{\boldsymbol{r}}, \infty)$; % *upper bound of* $\|A\tilde{\boldsymbol{x}} - \boldsymbol{b}\|_\infty$
> $e_{\text{abs}} = \alpha * \overline{r_{\text{norm}}}$;

Here, for given n-vector $\boldsymbol{v}$, the instruction **max**(**abs**$(\underline{\boldsymbol{v}})$, **abs**$(\overline{\boldsymbol{v}})$) produces the vector $\widehat{\boldsymbol{v}}$ with the i-th element $\widehat{v}_i = \max\{|\underline{v}_i|, |\overline{v}_i|\}$ and the instruction **norm**$(\boldsymbol{v}, \infty)$ is to calculate $\|\boldsymbol{v}\|_\infty$.

Thus, if we have a guaranteed upper bound of $\|A^{-1}\|_\infty$, we can obtain the error bound $\|\tilde{\boldsymbol{x}} - \boldsymbol{x}^*\|_\infty$ by calculating a guaranteed upper bound of $\|\boldsymbol{r}\|_\infty$. Although for a wide class of well posed problems the verification methods based on Eq. (6) work good, numerical experiments show that guaranteed error bounds obtained by such methods are overestimated in certain cases. The reason is mainly on the overestimation caused by estimating $\|A^{-1}\boldsymbol{r}\|_\infty$ through the formula $\|A^{-1}\|_\infty\|\boldsymbol{r}\|_\infty$.

In order to improve this situation, we shall propose to use the following theorem which is a refined version of the estimation (6).

Theorem 3. *Let A be an $n \times n$ real matrix and $\boldsymbol{b}$ an n-vector. Let $\boldsymbol{x}^*$ and $\tilde{\boldsymbol{x}}$ be an exact solution and an approximate solution of $A\boldsymbol{x} = \boldsymbol{b}$, respectively. Let also $\boldsymbol{r} = A\tilde{\boldsymbol{x}} - \boldsymbol{b}$ be a residual vector and $\tilde{\boldsymbol{z}}$ an approximate solution of $A\boldsymbol{z} = \boldsymbol{r}$. Then, the following inequality holds:*

$$\|\tilde{\boldsymbol{x}} - \boldsymbol{x}^*\|_\infty \leq \|\tilde{\boldsymbol{z}}\|_\infty + \|A^{-1}\|_\infty \|A\tilde{\boldsymbol{z}} - \boldsymbol{r}\|_\infty. \tag{11}$$

□

Proof:

$$\begin{aligned} \tilde{\boldsymbol{x}} - \boldsymbol{x}^* &= \tilde{\boldsymbol{x}} - A^{-1}\boldsymbol{b} = A^{-1}(A\tilde{\boldsymbol{x}} - \boldsymbol{b}) & (12) \\ &= A^{-1}\boldsymbol{r} + \tilde{\boldsymbol{z}} - \tilde{\boldsymbol{z}} \\ &= \tilde{\boldsymbol{z}} - A^{-1}(A\tilde{\boldsymbol{z}} - \boldsymbol{r}). & (13) \end{aligned}$$

From Eq. (13), it follows that

$$\begin{aligned} \|\tilde{\boldsymbol{x}} - \boldsymbol{x}^*\|_\infty &= \|\tilde{\boldsymbol{z}} - A^{-1}(A\tilde{\boldsymbol{z}} - \boldsymbol{r})\|_\infty \\ &\leq \|\tilde{\boldsymbol{z}}\|_\infty + \|A^{-1}(A\tilde{\boldsymbol{z}} - \boldsymbol{r})\|_\infty \\ &\leq \|\tilde{\boldsymbol{z}}\|_\infty + \|A^{-1}\|_\infty \|A\tilde{\boldsymbol{z}} - \boldsymbol{r}\|_\infty, & (14) \end{aligned}$$

which implies the desired formula (11). □

The following is a corollary of Theorem 3, which we shall use for verifying an accuracy of numerical solutions of $A\boldsymbol{x} = \boldsymbol{b}$.

Corollary 1. *Let A be an $n \times n$ real matrix and $\boldsymbol{b}$ an n-vector. Let $\boldsymbol{x}^*$ and $\tilde{\boldsymbol{x}}$ be an exact solution and an approximate solution of $A\boldsymbol{x} = \boldsymbol{b}$, respectively. Let also $\boldsymbol{r} = A\tilde{\boldsymbol{x}} - \boldsymbol{b}$ be a residual vector and let $\underline{\boldsymbol{r}}$ be a lower bound and $\overline{\boldsymbol{r}}$ an upper bound of $\boldsymbol{r}$ such that $\underline{\boldsymbol{r}} \leq \boldsymbol{r} \leq \overline{\boldsymbol{r}}$. Let further $\tilde{\boldsymbol{z}}$ be an approximate solution of $A\boldsymbol{z} = \underline{\boldsymbol{r}}$. Then, the following estimation holds:*

$$\|\tilde{\boldsymbol{x}} - \boldsymbol{x}^*\|_\infty \leq \|\tilde{\boldsymbol{z}}\|_\infty + \|A^{-1}\|_\infty (\|A\tilde{\boldsymbol{z}} - \underline{\boldsymbol{r}}\|_\infty + \|\overline{\boldsymbol{r}} - \underline{\boldsymbol{r}}\|_\infty). \tag{15}$$

□

Proof: Since $\underline{\boldsymbol{r}} \leq \boldsymbol{r} \leq \overline{\boldsymbol{r}}$, it holds that

$$\begin{aligned} \|A\tilde{\boldsymbol{z}} - \boldsymbol{r}\|_\infty &= \|A\tilde{\boldsymbol{z}} - \boldsymbol{r} + \underline{\boldsymbol{r}} - \underline{\boldsymbol{r}}\|_\infty \\ &\leq \|A\tilde{\boldsymbol{z}} - \underline{\boldsymbol{r}}\|_\infty + \|\boldsymbol{r} - \underline{\boldsymbol{r}}\|_\infty \\ &\leq \|A\tilde{\boldsymbol{z}} - \underline{\boldsymbol{r}}\|_\infty + \|\overline{\boldsymbol{r}} - \underline{\boldsymbol{r}}\|_\infty. & (16) \end{aligned}$$

From Eq. (11) and Eq. (16), we have the desired formula (15).

We here present a remark. Comparing the formula (15) with the formula (6), it is easily seen that if we want to have a better estimation using Eq. (15) than that for using Eq. (6), $\|\overline{r} - \underline{r}\|_\infty \ll \|r\|_\infty$ should be hold. For this reason, we shall propose to calculate $\underline{r}$ and $\overline{r}$ by making use of the high precision computations.

Based on Corollary 1, we now propose a method of verifying an accuracy of a given calculated solution $\tilde{x}$ of $Ax = b$, where $\tilde{x}$ is calculated by the double precision floating point arithmetic.

Verification method

Step 1: Calculate an upper bound of $\|A^{-1}\|_\infty$. (double precision).

Step 2: Calculate $\underline{r}$ and $\overline{r}$ such that $\underline{r} \le A\tilde{x} - b \le \overline{r}$. (high precision).

Step 3: Solve $Az = \underline{r}$. (double precision).

Step 4: Estimate $\|\tilde{x} - x^*\|_\infty$ applying Eq. (15). (double precision).

Here, we execute the steps 1, 3 and 4 using usual double precision computation and the rounding mode controlled computation. We use the high precision computation only for the step 2.

We now present Algorithm 6 for calculating an error bound $\|\tilde{x} - x^*\|_\infty$. In this algorithm, it is assumed that an upper bound α of $\|A^{-1}\|_\infty$ is given.

Algorithm 6. *Upper bound of $\|\tilde{x} - x^*\|_\infty$ using high precision computation*

$\textbf{function } e_{\text{abs}} = \textbf{errabs2}(A, b, \tilde{x}, \alpha)$
$[\underline{r}, \overline{r}] = \textbf{qincr}(A, b, \tilde{x});$ % *very sharp inclusion of* $A\tilde{x} - b$
$\tilde{z} = \textbf{solve}(Az = \underline{r});$ % *solving* $Az = \underline{r}$
$[\underline{t}, \overline{t}] = \textbf{dincr}(A, \underline{r}, \tilde{z});$ % *inclusion of* $A\tilde{z} - \underline{r}$
$\textbf{setround}(up);$
$\hat{t} = \textbf{max}(\textbf{abs}(\underline{t}), \textbf{abs}(\overline{t}));$
$\overline{d} = \overline{r} - \underline{r};$ % *upper bound of* $\overline{r} - \underline{r}$
$\overline{t_{\text{norm}}} = \textbf{norm}(\hat{t}, \infty);$ % *upper bound of* $\|A\tilde{z} - \underline{r}\|_\infty$
$\overline{d_{\text{norm}}} = \textbf{norm}(\overline{d}, \infty);$ % *upper bound of* $\|\overline{r} - \underline{r}\|_\infty$
$z_{\text{norm}} = \textbf{norm}(\tilde{z}, \infty);$
$e_{\text{abs}} = z_{\text{norm}} + \alpha * (\overline{t_{\text{norm}}} + \overline{d_{\text{norm}}});$

Here, **solve**$(Az = \underline{r})$ is an instruction of solving a matrix equation $Az = \underline{r}$ using the double precision computations. If Algorithm 6 terminates with success, then the following estimate holds:

$$\|\tilde{x} - x^*\|_\infty \le e_{\text{abs}}.$$

Furthermore, we present Algorithm 7 of calculating a relative error bound e_{rel} of $\|\tilde{x} - x^*\|_\infty / \|x^*\|_\infty$ from absolute error bound e_{abs} of $\|\tilde{x} - x^*\|_\infty$ based on Ref. [4].

Algorithm 7. *Rigorous upper bound of* $\|\tilde{\boldsymbol{x}} - \boldsymbol{x}^*\|_\infty / \|\boldsymbol{x}^*\|_\infty$

```
function e_rel = errrel(x̃, e_abs)
x_norm = norm(x̃, ∞);                  % ||x̃||_∞
setround(down);
d = x_norm − e_abs;
if d > 0                               % ||x̃||_∞ − e_abs > 0
  setround(up);
  e_rel = e_abs/d;                     % ||x̃ − x*||_∞/||x*||_∞ ≤ e_rel
else
  print("Verification failed.");
end
```

If Algorithm 7 terminates with success, then the following estimation holds:

$$\frac{\|\tilde{\boldsymbol{x}} - \boldsymbol{x}^*\|_\infty}{\|\boldsymbol{x}^*\|_\infty} \leq e_{\mathrm{rel}}.$$

4. Results of Verification

In this section, in order to illustrate that the proposed verification method gives a very sharp error bound for a numerical solution of $A\boldsymbol{x} = \boldsymbol{b}$, we shall report results of numerical experiments.

We will report two classes of results. One is done for a case of coefficient matrix A being a general dense matrix (Case 1) and the other is a case of A being a sparse monotone matrix (Case 2). Namely, we treated the following two cases:

Case 1 (general dense matrix)

- Coefficient matrix A is a random matrix.
- Matrix size of A is from 1000×1000 to 5000×5000.
- Right-hand side vector $\boldsymbol{b}$ is a random vector.
- An LU factorization of A is calculated.

Case 2 (monotone sparse matrix)

- Coefficient matrix A is a monotone matrix which is a discretized form of two-dimensional diffusion equation $\mathbf{div}\{-k \cdot \mathbf{grad}(u)\} = f$ with Dirichlet and Neumann boundary conditions by the finite element method using linear triangular elements.
- Matrix size of A is from 40000×40000 to 1000000×1000000.
- Right-hand side vector $\boldsymbol{b}$ also comes from the diffusion equation.
- The MICCG method is adopted as an iterative solver.

We have used the super computer HITACHI SR8000 (1 node; 8 processors, 8 GFLOPS, shared memory) with Fortran and C compilers. This computer environment satisfies the standard of IEEE 754. In addition, we have used quadruple precision equipped on SR8000 for the high precision computations. In order to illustrate the effect of proposed method, we have compared the results obtained by our method with those obtained by method proposed in Refs. [4] and [6] based on estimation (6).

In Case 1, the LU factorization has been applied to solve the matrix equation $A\boldsymbol{x} = \boldsymbol{b}$. Then, an upper bound of $\|A^{-1}\|_\infty$ has been calculated by the fast verification method proposed in Ref. [6] based on Theorem 1. By the way, in Case 1, we can also calculate the error bounds of computed solutions by the fast and very sharp verification method which is also proposed in Ref. [6] through the estimation (8)

$$\|\tilde{\boldsymbol{x}} - \boldsymbol{x}^*\|_\infty \leq \frac{\|R(A\tilde{\boldsymbol{x}} - \boldsymbol{b})\|_\infty}{1 - \|RA - I\|_\infty}.$$

In this paper, we call this method *mid-rad method*. Therefore, we also present results by the mid-rad method. Figure 2 shows that the present method gives relative error bounds of computed solutions much more sharply than the results obtained by the previous method and the mid-rad method gives good results comparable to the results obtained by the present method. Moreover, Table 1 shows that the elapsed time for step (C) of verifying a solution using the high precision computations is very short compared with that for step (A) of solving $A\boldsymbol{x} = \boldsymbol{b}$ and that for step (B) of verifying $\|A^{-1}\|_\infty$. This is because we can solve

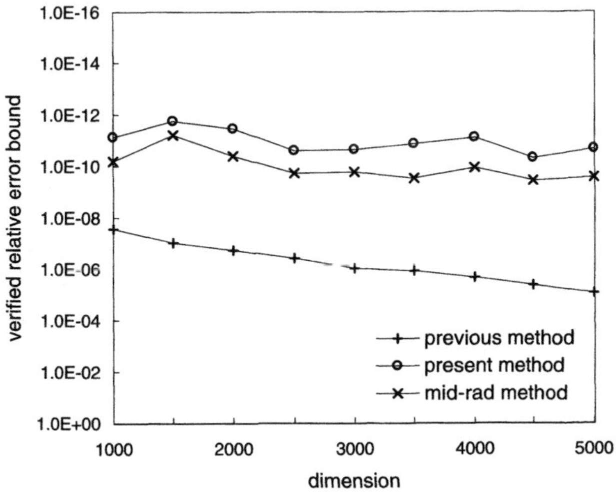

Fig. 2. Results of verification of $\|\tilde{\boldsymbol{x}} - \boldsymbol{x}^*\|_\infty / \|\boldsymbol{x}^*\|_\infty$ (Case 1)

Table 1. Results of elapsed time[sec] (Case 1) (A) Solving $Ax = b$. (B) Verifying $\|A^{-1}\|_\infty$. (C) Verifying $\|\tilde{x} - x^*\|_\infty / \|x^*\|_\infty$

Dimension	(A)	(B)	(C)	(B) + (C)	Total
1000	0.28	0.28	0.08	0.35	0.63
2000	1.53	1.43	0.29	1.72	3.25
3000	4.35	4.06	0.65	4.71	9.06
4000	9.33	9.15	1.14	10.29	19.62
5000	16.66	17.54	1.78	19.32	35.98

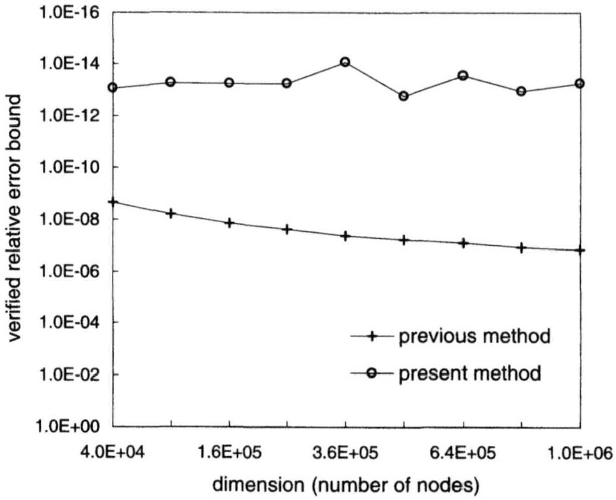

Fig. 3. Results of verification of $\|\tilde{x} - x^*\|_\infty / \|x^*\|_\infty$ (Case 2)

$Az = \underline{r}$ of Algorithm 6 by forward and backward substitutions only provided that the LU factorization of A is given.

In Case 2, we have calculated an upper bound of $\|A^{-1}\|_\infty$ by the fast verification method proposed in Ref. [4] based on Theorem 2. In this verification process, a matrix equation $Ay = e$ where $e = (1, 1, \dots, 1)^T$ must be solved. We have used the fast iterative solver MICCG(Modified Incomplete Cholesky Conjugate Gradient) method to solve the matrix equations $Ax = b$, $Ay = e$ and $Az = \underline{r}$. Moreover, we have chosen the stopping criteria for the iterations as follows: $\|A\tilde{x} - b\|_2 / \|b\|_2 < 10^{-12}$ for solving $Ax = b$, $\|A\tilde{y} - e\|_\infty < 10^{-3}$ for solving $Ay = e$ and $\alpha\|A\tilde{z} - \underline{r}\|_\infty < \|\tilde{z}\|_\infty$ for solving $Az = \underline{r}$. Here, α is an upper bound of $\|A^{-1}\|_\infty$. Figure 3 also shows that the present method gives relative error bounds of computed solutions much more sharply than the results obtained by the previous method. Moreover, Table 2 shows that the elapsed time for step (C) of verifying a solution becomes almost same as that for step (A) of solving $Ax = b$ and that for step (B) of verifying $\|A^{-1}\|_\infty$. This is due to the fact that we have to solve $Az = \underline{r}$ by an iterative solution method in verification process unlike the Case 1.

In conclusion, it turns out that we can get very sharp error bounds of numerical solutions of matrix equations by the proposed method.

Table 2. Results of elapsed time[sec] (Case 2) (A) Solving $Ax = \boldsymbol{b}$. (B) Verifying $\|A^{-1}\|_\infty$. (C) Verifying $\|\tilde{x} - x^*\|_\infty / \|x^*\|_\infty$

Dimension	(A)	(B)	(C)	(B) + (C)	Total
40000	0.97	0.54	0.65	1.19	2.16
160000	2.92	1.65	1.97	3.62	6.54
360000	20.36	10.54	15.99	26.53	46.89
640000	51.97	24.34	34.62	58.96	110.93
1000000	99.23	46.52	65.76	112.28	211.51

5. Residual Iteration and Verification

In this section, we shall show that our verification method for a numerical solution $\tilde{x}$ of $Ax = \boldsymbol{b}$ can be combined with the residual iteration naturally.

From Eq. (12) in Sect. 3, we have

$$x^* = \tilde{x} - A^{-1}(A\tilde{x} - \boldsymbol{b}). \tag{17}$$

We now present the following residual iteration procedures:

Residual iteration

Step 1: Compute a residual $A\tilde{x} - \boldsymbol{b}$ in high precision. (We denote the calculated residual as $\tilde{r}$.).

Step 2: Solve a matrix equation $Az = \tilde{r}$ in double precision. (We denote the obtained approximate solution as $\tilde{z}$.).

Step 3: Update $\tilde{x}$ by calculating $\tilde{x} - \tilde{z}$ in double precision.

In case where one has already verified the accuracy of $\tilde{x}$ using Algorithm 6, the steps 1 and 2 have been finished, thus in such a case one has only to execute Step 3.

We shall now report some numerical results for illustrating the effect of residual iteration. We also treat the examples considered in Sect. 4. Tables 3 and 4 show results of residual iteration and verification of solutions in Case 1 and Case 2, respectively. We have also compared the verification results obtained by the previous verification method proposed in Refs. [6] and [4] with the verification results obtained by the present verification method proposed in Sect. 3. Moreover, these tables show upper bounds of residual norm $\|A\tilde{x} - \boldsymbol{b}\|_\infty$.

Table 3. Results of verification with residual iteration (Case 1) Matrix size of A: 5000×5000. $\mathrm{cond}_\infty(A) \approx 10^8$. (R) Upper bound of $\|A\tilde{x} - \boldsymbol{b}\|_\infty$. (V) Verified relative error bound of solution

Number of residual iterations	Previous method		Present method	
	(R)	(V)	(R)	(V)
0	5.8×10^{-10}	7.7×10^{-5}	5.3×10^{-11}	1.7×10^{-11}
1	5.5×10^{-10}	7.3×10^{-5}	5.1×10^{-15}	1.2×10^{-16}
2	5.5×10^{-10}	7.3×10^{-5}	5.1×10^{-15}	1.2×10^{-16}

Table 4. Results of verification with residual iteration (Case 2) Matrix size of A: 160000×160000. $\mathrm{cond}_\infty(A) \approx 10^{12}$. (R) Upper bound of $\|A\tilde{x} - \boldsymbol{b}\|_\infty$. (V) Verified relative error bound of solution

Number of residual iterations	Previous method		Present method	
	(R)	(V)	(R)	(V)
0	7.9×10^{-5}	1.7×10^{-1}	7.9×10^{-5}	1.4×10^{-7}
1	5.5×10^{-6}	1.1×10^{-2}	2.9×10^{-6}	5.8×10^{-14}
2	5.4×10^{-6}	1.0×10^{-2}	2.9×10^{-6}	1.2×10^{-16}

These results show that if we apply the present method after residual iteration, then we can get much better results than the results obtained by applying the previous method after residual iteration. It turns out that we can get solutions of $A\boldsymbol{x} = \boldsymbol{b}$ which have almost the best accuracy for double precision by one or two residual iteration(s).

Acknowledgements

This research was partially supported by Grant-in-Aid for Scientific Research (B) No. 13440035, (C) No. 12650399 from the Ministry of Education, Science, Sports and Culture, Japan and Waseda University Grant for Special Research Projects No. 2001A-844, 2001B-014.

References

[1] Golub, G. H., Van Loan, C. F.: Matrix computations, 3rd ed. Baltimore London: The Johns Hopkins University Press 1996.
[2] Kulisch, U. W., Miranker, W. L.: Computer arithmetic in theory and practice. New York: Academic Press 1981.
[3] Kulisch, U. W., Miranker, W. L.: The arithmetic of the digital computer: A new approach. SIAM Rev. *28*, 1–40 (1986).
[4] Ogita, T., Oishi, S., Ushiro, Y.: Fast verification of solutions for sparse monotone matrix equations. Comput. Suppl. *15*, 175–187 (2001).
[5] Oishi, S.: Fast enclosure of matrix eigenvalues and singular values via rounding mode controlled computation. Linear Algebra Appl. *324*, 133–146 (2001).
[6] Oishi, S., Rump, S. M.: Fast verification of solutions of matrix equations. Numer. Math. *90*, 755–773 (2002).

Takeshi Ogita
Department of Mathematics
School of Education
Waseda University
Tokyo 169-8050
Japan
e-mail: ogita@aoni.waseda.jp

Shin'ichi Oishi
Department of Computer and Information Science
School of Science and Engineering
Waseda University
Tokyo 169-8555
Japan
e-mail: oishi@oishi.info.waseda.ac.jp

Yasunori Ushiro
Enterprise Server Division
Hitachi Ltd.
Kanagawa 243-0435
Japan
e-mail: yasunori.ushiro@itg.hitachi.co.jp

Computing [Suppl] 16, 185–197 (2002)

Computing

Schröder-like Methods for the Simultaneous Inclusion of Polynomial Zeros*

M. S. Petković and **L. D. Petković**

Abstract

The subject of this paper is the study of Schröder-like methods for the simultaneous inclusion of multiple complex zeros of a polynomial in circular complex arithmetic. These methods are based on the fundamental work of Gargantini and Henrici [7] and belong to the class of the most efficient inclusion methods. Apart from a review of the existing methods of Schröder's type, together with their convergence properties, a new, accelerated method in serial (single-step) mode with Schröder's correction is proposed. The comparison of the considered methods based on their computational efficiency and numerical results are given.

AMS Subject Classifications: 65H05, 65G20, 30C15.

Keywords: Polynomial zeros, simultaneous methods, inclusion of zeros, accelerated convergence, interval arithmetic.

1. Introduction

During the last thirty years various techniques were developed for *a posteriori* error estimates for the approximations of polynomial zeros. A new and very efficient approach to error estimates for a given set of approximate zeros uses circular arithmetic, as presented by Gargantini and Henrici [7] in 1972. After this fundamental paper a number of iterative methods for the simultaneous inclusion of polynomial zeros was constructed. The list of these methods and their characteristics (convergence properties, computational efficiency, numerical experiments) were presented in details in the book [11] and references cited therein. In general, inclusion methods, realized in circular complex arithmetic, produce resulting disks containing complex zeros. In this manner, the upper error bounds, given by the radii of disks, are obtained automatically. This very useful property of self-validated results, together with the ability to incorporate rounding errors without altering the fundamental structure of the iterative formula make frequent implementation of inclusion methods in many problems of applied mathematics and technical disciplines.

The purpose of this paper is to present some iterative methods of Schröder's type for the simultaneous inclusion of all simple or multiple zeros of a (complex)

*Dedicated to Prof. Dr. Jürgen Herzberger on his 60th birthday.

polynomial using circular complex interval arithmetic. The motivation for this study is the fact that these methods are the most efficient ones, as shown in [11, chap. 6]. First we give a review of the existing methods of Schröder's type together with their convergence features, and then, we propose a new, accelerated method in serial (single-step) mode with Schröder's correction. The comparison of the considered methods based on their computational efficiency and numerical results concludes the paper.

For the reader's convenience we briefly review the basic properties of circular interval arithmetic. For more details see the books [3] and [14].

For a (closed) disk Z with the center $z = \text{mid}Z$ and the radius $r = \text{rad}\, Z$, we shall adopt the following parametric notation $Z = \{z; r\}$. If $Z_k := \{c_k; r_k\} (k = 1, 2)$, then

$$Z_1 \pm Z_2 = \{c_1 \pm c_2; r_1 + r_2\},$$
$$Z_1 \cdot Z_2 = \{c_1 c_2; |c_1| r_2 + |c_2| r_1 + r_1 r_2\}.$$

If a disk $Z = \{z; r\}$ does not contain the origin (that is, $|z| > r$), then we can define the two types of inverse disks:

$$Z^{I_1} := Z^{-1} = \{z; r\}^{I_1} := \left\{ \frac{1}{z \cdot \left(1 - r^2/|z|^2\right)}; \frac{r}{|z|^2 - r^2} \right\} \quad \text{(the exact inversion)},$$
$$Z^{I_2} := Z^{I_c} = \{z; r\}^{I_2} := \left\{ \frac{1}{z}; \frac{r}{|z| \cdot (|z| - r)} \right\} \quad \text{(the centered inversion)}.$$

The inversion $\{z; r\}^{I_1}$ is an exact operation, that is, $\{z; r\}^{I_1} = \{z^{-1} | z \in Z\}$, but in general $\text{mid}\, Z^{-1} \neq (\text{mid}\, Z)^{-1}$. We emphasize that the two cases of inversion $()^{I_1} = ()^{-1}$ and $()^{I_2} = ()^{I_c}$, respectively, are handled simultaneously in this paper. Besides, the inversion of the form $1/()$ always assumes the exact inversion $()^{-1}$ without being noted. Sometimes we will write INV_1 and INV_2 instead of $()^{I_1}$ and $()^{I_2}$.

The centered inversion $()^{I_2}$ requires additional operations since the evaluation of $|z| = |z + iy| = \sqrt{x^2 + y^2}$ (appearing in $()^{I_2}$) needs extra operations. As proposed in [4], this disadvantage can be overcome by introducing a new type of centered inversion by

$$\{z; r\}^{I_2^*} := \left\{ \frac{1}{z}; \frac{2r}{|z|^2 - r^2} \right\}$$

since $|z|^2 = x^2 + y^2$ does not require the calculation of a square root. It is easy to show that

$$\{z; r\}^{I_2^*} \supset \{z; r\}^{I_2} \supset \{z; r\}^{-1}.$$

The radius of the inverse disk $\{z;r\}^{I_2^*}$ is two times larger compared to the exact inversion, but this is not a big disadvantage if r is very small (for instance, after the second or third iteration when very fast method is applied).

Following the definition of multiplication of disks and inversion of a disk which does not contain the origin, the division of disks is defined as

$$Z_1/Z_2 := Z_1 \cdot Z_2^{I_k} \quad (0 \notin Z_2) \quad (k \in \{1,2\}).$$

Throughout this paper disks in the complex plane will be denoted by capital letters.

2. Schröder-like Inclusion Methods without Corrections

Let P be a monic polynomial of degree $n \geq 3$ with distinct zeros $\zeta_1, \ldots, \zeta_\nu$ $(\nu \leq n)$ of the multiplicities $\mu_1, \ldots, \mu_\nu \geq 1$ $(\mu_1 + \cdots + \mu_\nu = n)$, that is

$$P(z) = z^n + a_1 z^{n-1} + \cdots + a_{n-1}z + a_n = \prod_{j=1}^{\nu} (z - \zeta_j)^{\mu_j}.$$

Applying the logarithmic derivative we find

$$\frac{P'(z)}{P(z)} = \sum_{j=1}^{\nu} \mu_j (z - \zeta_j)^{-1},$$

wherefrom

$$\zeta_i = z - \frac{1}{\frac{1}{N(z,\mu_i)} - \frac{1}{\mu_i} \sum_{\substack{j=1 \\ j \neq i}}^{\nu} \mu_j (z - \zeta_j)^{-1}} \quad (i = 1, \ldots, \nu). \tag{1}$$

Here $N(z, \mu_i) := \mu_i P(z)/P'(z)$ is Schröder's correction appearing in the Schröder iterative formula

$$\hat{z} = z - \mu \frac{P(z)}{P'(z)}$$

for finding multiple zeros, where $\hat{z}$ is a new approximation. In the sequel, when $z = z_i$, we will write $N(z_i)$ instead of $N(z_i, \mu_i)$ for simplicity.

The fixed point relation (1) is the basis for the construction of iterative methods of Schröder's type for the simultaneous inclusion of multiple zeros of a polynomial. To compute circular approximations for the disjoint zeros $\zeta_1, \ldots, \zeta_\nu$ of a polynomial P simultaneously, we assume that $\nu \geq 2$ and initial disjoint disks $Z_1^{(0)}, \ldots, Z_\nu^{(0)}$, containing the zeros $\zeta_1, \ldots, \zeta_\nu$, are known. We write mid $Z_i^{(m)} = z_i^{(m)}$ and rad $Z_i^{(m)} = r_i^{(m)}$ for the center and the radius of the disk $Z_i^{(m)}$ in the m-th iteration step. Sometimes we will omit the iteration indices for the quantities in the m-th iteration and use the symbol $\widehat{\ }$ for the quantities in the $(m+1)$-st iterative step. For example, we write $z_i, r_i, Z_i, \hat{z}_i, \hat{r}_i, \widehat{Z}_i$ instead of $z_i^{(m)}, r_i^{(m)}, Z_i^{(m)}, z_i^{(m+1)}, r_i^{(m+1)}, Z_i^{(m+1)}$.

In this paper, we will deal with the following disk approximations:

$$\begin{aligned}
&\boldsymbol{Z} = (Z_1, \ldots, Z_\nu) \quad \text{(current approximation)},\\
&\widehat{\boldsymbol{Z}} = (\widehat{Z}_1, \ldots, \widehat{Z}_\nu) \quad \text{(new approximation)},\\
&\boldsymbol{Z}_N = (Z_1 - N(z_1), \ldots, Z_\nu - N(z_\nu)), \quad z_i = \operatorname{mid} Z_i.\\
&\qquad \text{(quasi-Schröder's approximation)}.
\end{aligned}$$

Following the notation for the two types of inversion of a disk, let us introduce the abbreviation

$$\Sigma_{i,\mathrm{INV}_k}(\boldsymbol{A}, \boldsymbol{B}) = \sum_{j=1}^{i-1} \mu_j (z_i - A_i)^{I_k} + \sum_{j=i+1}^{\nu} \mu_j (z_i - B_i)^{I_k} \quad (k \in \{1, 2\}), \tag{2}$$

where $\boldsymbol{A} = (A_1, \ldots, A_\nu)$ and $\boldsymbol{B} = (B_1, \ldots, B_\nu)$ are some vectors of approximations in the form of disks, and INV_k points to the type of disk inversion.

Starting from (1) and using circular interval operations, Gargantini established in [6] the Schröder-like algorithm for the simultaneous inclusion of all zeros of P,

$$\widehat{Z}_i = z_i - \frac{1}{\frac{1}{N(z_i)} - \frac{1}{\mu_i} \sum_{\substack{j=1\\ j\neq i}}^{\nu} \mu_j (z_i - Z_j)^{-1}} \quad (i = 1, \ldots, \nu), \tag{3}$$

where $z_i = \operatorname{mid} Z_i$. The inclusion method (3) has a *cubic* convergence. In the case of simple zeros, the iteration method (3) reduces to the inclusion method proposed in [7]. Using the abbreviation (2), the iterative formula (3) can be written in the form

$$\widehat{Z}_i = z_i - \left(\frac{1}{N(z_i)} - \frac{1}{\mu_i} \Sigma_{i,\mathrm{INV}_1}(\boldsymbol{Z}, \boldsymbol{Z}) \right)^{-1} \quad (i = 1, \ldots, \nu). \tag{4}$$

For $m = 0, 1, \ldots$ let us introduce the abbreviations

$$\begin{aligned}
&r_m = \max_{1 \le i \le n} r_i^{(m)}, \quad \rho_m = \min_{\substack{i,j\\ i\neq j}} \left\{ |z_i^{(m)} - z_{(m)}^j| - r_{(m)}^j \right\},\\
&\varepsilon_i^{(m)} = z_i^{(m)} - \zeta_i, \quad \varepsilon_m = \max_{1 \le i \le n} |\varepsilon_i^{(m)}|.
\end{aligned}$$

As a special case of a general convergence theorem, the following assertion was proved in [10]:

Theorem 1. *Let* $(Z_1,\ldots,Z_\nu) =: \left(Z_1^{(0)},\ldots,Z_\nu^{(0)}\right)$ *be initial disks such that* $\zeta_i \in Z_i$ $(i=1,\ldots,\nu)$. *Then, under the initial condition*

$$\rho_0 > 2nr_0,$$

the interval method (4) *is convergent with the order three, and* $\zeta_i \in Z_i^{(m)}$ *holds for each* $i=1,\ldots,\nu$ *and* $m=1,2,\ldots$.

The rate of convergence of the total-step method (4) can be accelerated by using the already calculated disks as soon as they are available (the so-called Gauss-Seidel approach or single-step mode). In this way the following single-step method is obtained

$$\widehat{Z}_i = z_i - \left(\frac{1}{N(z_i)} - \frac{1}{\mu_i}\Sigma_{i,\mathrm{INV}_1}(\widehat{\boldsymbol{Z}},\boldsymbol{Z})\right)^{-1} \quad (i=1,\ldots,\nu). \tag{5}$$

Using the technique presented in [2] and the notion of the R-order of convergence introduced by Ortega and Rheinboldt [9], the following assertion was proved in [11]:

Theorem 2. *Let* $(Z_1,\ldots,Z_\nu) =: \left(Z_1^{(0)},\ldots,Z_\nu^{(0)}\right)$ *be initial disks such that* $\zeta_i \in Z_i$ $(i=1,\ldots,\nu)$. *Then, under the initial condition*

$$\rho_0 > 2nr_0,$$

the single-step interval method (5) *is convergent with the R-order at least* $2+\tau_\nu > 3$, *where* τ_ν *is the unique positive zero of the polynomial* $g_\nu(\tau) = \tau^\nu - \tau - 2$, *and* $\zeta_i \in Z_i^{(m)}$ *holds for each* $i=1,\ldots,\nu$ *and* $m=1,2,\ldots$.

3. Schröder-like Inclusion Method with Correction

The centers of disks produced by the Schröder-like inclusion method (3) behave according to the formula

$$\hat{z}_i = z_i - \frac{1}{\frac{P'(z_i)}{P(z_i)} - \frac{1}{\mu_i}\sum_{\substack{j=1\\ j\neq i}}^{\nu}\mu_j(z_i - z_j)^{-1}} \quad (i=1,\ldots,\nu).$$

In fact, this formula defines an iterative method in ordinary complex arithmetic with the cubic convergence. In the case of simple zeros the above formula reduces to the well known Ehrlich-Aberth method considered by Ehrlich [5] and Aberth [1]. Using Newton's correction $P(z_i)/P'(z_i)$, Nourein [8] improved the Ehrlich-Aberth method increasing the convergence order from three to four, without additional calculations. Analogously, using Nourein's approach, one can state the following iteration formula for multiple zeros

$$\hat{z}_i = z_i - \frac{1}{\frac{1}{N(z_i)} - \frac{1}{\mu_i}\sum_{\substack{j=1\\ j\neq i}}^{\nu}\mu_j(z_i - z_j + N(z_j))^{-1}} \quad (i=1,\ldots,\nu), \tag{6}$$

with the order of convergence equal to *four*, where $N(z_i) = \mu_i P(z_i)/P'(z_i)$ is the Schröder correction.

Further improvement of the convergence rate of Schröder-like inclusion methods has been attained by using Schröder's correction in the similar way as in (6). The convergence of the midpoints and the convergence of the radii are coupled so that a further improvement of the midpoints via (5) actually improves the convergence of the radii. The construction of inclusion methods requires suitable initial conditions which enable the implication

$$\zeta_i \in Z_i^{(0)} \quad \Rightarrow \quad \zeta_i \in Z_i^{(0)} - N(\mathrm{mid}\, Z_i^{(0)}).$$

As shown in [4] such conditions for the Schröder-like methods with correction are given by the inequality (9). Then, substituting the inclusion disks $Z_j^{(0)} - N(\mathrm{mid}\ Z_j^{(0)})$ instead of the exact zeros ζ_j in (1), one obtains the Schröder-like inclusion method with correction

$$\widehat{Z}_i = z_i - \left(\frac{1}{N(z_i)} - \frac{1}{\mu_i}\Sigma_{i,\mathrm{INV}_k}(\boldsymbol{Z}_N, \boldsymbol{Z}_N)\right)^{-1} \quad (i = 1, \ldots, \nu;\ k \in \{1, 2\}). \tag{7}$$

In (7) INV_1 and INV_2 denote inversions of a disk defined above, that is, $\mathrm{INV}_k = I_k \in \{()^{I_1}, ()^{I_2}\}$; thus, (7) describes two different methods simultaneously. To emphasize which of two inversions $()^{I_1}$ or $()^{I_2}$ is applied in (7), we will denote the two above methods with $(7)_{I_1}$ and $(7)_{I_2}$, or sometimes $(7(\mathrm{I}_1))$ and $(7(\mathrm{I}_2))$.

We note that another inclusion methods with correction have been considered in [12] and [13] (see, also, the book [14, chap. 6]).

To estimate the order of convergence of iteration methods one usually introduces the error-sequences

$$h_i^{(m)} = |z_i^{(m)} - \zeta_i| \quad (i = 1, \ldots, \nu),$$

where $z_i^{(m)}$ is an approximation to the zero ζ_i produced in the mth step. In the convergence analysis of inclusion methods with corrections we deal with mutually dependent sequences of centers and radii of disks. To resolve this problem we use the following result given by J. W. Schmidt in [15].

Theorem 3. *Given the error-recursion*

$$h_i^{(m+1)} \leq \alpha_i \prod_{j=1}^{\nu} \left(h_j^{(m)}\right)^{t_{ij}} \quad (i = 1, \ldots, \nu;\ m \geq 0), \tag{8}$$

where $t_{ij} \geq 0$, $\alpha_i > 0$, $1 \leq i, j \leq \nu$. *Denote the matrix of exponents appearing in* (8) *with* T_ν, *that is* $T_\nu = [t_{ij}]_{\nu \times \nu}$. *Assume that the matrix* T_ν *has the spectral radius*

$\rho(T_\nu) > 1$ and the corresponding eigenvector $\boldsymbol{x}_\rho = (x_1, \ldots, x_\nu)$ is positive, i.e. $x_1 > 0, \ldots, x_\nu > 0$. Then all the sequences $\{h_i^{(m)}\}$ $(i = 1, \ldots, \nu)$ have the R-order at least $\rho(T_\nu)$.

Following the notion of the R-order of convergence, the matrix $T_\nu = [p_{ij}]$ will be called the *R-matrix* in the sequel. If the eigenvector $\boldsymbol{x}_\rho$ corresponding to the spectral radius $\rho(T_\nu)$ is positive then we write $\boldsymbol{x}_\rho > 0$. The R-order of convergence of the iterative method IM will be denoted by $O_R(\mathrm{IM})$.

In what follows we will write $w_1 \sim w_2$ or $w_1 = O_M(w_2)$ (the same order of magnitude) for two complex numbers w_1 and w_2 that satisfy $|w_1| = O(|w_2|)$. Using the above notation, the following relations have been derived in [4] for the inclusion method (7):

$$\varepsilon_{m+1} = O_M(\varepsilon_m^2 r_m^2), \quad r_{m+1} = O_M(|\varepsilon_m|^2 r_m) \quad \text{for the inversion } \mathrm{INV}_1 = ()^{-1},$$

$$\varepsilon_{m+1} = O_M(\varepsilon_m^4), \quad r_{m+1} = O_M(|\varepsilon_m|^2 r_m) \quad \text{for the inversion } \mathrm{INV}_2 = ()^{I_2}.$$

According to Theorem 3, from the above relations we form the following R-matrices:

(i) in the case of the exact inversion $\mathrm{INV}_1 = ()^{-1}$ we have

$$T_2^{(1)} = \begin{bmatrix} 2 & 2 \\ 2 & 1 \end{bmatrix}$$

with the spectral radius $\rho(T_2^{(1)}) = \frac{1}{2}(3 + \sqrt{17}) \approx 3.562$ and the corresponding vector $\boldsymbol{x}_\rho = \left(\frac{1}{4}(1 + \sqrt{17}), 1\right) > 0$;

(ii) in the case of the centered inversion $\mathrm{INV}_2 = ()^{I_2}$ we have

$$T_2^{(2)} = \begin{bmatrix} 4 & 0 \\ 2 & 1 \end{bmatrix}$$

with the spectral radius $\rho(T_2^{(2)}) = 4$ and the corresponding vector $\boldsymbol{x}_\rho = (3, 2) > 0$; The superscript indices denote the type of the employed inversion, INV_1 or INV_2.

In view of Theorem 3 and the conditions for the convergence considered in [4], we have the following assertion:

Theorem 4. *Let $(Z_1, \ldots, Z_\nu) =: (Z_1^{(0)}, \ldots, Z_\nu^{(0)})$ be initial disks such that $\zeta_i \in Z_i$ $(i = 1, \ldots, \nu)$. Let $O_R((7))$ denote the R-order of convergence of the radii for the method* (7), *where* $\mathrm{INV}_k \in \{()^{I_1}, ()^{I_2}\}$. *If*

$$\rho_0 > 4nr_0, \tag{9}$$

then the interval method (7) is convergent with the R-order given by

$$O_R((7)) \geq \begin{cases} \frac{1}{2}(3+\sqrt{17}) \cong 3.562, & \textit{if} \quad \mathrm{INV}_k = ()^{I_1}, \\ 4, & \textit{if} \quad \mathrm{INV}_k = ()^{I_2}, \end{cases}.$$

and $\zeta_i \in Z_i^{(m)}$ *for each* $i = 1, \ldots, \nu$ *and* $m = 1, 2, \ldots$.

We note that the order of convergence of the inclusion methods $(7)_{I_1}$ and $(7)_{I_2}$ was obtained in a different (and more complicated) manner in [4].

4. Single-step Method with Schröder's Correction

In this section we propose a new accelerated inclusion method of Schröder's type. Continuing to develop inclusion methods with corrections we notice that further acceleration of the convergence of Schröder-like method with correction (7) may be attained using the Gauss-Seidel approach. In this way we can construct the following single-step inclusion method with Schröder's correction:

$$\widehat{Z}_i = z_i - \left(\frac{1}{N(z_i)} - \frac{1}{\mu_i}\Sigma_{i,\mathrm{INV}_1}(\widehat{\mathbf{Z}}, \mathbf{Z}_N)\right)^{-1} \quad (i = 1, \ldots, \nu;\ k \in \{1, 2\}). \tag{10}$$

Thus, we use the circular approximations already calculated in the same iteration.

It is very difficult to find the R-order of convergence of this method. Aprat from the very complicated mutual dependence of even 2ν sequences of centres and radii of produced disks, the number of different zeros ν appears as a parameter. For this reason, we will use the denotation $O_R(\mathrm{IM}, \nu)$ for the R-order. We note that the determination of the bounds of the R-order for an arbitrary ν requires an enormous labor and tedious work. However, we can estimate easily the limit bounds of the R-order taking the limit cases $\nu = 2$ and very large ν.

First, since the convergence rate of the considered single-step method becomes almost the same as the one of the corresponding total-step method when the polynomial degree is very large, according to Theorem 4 for very large ν we have $O_R((10)_{I_1}, \nu) \geq 2 + \sqrt{7} \cong 3.562$ and $O_R((10)_{I_2}, \nu) \geq 4$.

Consider now the single-step method (10) for $\nu = 2$ and the centered inversion $()^{-1}$. Assuming that $1 > |\varepsilon_1^{(0)}| = |\varepsilon_2^{(0)}| = r_1^{(0)} = r_2^{(0)}$ (the "worst case" model), after an extensive but elementary calculation we derive the following relations for the two successive iterations

$$\hat{\varepsilon}_1 = O_M(\varepsilon_1^2 r_2^2),\quad \hat{\varepsilon}_2 = O_M(\varepsilon_1^2 \varepsilon_2^2 r_2^2),\quad \hat{r}_1 = O(|\varepsilon_1|^2 r_2),\quad \hat{r}_2 = O(|\varepsilon_1|^2 |\varepsilon_2|^2 r_2).$$

The corresponding R-matrix has the form

$$T_4^{(1)} = \begin{bmatrix} 2 & 0 & 0 & 2 \\ 2 & 2 & 0 & 2 \\ 2 & 0 & 0 & 1 \\ 2 & 2 & 0 & 1 \end{bmatrix}$$

with $\rho(T_4^{(1)}) = 4.828$, $\boldsymbol{x}_\rho = (\sqrt{2}/2, \frac{1}{2}(1+\sqrt{2}), 1/2, 1) > 0$. In regard to Theorem 3 we obtain

$$O_R((10)_{I_1}, 2) \geq \rho(T_4^{(1)}) = 4.828.$$

Consider now the centered form of the inverse disk. Using this type of inversion we have already proved that $O_R((10)_{I_2}) \geq 4$ in the case of the total-step mode and the presence of Schröder's correction terms. If we apply the inverse $()^{I_2}$ in the iteration formula (10), the corresponding single-step method $(10)_{I_2}$ will have a greater R-order of convergence than $(10)_{I_1}$. More precisely, considering again the case $v = 2$, for the iteration method $(10)_{I_2}$ we find the following relations for the two successive iterations

$$\hat{\varepsilon}_1 = O_M(\varepsilon_1^2\varepsilon_2^2), \quad \hat{\varepsilon}_2 = O_M(\varepsilon_1^2\varepsilon_2^4), \quad \hat{r}_1 = O(|\varepsilon_1|^2 r_2), \quad \hat{r}_2 = O(|\varepsilon_1|^2|\varepsilon_2|^2 r_2).$$

The corresponding R-matrix is given by

$$T_4^{(2)} = \begin{bmatrix} 2 & 2 & 0 & 0 \\ 2 & 4 & 0 & 0 \\ 2 & 0 & 0 & 1 \\ 2 & 2 & 0 & 1 \end{bmatrix}$$

with $\rho(T_4^{(2)}) = 5.236$, $bsx_\rho = (\frac{1}{4}(1+\sqrt{5}), \frac{1}{4}(3+\sqrt{5}), 1/2, 1) > 0$. By Theorem 3 one obtains

$$O_R((10)_{I_2}, 2) \geq \rho(T_4^{(2)}) = 5.236.$$

Let $\Omega_4, \Omega_5, \Omega_{7(I_1)}, \Omega_{7(I_2)}, \Omega_{10(I_1)}, \Omega_{10(I_2)}$ denote the ranges of the lower bounds of the R-order of convergence of the considered inclusion methods of Schröder's type. The lower bound of the range Ω_k is the R-order for very large v, and the upper bound is the R-order for $v = 2$. Taking into account the convergence analysis of the total-step method with Schröder's correction (7) and the above results, we can state the following convergence theorem for the improved inclusion method (10):

Theorem 5. *Let* $(Z_1, \ldots, Z_v) =: (Z_1^{(0)}, \ldots, Z_v^{(0)})$ *be initial disks such that* $\zeta_i \in Z_i$ $(i = 1, \ldots, v)$. *If the inequality* (9) *holds, then the interval method* (10) *is convergent with the R-order in the interval*

$$\begin{aligned} \Omega_{10(I_1)} &\in (3.562, 4.828), \quad && \text{if INV}_k = ()^{I_1}, \\ \Omega_{10(I_2)} &\in (4, 5.236), && \text{if INV}_k = ()^{I_2}, \end{aligned}$$

and $\zeta_i \in Z_i^{(m)}$ *for each* $i = 1, \ldots, v$ *and* $m = 1, 2, \ldots$.

According to the previous results and the bounds $2 + \tau_\infty$ and $2 + \tau_2$ for the single-step method (5), we obtain

$$\Omega_4 = 3, \quad \Omega_5 = (3,4), \quad \Omega_{7(I_1)} = 3.562, \quad \Omega_{7(I_2)} = 4,$$
$$\Omega_{10(I_1)} = (3.562, 4.828), \quad \Omega_{10(I_2)} = (4, 5.236).$$

The ranges of the R-order of the improved inclusion methods point to the higher order of convergence of the methods with corrections. Since the increased convergence is attained with a negligible number of additional calculations (because Schröder's correction $\mu_i P(z_i)/P'(z_i)$ is already calculated), we conclude that the improved inclusion methods possess a great computational efficiency.

5. Computational Results

We have already mentioned that the methods (4) and (5) possess the highest computational efficiency among interval methods for the simultaneous inclusion of all polynomial zeros ([11, chap. 6]). In this section, we will mutually compare the six presented methods of Schröder's type with each other. For simplicity, only the case of simple complex zeros will be treated. As usual, we will use for this purpose the *coefficient of efficiency* for an interval method IM given by

$$E(\mathrm{IM}, n) = R(n)^{c/\theta(n)}, \tag{11}$$

where n is the polynomial degree, $R(n)$ is the R-order of convergence, c is a normalization contant which also involves characteristics of arithmetic units of the employed computer machine and $\theta(n)$ is the computational cost of evaluation (for all n zeros) per iteration expressed by the number of *flops* where one flop is defined as one multiplication plus one addition. We note that for the considered methods without corrections and with the exact inversion $()^{-1}$ this number is $\theta(n) = 36n^2 + O(n)$ flops and about $\theta(n) = 40n^2 + O(n)$ flops for the methods $(7)_{I_2}$ and $(10)_{I_2}$ which use the centered inversion. Additional operations in this last case are required since the evaluation of $|z| = |z + iy| = \sqrt{x^2 + y^2}$ (appearing in $()^{I_c}$) needs extra operations, as mentioned in the introduction.

This disadvantage can be overcome by using a new type of centered inversion $()^{I_2^*}$ given in introduction. The convergence behavior of the inclusion methods that use this type of inversion is almost the same as in the case of $()^{I_2}$ under suitable initial conditions. However, numerical examples have shown that the increased radius $\mathrm{rad}()^{I_2^*}$ (compared to the radii of the inversions $()^{I_1}$ and $()^{I_2}$) can cause certain difficulties (slower convergence and larger resulting disks, see the fifth column in Table 2) if the initial disks are not sufficiently small. On the other hand, the corresponding total-step and single step methods with corrections which use the new centered inversion $()^{I_2^*}$, denoted by $(7)_{I_2^*}$ and $(10)_{I_2^*}$, have even higher computational efficiency than the other methods of Schröder's type (see Table 1). This is a consequence of the very high order $R(n)$ of convergence of these methods. But, theoretical value of the order $R(n)$ is obtained in a limiting process and it is lower in practice so that the computational efficiency of the methods $(7)_{I_2^*}$ and $(10)_{I_2^*}$ is overestimated. Having in mind this fact, as well as the mentioned disadvantage

Table 1. The values of $E(\mathrm{IM}, \nu)$

Methods	3	4	5	6	7	8	9	10
(4)	2.350	1.617	1.360	1.238	1.170	1.128	1.100	1.080
(5)	2.662	1.698	1.393	1.255	1.180	1.134	1.104	1.083
(7)(I_1)	2.686	1.743	1.427	1.280	1.199	1.149	1.116	1.093
(7)(I_2)	2.639	1.770	1.432	1.275	1.197	1.146	1.114	1.091
(7)(I_2^*)	2.939	1.834	1.474	1.309	1.219	1.164	1.127	1.102
10(I_1)	3.081	1.842	1.468	1.301	1.211	1.157	1.122	1.097
10(I_2)	2.932	1.848	1.463	1.289	1.205	1.152	1.117	1.094
10(I_2^*)	3.304	1.920	1.508	1.326	1.229	1.170	1.131	1.105

Table 2. Maximal radii of inclusion disks in three iterations. $A(-h)$ means $A \times 10^{-h}$

Methods	(4)	(5)	$(7)_{I_1}$	$(7)_{I_2}$	$(7)_{I_2^*}$	$(10)_{I_1}$	$(10)_{I_2}$
$\max r_i^{(1)}$	1.16(−1)	1.16(−1)	1.25(−1)	2.44(−1)	3.33(−1)	4.53(−2)	5.97(−2)
$\max r_i^{(2)}$	9.55(−4)	2.38(−4)	3.78(−5)	5.20(−4)	3.54(−3)	3.93(−6)	1.38(−6)
$\max r_i^{(3)}$	4.35(−13)	1.41(−17)	3.61(−17)	5.18(−16)	1.25(−12)	3.43(−24)	4.57(−23)

consisting in producing larger inclusion disks compared to the other methods, we cannot assert that these methods are the most efficient in spite of their highest coefficient of efficiency. As proposed in [4], a good approach to construct a very efficient inclusion algorithm is to start with a method with the inversion $()^{-1}$ only in the first iterative step and then apply one of the inclusion methods $(7)_{I_2^*}$ or $(10)_{I_2^*}$.

According to the entries of the computational efficiency displayed in Table 1 the following conclusion may be deduced:

Among Schröder-like inclusion methods a new single step method with the exact inversion $(10)_{I_1}$ is the most efficient. Although its convergence rate is slower compared to the other single step method $(10)_{I_2}$, this method with the exact inversion requires less numerical operations, as mentioned above. The advantage of the method $(10)_{I_1}$ in reference to $(10)_{I_2}$ is even greater taking into account the following facts of practical importance. The values of the convergence order given by the ranges $\Omega_{10(I_1)} = (3.562, 4.828)$ and $\Omega_{10(I_2)} = (4, 5.236)$ should be regarded as asymptotical ones meaning that the notified speed of convergence can be realized after several iterative steps. Numerical experiments have shown that such situation most frequently begins after the second or third iteration. Since the considered methods possess very fast convergence, the wanted accuracy of approximations is attained applying only a few iterative steps. For these reasons, considering the first, say, three iterations, the convergence speed of the methods with centered inversion $(7)_{I_2}$ and $(10)_{I_2}$ is somewhat smaller than the theoretical one. The same conclusion holds for the total step methods $(7)_{I_1}$ and $(7)_{I_2}$ meaning that $(7)_{I_1}$ has an advantage in reference to $(7)_{I_2}$. A discussion for these two methods can also be found in [4].

According to the previous comments and the entries given in Table 1 we have the following rating of the considered inclusion methods:

1.	2.	3.	4.	5.	6.
$(10)_{I_1}$	$(10)_{I_2}$	$(7)_{I_1}$	$(7)_{I_2}$	(5)	(4)

Finally, we note that this "rating list" of the considered iterative methods shows a considerable agreement with the rating list of methods based on the measured total CPU time which is necessary for the calculation of all zeros.

Theoretical results which give the convergence order of the considered Schröder-like methods mainly well coincide with the convergence behavior of these methods in practice, especially when the number of iterative steps increases. These inclusion methods have been tested in solving many polynomial equations. Apart from the six considered methods presented in the previous sections, we have also tested the inclusion method $(7)_{I_2^*}$ with the modified centered inversion $()^{I_2^*}$ to illustrate the above discussion. For demonstration, we give numerical results obtained in finding the zeros of the polynomial

$$\begin{aligned} P(z) &= z^9 + (-2+3i)z^8 + (48-6i)z^7 + (-94+152i)z^6 + (522-298i)z^5 \\ &\quad + (-950+1974i)z^4 - (1400+3650i)z^3 + (3750+1200i)z^2 \\ &\quad + (-1875+1250i)z - 625i. \end{aligned}$$

The exact zeros of P are $\zeta_1 = 1, \zeta_2 = -i, \zeta_3 = -5i$ and $\zeta_4 = -5i$ with the multiplicities $\mu_1 = 2, \mu_2 = 3, \mu_3 = 2$, and $\mu_4 = 2$. For the initial inclusion disks we have taken the circular regions $Z_k^{(0)}$ $(k = 1,2,3,4)$ with the radii $r_k^{(0)} = 0.9$ and the centers

$$z_1^{(0)} = 1.1 + 0.2i,\ z_2^{(0)} = 0.2 - 0.8i,\ z_3^{(0)} = -0.6 - 4.4i,\ z_4^{(0)} = -0.6 + 4.4i.$$

The programs have been realized using multiple precision arithmetic. The maximal radii $r^{(m)} = \max_{1\le 1\le \nu} r_i^{(m)}$ $(m = 1,2,3)$ for all tested methods are shown in Table 2.

From Table 2, we see that the proposed single step methods with corrections $(10)_{I_1}$ and $(10)_{I_2}$ produce very small disks, especially in the third iterative step. The inclusion method $(7)_{I_2^*}$ has produced somewhat larger disks, which coincides with the above comment. A number of tested examples have shown similar behavior. According to this, and taking into account the above ranking list based on the computational efficiency, we can conclude that the proposed single step methods become the most efficient methods for the simultaneous inclusion of polynomial zeros.

References

[1] Aberth, O.: Iteration methods for finding all zeros of a polynomial simultaneously. Math. Comp. *27*, 339–344 (1973).
[2] Alefeld, G., Herzberger, J.: On the convergence speed of some algorithms for the simultaneous approximation of polynomial zeros. SIAM. Anal. *11*, 237–243 (1974).

[3] Alefeld, G., Herzberger, J.: Introduction to interval computation. New York: Academic Press 1983.
[4] Carstensen, C., Petković, M. S.: An improvement of Gargantini's simultaneous inclusion method for polynomial roots by Schroeder's correction. Appl. Numer. Math. *25*, 59–67 (1993).
[5] Ehrlich, L. W.: A modified Newton method for polynomials. Comm. ACM *10*, 107–108 (1967).
[6] Gargantini, I.: Further applications of circular arithmetic: Schroeder-like algorithms with error bounds for finding zeros of polynomials. SIAM J. Numer. Anal. *15*, 497–510 (1979).
[7] Gargantini, I., Henrici, P.: Circular arithmetic and the determination of polynomial zeros. Numer. Math. *18*, 305–320 (1972).
[8] Nourein, A. W. M.: An improvement on two iteration methods for simultaneously determination of the zeros of a polynomial. Inter. J. Comput. Math. *6*, 241–252 (1977).
[9] Ortega, J. M., Rheinboldt, W. C.: Iterative solution of nonlinear equations in several variables. New York: Academic Press 1970.
[10] Petković, M. S.: On a generalization of the root iterations for polynomial complex zeros in circular interval arithmetic. Comput. *27*, 37–55 (1981).
[11] Petković, M. S.: Iterative methods for simultaneous inclusion of polynomial zeros. Berlin: Springer 1989.
[12] Petković, M. S.: Halley-like method with corrections for the inclusion of polynomial complex zeros. Comput. *62*, 69–88 (1999).
[13] Petković, M. S., Carstensen, C.: On some improved inclusion methods for polynomial roots with Weierstrass' correction. Comput. Math. Appl. *25*, 59–67 (1993).
[14] Petković, M. S., Petković, L. D.: Complex interval arithmetic and its applications. Berlin: Wiley VCH 1998.
[15] Schmidt, J. W.: On the R-order of coupled sequences. Comput. *26*, 333–342 (1981).

M. S. Petković
Faculty of Electronic Engineering
University of Niš
18 000 Niš
Yugoslavia
e-mail: msp@junis.ni.ac.yu

L. D. Petković
Faculty of Mechanical Engineering
University of Niš
18 000 Niš
Yugoslavia

Computing [Suppl] 16, 199–211 (2002)

Computing

Interval Root-finding Methods of Laguerre's Type*

L. D. Petković, M. S. Petković
and **D. Živković**

Abstract

In this paper, several algorithms of Laguerre's type for the inclusion of simple complex zeros of a polynomial are presented. These methods are realized in complex circular arithmetic and have the convergence order ≥ 4. The proposed algorithms possess a great computational efficiency since the acceleration of the convergence is attained with few additional calculations. High convergence speed is demonstrated on numerical examples.

AMS Subject Classifications: 65H05, 65G20, 30C15.

Keywords: Zeros of polynomials, simultaneous methods, convergence rate, circular arithmetic.

1. Introduction

Interval methods, developed in the sixties of the last century, appeared as a new and very efficient approach for solving and analyzing a wide class of linear and nonlinear problems, with the remarkable advantage of self-validated results. In this paper we present several new iterative methods of Laguerre's type for the simultaneous inclusion of all simple complex zeros of a polynomial in circular complex arithmetic. Such an approach provides a posteriori error estimates for the resulting approximations to the polynomial zeros and an efficient work in the presence of uncertain data. More details on inclusion simultaneous methods can be found in the books [1], [8] and [10] and many papers cited therein.

The proposed methods are realized in circular complex arithmetic and have the order of convergence ≥ 4. The modified methods which improve the convergence rate are obtained with negligible number of additional operations so that these methods possess a high computational efficiency. The purpose of this paper is to give a review of new algorithms of Laguerre's type, without convergence analysis and stating initial conditions that provide both the guaranteed convergence and the inclusion property $\zeta_i \in Z_i^{(m)}$ for the inclusion disks $Z_1^{(m)}, \ldots, Z_n^{(m)}$ produced at each iterative step $m = 1, 2, \ldots$, where $\zeta_1, \ldots, \zeta_n$ are the zeros of a given polynomial P. These important questions will be considered in forthcoming papers. Instead of that we give quantitative data on the convergence order of the

---

* Dedicated to Prof. Dr. Jürgen Herzberger on his 60th birthday.

presented methods and demonstrate the convergence speed of these methods on numerical examples.

In Sect. 2, we give some preliminary and axillary results. New algorithms and their convergence rate are considered in Sect. 3. Results of numerical experiments are given in Sect. 4. The presented algorithms are realized in terms of circular regions – disks. Basic definitions and properties of circular arithmetic can be found in details in the books [1], [8] and [10]. Here we will only give the definitions of the square root and the inversion of a nonzero circular region which are of essential importance in developing proposed methods.

The square root of a disk $\{c; r\}$, where $c = |c|e^{i\theta}$, $|c| > r$ (that is, $0 \notin \{c; r\}$) is defined as the union of two disks:

$$\{c; r\}^{1/2} := \left\{\sqrt{|c|}e^{i\theta/2}; \sqrt{|c|} - \sqrt{|c| - r}\right\} \bigcup \left\{-\sqrt{|c|}e^{i\theta/2}; \sqrt{|c|} - \sqrt{|c| - r}\right\}. \quad (1)$$

We use two types of inversion of a disk not containing 0:

the exact inversion:

$$Z^{-1} = \{z; r\}^{I_e} := \left\{\frac{1}{z \cdot \left(1 - r^2/|z|^2\right)}; \frac{r}{|z|^2 - r^2}\right\} = \{z^{-1} | z \in Z\};$$

the centered inversion:

$$Z^{I_c} = \{z; r\}^{I_2} := \left\{\frac{1}{z}; \frac{r}{|z| \cdot (|z| - r)}\right\}.$$

We note that the inclusion $Z^{I_e} \subset Z^{I_c}$ holds for any Z $(0 \notin Z)$. In the sequel the abbreviation INV will be used to denote one of the above inversions of a disk not containing 0, that is, INV$\in \{()^{I_e}, ()^{I_c}\}$. Sometimes, we will use the superscript indices (e) and (c) to indicate the exact and centered inversion, respectively.

To construct an algorithm for the inclusion of only one zero, we need the inversion of a closed exterior of a circle. Let $W = \{w | |w - c| \geq r\}$ be a closed exterior of a circle such that $0 \notin W$, that is, $|c| < r$. Applying the theory of conformal mapping we find that the inverse of W^{-1} is a closed disk given by

$$W^{-1} = \left\{\frac{-\bar{c}}{r^2 - |c|^2}; \frac{r}{r^2 - |c|^2}\right\}. \quad (2)$$

2. Some Preliminary Results

Laguerre-like Fixed Point Relation

Let $P(z) = z^n + a_{n-1}z^{n-1} + \cdots + a_1 z + a_0 = \prod_{j=1}^{n}(z - \zeta_j)$ be a monic polynomial of degree n with simple zeros $\zeta_1, \ldots, \zeta_n$, and let $I_n := \{1, \ldots, n\}$ be the index set. We will deal with the functions

$$\delta_1(z) = \frac{P'(z)}{P(z)}, \quad \delta_2(z) = \frac{P'(z)^2 - P(z)P''(z)}{P(z)^2}.$$

For the point $z = z_i$ $(i \in I_n)$ let us introduce the abbreviations

$$\Sigma_{k,i} = \sum_{\substack{j=1 \\ j \neq i}}^{n} \frac{1}{(z_i - \zeta_j)^k} \quad (k = 1, 2), \quad q_i = n\Sigma_{2,i} - \frac{n}{n-1}\Sigma_{1,i}^2,$$

$$\delta_{1,i} = \delta_1(z_i), \quad \delta_{2,i} = \delta_2(z_i), \quad \varepsilon_i = z_i - \zeta_i.$$

In [4] the following fixed point relation

$$\zeta_i = z_i - \frac{n}{\delta_{1,i} \pm \sqrt{(n-1)(n\delta_{2,i} - \delta_{1,i}^2 - q_i)}} \quad (i \in I_n) \tag{3}$$

has been derived. This relation is the base for the construction of inclusion root-finding methods of Laguerre's type.

Assume that we have found n disjoint inclusion disks $Z_1, \ldots, Z_n$, $\zeta_i \in Z_i$ $(i \in I_n)$, and let $z_i = \operatorname{mid} Z_i$. The substitution of the zero ζ_j by its inclusion disk Z_j in q_i gives a circular interval extension Q_i of q_i, that is,

$$q_i \in Q_i = n \sum_{j \neq i} \left(\frac{1}{z_i - Z_j}\right)^2 - \frac{n}{n-1}\left(\sum_{j \neq i} \frac{1}{z_i - Z_j}\right)^2.$$

Using the inclusion isotonicity property, from the fixed point relation (3) we obtain

$$\zeta_i \in \hat{Z}_i := z_i - \frac{n}{\delta_{1,i} \pm \sqrt{(n-1)(n\delta_{2,i} - \delta_{1,i}^2 - Q_i)}} \quad (i \in I_n). \tag{4}$$

If the denominator in (4) is not a disk containing 0, then $\hat{Z}_i$ is a new outer circular approximation to the zero ζ_i, that is, $\zeta_i \in \hat{Z}_i$.

Remark 1. Neglecting the term Q_i in (4), we obtain the classical Laguerre method of the third order for finding a single zero,

$$\hat{z}_i = z_i - \frac{n}{\delta_{1,i} \pm \sqrt{(n-1)\left(n\delta_{2,i} - \delta_{1,i}^2\right)}}.$$

For this reason, all algorithms presented in this paper will be referred to as Laguerre-like algorithms.

Selection of a Proper Disk

In formula (4) and the subsequent modified formulas of this type we have to choose the sign in front of the square root in the denominator. If inclusion disks $Z_1, \dots, Z_n$ are sufficiently small, then $W_i := (n-1)(n\delta_{2,i} - \delta_{1,i}^2 - Q_i)$ does not contain the origin and, according to (1), $W_i^{1/2}$ is the union of two disjoint disks, say, $W_i^{1/2} = U_{1,i} \cup U_{i,2}$ with centers $u_{1,i} = \text{mid } U_{1,i}$ and $u_{i,2} = \text{mid } U_{2,i}$, where $u_{i,1} = -u_{i,2}$. The following criterion for the choice of a proper disk of a square root (between two) can be stated (see, e.g., [3]):

If the disks $Z_1, \dots, Z_n$ are reasonably small, then we have to choose that disk (between $U_{1,i}$ and $U_{2,i}$) whose center maximizes $|\delta_{1,i} + u_{i,k}|$ $(k = 1, 2)$.

In the sequel, the selection of a proper disk will be indicated by the symbol $*$ and the proper disk will be denoted by $[W_i]_*^{1/2}$.

Types of Circular Approximations

Let $Z_1^{(m)}, \dots, Z_n^{(m)}$ $(m = 1, 2, \dots)$ be circular approximations to the zeros $\zeta_1, \dots, \zeta_n$ of P, obtained in the mth iterative step. Let $z_i^{(m)} = \text{mid } Z_i^{(m)}$ and let $\delta_{1,i}^{(m)}, \delta_{2,i}^{(m)}, Q_i^{(m)}$ be the quantities evaluated at $z_1^{(m)}, \dots, z_n^{(m)}$. For simplicity, we will omit the iteration index m and denote the quantities in the next $(m+1)$-th iteration by the symbol $\widehat{\ }$ (*hat*).

Let $\boldsymbol{A} = (A_1, \dots, A_n)$ and $\boldsymbol{B} = (B_1, \dots, B_n)$ be some vectors of disks. We deal with the disk

$$Q_i(\boldsymbol{A}, \boldsymbol{B}) = n \sum_{j=1}^{i-1} \left(\text{INV}(z_i - A_j)\right)^2 + n \sum_{j=i+1}^{n} \left(\text{INV}(z_i - B_j)\right)^2 - \frac{n}{n-1} \left(\sum_{j=1}^{i-1} \text{INV}(z_i - A_j) \right)^2 - \frac{n}{n-1} \left(\sum_{j=i+1}^{n} \text{INV}(z_i - B_j) \right)^2,$$

where $\text{INV} \in \{()^{I_e}, ()^{I_c}\}$, and the following circular approximations:

$\boldsymbol{Z} = (Z_1, \dots, Z_n)$ (current approximations),
$\widehat{\boldsymbol{Z}} = (\widehat{Z}_1, \dots, \widehat{Z}_n)$ (new approximations),
$\boldsymbol{Z}_N = (Z_1 - N(z_1), \dots, Z_n - N(z_n))$, $z_i = \text{mid } Z_i$ (Newton's approximations).

Convergence of Mutually Dependent Sequences

The determination of the order of convergence of the interval methods usually requires some manipulations with the error-sequences

$$\varepsilon_i^{(m)} = |\text{mid } Z_i^{(m)} - \zeta_i|, \quad r_i^{(m)} = \text{rad } Z_i^{(m)} \quad (i = 1, \dots, n),$$

where $Z_i^{(m)}$ is an outer approximation to the zero ζ_i produced at the mth step. As shown in [2] and [9], in the convergence analysis of inclusion methods with corrections we deal with mutually dependent sequences of centers and radii of disks. In the analysis of an iterative method (IM) we use the concept of the R-order of convergence, denoted by $O_R(IM)$ (see [6]), and the following result given in [11] and [5].

Theorem A. *Given the error-recursion*

$$u_i^{(m+1)} \leq \alpha_i \prod_{j=1}^{n} \left(u_j^{(m)}\right)^{t_{ij}}, \quad (i = 1, \ldots, n;\ m \geq 0), \tag{5}$$

where $t_{ij} \geq 0,\ \alpha_i > 0,\ 1 \leq i, j \leq n$. *Denote the matrix of exponents appearing in* (5) *with* T, *that is* $T = [t_{ij}]_{n \times n}$. *If the nonnegative matrix* T *has the spectral radius* $\rho(T) > 1$ *and a corresponding eigenvector* $\boldsymbol{x}_\rho > 0$, *then the R-order of convergence of all sequences* $\{u_i^{(m)}\}$ $(i = 1, \ldots, n)$ *is at least* $\rho(T)$.

We call matrix $T = [t_{ij}]$ the *R-matrix* because of its connection to the R-order of convergence. Further, we write $w_1 \sim w_2$ or $w_1 = O_M(w_2)$ (the same order of magnitude) for two complex numbers w_1 and w_2 satisfying $|w_1| = O(|w_2|)$. In the convergence analysis of inclusion methods it is adopted $1 > |\varepsilon^{(0)}| = r^{(0)} > 0$ (the "worst case" model). This assumption has no influence to the final result of the limit process which we apply in order to obtain the lower bound of the R-order of convergence.

3. New Algorithms and their Convergence

The fixed point relation (3) and formula (4) suggest the construction of new algorithms of Laguerre's type for the simultaneous inclusion of all simple zeros of a given polynomial P. Depending on the choice of the type of circular approximations given in the previous section, we can construct various methods for the simultaneous inclusion of polynomial zeros. For these methods we give convergence theorems based on Theorem A. Due to the lack of space, we omit the proof and only present the corresponding T-matrices appearing in Theorem A together with their spectral radii, which give the R-order of convergence.

Total Step Method

First, following (4), we state the total-step method of Laguerre's type:

Algorithm (TS) – *Total step method:*

$$\hat{Z}_i = z_i - \frac{n}{\delta_{1,i} + \left[(n-1)\left(n\delta_{2,i} - \delta_{1,i}^2 - Q_i(\boldsymbol{Z}, \boldsymbol{Z})\right)\right]_*^{1/2}} \quad (i \in I_n).$$

The symbol $*$ indicates the proper square root of a disk.

Using circular arithmetic operations, in the similar way as in [3] we can show that $\hat{r}_i \sim r_i^4$ for every $i \in I_n$. Hence the following theorem can be stated:

Theorem 1. *If the radii* $r_1^{(0)}, \ldots, r_n^{(0)}$ *of the inclusion disks* $Z_1^{(0)}, \ldots, Z_n^{(0)}$ *are sufficiently small, then the interval method (TS) has the convergence order four.*

Remark 2. We note that the term "reasonably small disks" appearing in the above theorem and the next convergence theorems is rather indefinite and imprecise. This shortcoming can be overcame by using the initial conditions of the form

$$\min_{\substack{i,j\\ i\neq j}}\left\{|z_i^{(0)} - z_j^{(0)}| - r_j^{(0)}\right\} > C_n \max_{1\le j\le n} |r_j^{(0)}|, \quad r_j^{(0)} = \operatorname{rad} Z_j^{(0)},\ z_j^{(0)} = \operatorname{mid} Z_j^{(0)},$$

where C_n is suitably chosen constant depending only on the polynomial degree n (see [8] and references cited therein). The condition in the form of the last inequality that provides the guaranteed convergence and the inclusion $\zeta_i \in Z_i^{(m)}$ $(i \in I_n;\ m = 1, 2, \ldots)$ will be considered in the forthcoming paper.

Single Step Method

The convergence order of the total-step method (TS) can be increased by applying the Gauss-Seidel approach, that is, by the use of already calculated disks in the current iteration (single step mode). This approach leads to the single step algorithms:

Algorithm (SS) – *Single step method:*

$$\hat{Z}_i = z_i - \frac{n}{\delta_{1,i} + \left[(n-1)\left(n\delta_{2,i} - \delta_{1,i}^2 - Q_i(\hat{\boldsymbol{Z}}, \boldsymbol{Z})\right)\right]_*^{1/2}} \qquad (i \in I_n).$$

The lower bound of the R-order of convergence of Algorithm (SS) is given in the following theorem and it can be determined in a similar way as in [8, chap. 2].

Theorem 2. *The R-order of (SS) is* $O_R(SS) \ge 3 + \sigma_n$, *where* $\sigma_n > 1$ *is the unique positive root of the equation*

$$\sigma^n - \sigma - 3 = 0.$$

Total Step Method with Newton's Correction

Following the idea presented in [2] and [9], we are able to attain further acceleration of Laguerre-like algorithms. This approach is based on the use of corrections which appear in root-finding methods of the second order, most frequently Newton's correction $N(z_i) = P(z_i)/P'(z_i)$ or Weierstrass' correction $W(z_i) = P(z_i)/\prod_{j\neq i}(z_i - z_j)$. In this paper we apply Newton's correction in such a way that we use inclusion disks $Z_j - N(\text{mid}Z_j)$ (Newton's approximations) instead of Z_j. Naturally, to keep the enclosure of zeros it is necessary to perform such a choice of initial disks $Z_1^{(0)}, \ldots, Z_n^{(0)}$ that provides the implication $\zeta_i \in Z_i^{(0)} \Rightarrow \zeta_i \in Z_i - N(\text{mid}\, Z_i^{(0)})$ for every $i \in I_n$. As explained in [2] and [9], the forced acceleration of the convergence of the centers of inclusion disks improves the convergence of the radii of these disks.

First we present the total step method of Laguerre's type with Newton's corrections:

Algorithm (TSN) – *Total step method with Newton's correction:*

$$\hat{Z}_i = z_i - \frac{n}{\delta_{1,i} + \left[(n-1)\left(n\delta_{2,i} - \delta_{1,i}^2 - Q_i(\boldsymbol{Z}_N, \boldsymbol{Z}_N)\right)\right]_*^{1/2}} \quad (i \in I_n).$$

Let us introduce

$$|\varepsilon| = \max_{1\leq j\leq n} |\varepsilon_j|, \quad |\hat{\varepsilon}| = \max_{1\leq j\leq n} |\hat{\varepsilon}_j|, \quad r = \max_{1\leq j\leq n} r_j, \quad \hat{r} = \max_{1\leq j\leq n} \hat{r}_j.$$

In a similar way as in [2] and [9] we can derive the following relations for the method (TSN):

(i) $|\hat{\varepsilon}| \sim |\varepsilon|^3 r^2,\ \hat{r} \sim |\varepsilon|^3 r$ *if* $\text{INV} = ()^{I_e}$;

(ii) $|\hat{\varepsilon}| \sim |\varepsilon|^5,\ \hat{r} \sim |\varepsilon|^3 r$ *if* $\text{INV} = ()^{I_c}$.

According to these relations and Theorem A we form the R-matrices and find their spectral radii and eigenvectors:

$$\text{the case (i)}: \ T_2^{(e)} = \begin{bmatrix} 3 & 2 \\ 3 & 1 \end{bmatrix}, \quad \rho(T_2^{(e)}) = 2 + \sqrt{7}, \ \boldsymbol{x}_\rho = \left(\frac{1+\sqrt{7}}{3}, 1\right) > 0;$$

$$\text{the case (ii)}: \ T_2^{(c)} = \begin{bmatrix} 5 & 0 \\ 3 & 1 \end{bmatrix}, \quad \rho(T_2^{(c)}) = 5, \ \boldsymbol{x}_\rho = \left(\frac{4}{3}, 1\right) > 0.$$

The superscript indices (e) and (c) are relating to the exact and centered inversion, respectively.

Using the previous results and Theorem A we have the following assertion:

Theorem 3. *The lower bound of the R-order of convergence of the total-step method (TSN) is given by*

$$O_R((TSN)) \geq \begin{cases} 2+\sqrt{7} \cong 4.646, & \textit{if } \mathrm{INV}_k = ()^{I_e}, \\ 5, & \textit{if } \mathrm{INV}_k = ()^{I_c}. \end{cases}$$

Single Step Method with Newton's Correction

Applying Gauss-Seidel approach to the total-step method (TSN) we state the single step method with Newton's correction which presents further improvement of the previous methods:

Algorithm (SSN) – *Single step method with Newton's correction:*

$$\hat{Z}_i = z_i - \frac{n}{\delta_{1,i} + \left[(n-1)\left(n\delta_{2,i} - \delta_{1,i}^2 - Q_i(\widehat{\mathbf{Z}}, \mathbf{Z}_N)\right)\right]_*^{1/2}} \quad (i \in I_n).$$

The determination of the R-order of convergence of the single step method (SSN) is very difficult task since we should deal with $2n$ mutually dependent sequences of centers and radii of produced disks. Instead of that, we are able to find the lower and upper bounds of the R-order of convergence taking into account the following facts:

(1) $O_R((SSN, n)) \cong O_R(TSN)$ for very large n and, therefore, we can take $O_R(TSN)$ for (approximate) lower bounds of $O_R((SSN, n))$.

(2) The upper bound of $O_R((SSN, n))$ is given by $O_R((SSN, 2))$, noting that the analysis for the concrete case $n = 2$ is not too difficult.

According to the previous facts and Theorem 3 we first find the lower bound of $O_R((SSN, n))$ as

$$O_R((SSN, n)) \cong O_R(TSN) \geq \begin{cases} 2+\sqrt{7} \cong 4.646, & \text{if } \mathrm{INV} = ()^{I_e}, \\ 5, & \text{if } \mathrm{INV} = ()^{I_c}. \end{cases}$$

Second, for $n = 2$ after extensive but elementary calculations we derive the following relations:

$$|\hat{\varepsilon}_1| \sim |\varepsilon_1|^3 r_2^2, \quad |\hat{\varepsilon}_2| \sim |\varepsilon_1|^3 |\varepsilon_2|^3 r_2^2, \quad \hat{r}_1 \sim |\varepsilon_1|^3 r_2, \quad \hat{r}_2 \sim |\varepsilon_1|^3 |\varepsilon_2|^3 r_2 \quad \text{if } \mathrm{INV} = ()^{I_e}.$$

$$|\hat{\varepsilon}_1| \sim |\varepsilon_1|^3 |\varepsilon_2|^2, \quad |\hat{\varepsilon}_2| \sim |\varepsilon_1|^3 |\varepsilon_2|^5, \quad \hat{r}_1 \sim |\varepsilon_1|^3 r_2, \quad \hat{r}_2 \sim |\varepsilon_1|^3 |\varepsilon_2|^3 r_2 \quad \text{if } \mathrm{INV} = ()^{I_c}.$$

Taking into account the above relations and Theorem A, we form the R-matrices and find their spectral radii and eigenvectors:

$$T_4^{(e)} = \begin{bmatrix} 3 & 0 & 0 & 2 \\ 3 & 3 & 0 & 2 \\ 3 & 0 & 0 & 1 \\ 3 & 3 & 0 & 1 \end{bmatrix}, \quad \rho(T_4^{(e)}) = 6.29654, \; \boldsymbol{x}_\rho = (1, 1.91, 0.7382, 1.6483) > 0,$$

and

$$T_4^{(c)} = \begin{bmatrix} 3 & 2 & 0 & 0 \\ 3 & 5 & 0 & 0 \\ 3 & 0 & 0 & 1 \\ 3 & 3 & 0 & 1 \end{bmatrix}, \quad \rho(T_4^{(c)}) = 6.64575, \;\; \boldsymbol{x}_\rho = (1, 1.8229, 0.6771, 1.5) > 0.$$

Therefore, according to Theorem A, the upper bound of the R-order of convergence is determined by

$$O_R((SSN, n=2)) \geq \begin{cases} 6.296, & \textit{if} \;\; \mathrm{INV}_k = ()^{I_e}, \\ 6.646, & \textit{if} \;\; \mathrm{INV}_k = ()^{I_c}. \end{cases}$$

With regard to the last result and Theorem 3 we find the ranges of the lower bounds of the R-order of convergence of (SSN):

$$O_R^{(e)}(SSN) \in (4.646, 6.297) \quad \text{if INV} = ()^{I_e},$$
$$O_R^{(c)}(SSN) \in (5, 6.646) \quad \text{if INV} = ()^{I_c}.$$

Method for a Single Zero

In some cases it is necessary to find sufficiently small disk containing only one isolated complex zero of a polynomial. Let $\zeta_1 = \zeta \neq 0$ be the sought zero of a polynomial and let $Z^{(0)} = \{a; R\}$ be a disk such that it contains only that zero ζ. The remaining zeros $\zeta_2, \ldots, \zeta_n$ belong to the open set $V^{(0)} = \{z | z \notin Z^{(0)}\} = \operatorname{ext} Z^{(0)}$. Let $z \in Z^{(0)}$. Since $\zeta_j \in V^{(0)} = \operatorname{ext} Z^{(0)}$ $(j = 2, \ldots, n)$, according to the inclusion isotonicity we obtain

$$\left(\frac{1}{z - \zeta_j}\right)^k \in \left(\frac{1}{z - V^{(0)}}\right)^k \quad (j = 2, \ldots, n; k = 1, 2).$$

Since $z \notin V^{(0)}$, that is $|z - a| < R$, the exact inversion of the region

$$z - V^{(0)} = \{v | v - (z - a)| > R\}$$

is a closed interior of a circle given by

$$X^{(0)} = (z - V^{(0)})^{-1} = \left\{ v \left| \left| v + \frac{\bar{z} - \bar{a}}{R^2 - |z - a|^2} \right| \leq \frac{R}{R^2 - |z - a|^2} \right. \right\} = \{h; d\}$$

(see (2)), where

$$h = \frac{\bar{a} - \bar{z}}{R^2 - |z - a|^2}, \quad d = \frac{R}{R^2 - |z - a|^2}.$$

Using the notation from Sect. 2, we calculate

$$Q(\text{ext } Z^{(0)}) = n\sum_{j=2}^{n}\{h;d\}^2 - \frac{n}{n-1}\left(\sum_{j=2}^{n}\{h;d\}\right)^2 = n(n-1)d\{0;4|h|+d\}.$$

Now, from the fixed point relation (3) we obtain

$$\zeta \in z - \frac{n}{\delta_{1,i} \pm \sqrt{(n-1)(n\delta_{2,i} - \delta_{1,i}^2 - Q(\text{ext } Z^{(0)}))}}. \tag{6}$$

Let

$$Z^{(m)} = \{z^{(m)}; r^{(m)}\} \quad \text{and} \quad X^{(m)} = \left(z^{(m)} - \text{ext } Z^{(m)}\right)^{-1} = \{h^{(m)}; d^{(m)}\},$$

where

$$h^{(m)} = \frac{\bar{a} - \bar{z}^{(m)}}{R^2 - |z^{(m)} - a|^2}, \qquad d^{(m)} = \frac{R}{R^2 - |z^{(m)} - a|^2}.$$

Denoting

$$\delta_k^{(m)} = \delta_{k,i}(z^{(m)}) \ (k = 1,2), \quad Q(\text{ext } Z^{(m)}) = n(n-1)d^{(m)}\{0; 4|h^{(m)}| + d^{(m)}\},$$

and using the previous consideration and (6), we can state the following algorithm for the inclusion of a simple complex zero ζ of a polynomial P:

Algorithm (One) – *Method for a single zero:*

$$Z^{(m+1)} = z^{(m)} - \frac{n}{\delta_1^{(m)} + \left[(n-1)\left(n\delta_2^{(m)} - (\delta_1^{(m)})^2 - Q(\text{ext } Z^{(m)})\right]_*^{1/2}}$$

$$(i \in I_n; \ m = 0, 1, \ldots).$$

In the implementation of Algorithm (One) it is necessary to find such initial region $Z^{(0)} = \{a; R\}$ that enables the inclusion $Z^{(m)} \subset Z^{(0)}$ for each $m = 1, 2, \ldots$, which will provide that the inverse disk $X^{(m)}$ is well defined in each iteration.

In a similar way as in [7] we can prove the following assertion:

Theorem 4. *The order of convergence of Algorithm (One) is three.*

The decrease of the convergence rate of Algorithm (One) related to the preceding algorithms comes from the fact that $\text{rad}Q(\text{ext } Z^{(m)}) = O(1)$, while for the disk

$Q_i(\cdot, \cdot)$ appearing in the algorithms (TS), (SS), (TSN) and (SSN) we have rad $Q_i(\cdot, \cdot) = O(r^{(m)})$, which preserves the convergence rate at least four.

4. Numerical Results

The presented algorithms have been tested in solving many polynomial equations. To provide the enclosure of the zeros in the second and third iteration that produce very small disks, we used the programming package *Mathematica 4.0* with great number of significant decimal digits. For demonstration, we give two numerical examples. The results of the third iteration are included to illustrate very fast convergence of the proposed algorithms although, at present time, the produced (very small) disks are pointless from a practical point of view.

Example 1. To find the circular inclusion approximations to the zeros of the polynomial

$$P(z) = z^9 + 3z^8 - 3z^7 - 9z^6 + 3z^5 + 9z^4 + 99z^3 + 297z^2 - 100z - 300,$$

Table 1. The radii of inclusion disks in three iterations. $A(-h)$ means $A \times 10^{-h}$

i	(TS)	(SS)	(TSN)	(SSN)
		$m = 1$		
1	1.49(−2)	1.49(−2)	1.67(−2)	1.67(−2)
2	8.95(−2)	8.57(−2)	1.77(−2)	1.49(−2)
3	2.57(−2)	2.44(−2)	2.61(−2)	2.37(−2)
4	7.93(−3)	5.09(−3)	7.49(−3)	4.23(−3)
5	4.92(−3)	2.77(−3)	5.09(−3)	2.41(−3)
6	8.59(−3)	2.11(−3)	9.19(−3)	1.12(−3)
7	9.25(−2)	2.69(−3)	3.20(−2)	1.15(−3)
8	2.33(−3)	3.61(−4)	1.46(−3)	2.81(−4)
9	2.03(−3)	6.65(−5)	2.08(−3)	5.08(−5)
		$m = 2$		
1	2.97(−10)	8.44(−11)	9.93(−11)	2.00(−11)
2	3.72(−8)	4.87(−9)	1.73(−9)	2.76(−10)
3	1.58(−9)	8.69(−11)	8.06(−14)	5.76(−15)
4	1.53(−12)	1.34(−14)	3.31(−13)	1.09(−14)
5	2.76(−12)	1.89(−15)	4.93(−15)	2.04(−17)
6	2.41(−12)	6.43(−15)	6.07(−12)	4.74(−15)
7	2.06(−8)	1.58(−17)	3.31(−11)	1.78(−18)
8	9.27(−14)	7.77(−19)	3.97(−15)	6.16(−20)
9	4.05(−13)	1.06(−26)	1.59(−15)	4.35(−31)
		$m = 3$		
1	1.06(−41)	4.46(−45)	4.27(−54)	2.39(−58)
2	2.34(−35)	1.79(−39)	1.15(−49)	2.80(−56)
3	1.63(−39)	1.78(−51)	1.74(−64)	1.11(−72)
4	4.65(−49)	8.57(−64)	6.94(−63)	1.21(−74)
5	4.92(−50)	2.38(−64)	1.73(−68)	6.49(−79)
6	6.13(−48)	3.72(−66)	1.61(−55)	6.97(−79)
7	1.08(−35)	1.12(−75)	3.63(−53)	1.50(−93)
8	1.17(−52)	1.56(−86)	9.31(−72)	1.94(−115)
9	8.94(−51)	2.22(−124)	7.02(−73)	1.36(−167)

we implemented the interval methods (TS) and (SS) without corrections, and (TSN) and (SSN) with Newton's corrections and the centered inversion of disks. The exact zeros of P are

$$\zeta_1 = -3, \ \zeta_{2,3} = \pm 1, \ \zeta_{4,5} = \pm 2i, \ \zeta_{6,7} = -2 \pm i, \ \zeta_{8,9} = 2 \pm i.$$

The initial inclusion disks were chosen to be circular regions with the radii rad $Z_i^{(0)} = 0.4$. The radii of the inclusion disks produced in the first three iterative steps are given in Table 1.

Example 2. We considered the polynomial

$$P(z) = z^7 + z^5 - 10z^4 - z^3 - z + 10$$

with the zeros

$$\zeta_1 = 2, \ \zeta_{2,3} = \pm 1, \ \zeta_{4,5} = \pm i, \ \zeta_{6,7} = -1 \pm 2i.$$

The radii of the initial inclusion disks were $r_i = \text{rad } Z_i^{(0)} = 0.4$. The radii of resulting disks are given in Table 2.

From Tables 1 and 2 and a lot of numerical experiments we can conclude that theoretical results which give the convergence order of the considered Laguerre-

Table 2. The radii of inclusion disks in three iterations. $A(-h)$ means $A \times 10^{-h}$

i	(TS)	(SS)	(TSN)	(SSN)
		$m = 1$		
1	1.41(−2)	1.41(−2)	8.38(−3)	8.38(−3)
2	2.12(−2)	6.65(−3)	3.54(−2)	6.69(−3)
3	1.85(−2)	1.65(−2)	1.56(−2)	1.29(−2)
4	1.18(−2)	4.46(−3)	1.32(−2)	5.29(−3)
5	4.79(−3)	9.93(−4)	3.46(−3)	1.07(−3)
6	5.34(−3)	2.60(−4)	5.42(−3)	2.46(−4)
7	1.77(−3)	1.76(−5)	1.59(−3)	1.46(−5)
		$m = 2$		
1	1.77(−11)	5.87(−12)	1.18(−14)	2.41(−15)
2	1.46(−9)	2.01(−13)	8.89(−11)	1.08(−13)
3	1.13(−11)	2.25(−12)	3.41(−13)	8.60(−14)
4	1.14(−10)	4.41(−13)	2.29(−12)	1.49(−14)
5	1.57(−12)	1.29(−16)	1.38(−13)	4.57(−18)
6	3.93(−13)	3.10(−21)	1.01(−14)	1.14(−22)
7	2.58(−14)	5.55(−33)	1.38(−16)	6.52(−37)
		$m = 3$		
1	2.57(−45)	3.76(−52)	3.64(−66)	1.73(−73)
2	2.29(−41)	1.33(−54)	3.33(−59)	2.33(−68)
3	4.94(−47)	2.99(−51)	2.11(−65)	1.79(−68)
4	1.97(−43)	4.31(−58)	8.16(−60)	3.22(−78)
5	1.58(−49)	1.88(−77)	1.91(−63)	1.71(−98)
6	7.96(−53)	2.43(−103)	1.40(−69)	1.99(−135)
7	2.06(−57)	7.56(−154)	2.44(−75)	4.42(−194)

like methods mainly well coincide with the convergence behavior of these methods in practice, especially when the number of iterative steps increases. Some exceptions appear in the first iteration since Newton's corrections cannot improve resulting disks to a great extent when initial approximations (centers of initial disks) are not sufficiently close to the sought zeros.

References

[1] Alefeld, G., Herzberger, J.: Introduction to interval computations. New York: Academic Press 1983.
[2] Carstensen, C., Petković, M. S.: An improvement of Gargantini's simultaneous inclusion method for polynomial roots by Schroeder's correction. Appl. Numer. Math. *25*, 453–468 (1993).
[3] Gargantini, I.: Parallel Laguerre iterations: Complex case. Numer. Math. *26*, 317–323 (1976).
[4] Hansen, E., Patrick, M., Rusnak, J.: Some modifications of Laguerre's method. BIT *17*, 409–417 (1977).
[5] Herzberger, J., Metzner, L.: On the *Q*-order and *R*-order of convergence for coupled sequences arising in iterative numerical processes. In: Numerical methods and error bounds (G. Alefeld and J. Herzberger, eds.), Mathematical research, vol. 89. Berlin: Akademie Verlag 1996, pp. 120–131.
[6] Ortega, J. M., Rheinboldt, W. C.: Iterative solution of nonlinear equations in several variables. New York: Academic Press 1970.
[7] Petković, M. S.: Some interval iterations for finding a zero of a polynomial with error bounds. Comput. Math. Appl. *14*, 479–495 (1987).
[8] Petković, M. S.: Iterative methods for simultaneous inclusion of polynomial zeros. Berlin: Springer 1989.
[9] Petković, M. S., Carstensen, C.: On some improved inclusion methods for polynomial roots with Weierstrass' correction. Comput. Math. Appl. *25*, 59–67 (1993).
[10] Petković, M. S., Petković, L. D.: Complex interval arithmetic and its applications. Berlin: Wiley VCH 1998.
[11] Schmidt, J. W.: On the R-order of coupled sequences. Comput. *26*, 333–342 (1981).

L. D. Petković,
D. Živković
Faculty of Mechanical Engineering
University of Niš
18 000 Niš
Yugoslavia

M. S. Petković
Faculty of Electronic Engineering
University of Niš
18 000 Niš
Yugoslavia
e-mail: msp@junis.ni.ac.yu

Computing [Suppl] 16, 213–236 (2002)

Computing

Exact Behavior of Singularities of Protter's Problem for the 3-D Wave Equation*

N. Popivanov and **T. Popov**

Abstract

For the wave equation we study boundary value problems, which are three-dimensional analogues of Darboux-problems on the plane. It is shown that for n in $\mathbb{N}$ there exists a right hand side smooth function from $C^n(\bar{\Omega})$, for which the corresponding unique generalized solution belongs to $C^n(\bar{\Omega}\backslash O)$, but it has a strong power-type singularity at the point O. This singularity is isolated at the vertex O of the characteristic cone and does not propagate along the cone. The present article describes the exact behavior of the singular solutions at the point O. It states some exact a priori estimates for the solution.

Mathematics Subject Classification: Primary 35L05, 35L20, 35D05, 35A20.

Keywords: Wave equation, boundary value problems, generalized solution, singular solutions, propagation of singularities, special functions.

1. Introduction

We discuss some boundary value problems for the wave equation

$$\square u = u_{x_1x_1} + u_{x_2x_2} - u_{tt} = f \tag{1.1}$$

in a simply connected domain $\Omega \subset \mathbb{R}^3$. The domain

$$\Omega := \{(x_1,x_2,t) : 0 < t < 1/2, t < \sqrt{x_1^2+x_2^2} < 1-t\}$$

is bounded by two characteristic cones of (1.1)

$$S_1 = \{(x_1,x_2,t) : 0 < t < 1/2, \sqrt{x_1^2+x_2^2} = 1-t\},$$
$$S_2 = \{(x_1,x_2,t) : 0 < t < 1/2, \sqrt{x_1^2+x_2^2} = t\}$$

and the circle $S_0 = \{(x_1,x_2,t) : t = 0, x_1^2+x_2^2 < 1\}$. The following three-dimensional analogues of the plane Darboux problems are stated by M. Protter [22]:

*Dedicated to the 60th anniversary of Prof. Dr. Jürgen Herzberger.

Problem P1. To find a solution of the wave equation (1.1) in Ω which satisfies the boundary conditions

P1 $u|_{S_0} = 0$ and $u|_{S_1} = 0$.

The problem adjoint to $P1$ is:

Problem P1*. To find a solution of the wave equation (1.1) in Ω which satisfies the boundary conditions

P1* $u|_{S_0} = 0$ and $u|_{S_2} = 0$.

Protter [22] formulated and studied problems $P1$ and $P1^*$ in Ω as multi-dimensional analogues of the Darboux problem on the plane. For recent known results concerning Protter's Problems $P1$ and $P1^*$ see paper the [21] of Popivanov, Schneider and references therein. For further publications in this area see [1], [2], [8], [13], [16], [17]. Substituting the boundary condition $u|_{S_0} = 0$ by $[u_t + \alpha u]|_{S_0} = 0$, one obtains Problem P_α, for which we refer to [11] and references therein. In the case of the wave equation, involving either lower order terms or some other type perturbations, Problem P_α in Ω with $\alpha \equiv 0$ has been studied by Aldashev in [1–3]. On the other hand, Bazarbekov [5] gives another analogue of the classical Darboux problem in the same domain Ω. Some other statements of Darboux type problems can be found in [4], [6], [15] in bounded or unbounded domains different from Ω.

It is well known that the $3 - D$ Problem $P1$ is not well posed in contrast to the Darboux-problem on the plane. The reason is that the adjoint homogeneous Problem $P1^*$ has smooth solutions and their span is infinite dimensional as one could see in Tong Kwang-Chang [24], Popivanov and Schneider [20], and Khe Kan Cher [17]. More precisely, in polar coordinates $x_1 = \rho\cos\varphi, x_2 = \rho\sin\varphi, t$ we have the following lemma for Problem $P1^*$:

Lemma 1.1. *Let $n \in \mathbb{N}$, $n \geq 4$,*

$$H_k^n(\rho, t) := \sum_{i=0}^{k} A_i^k \frac{t(\rho^2 - t^2)^{n-3/2-k-i}}{\rho^{n-2i}}, \tag{1.2}$$

where the coefficients are

$$A_i^k := (-1)^i \frac{(k-i+1)_i (n-1/2-k-i)_i}{i!(n-i)_i}, \quad (a)_i := a(a+1)\dots(a+i-1).$$

Then the functions

$$V_k^{n,1}(\rho, t, \varphi) = H_k^n(\rho, t)\cos n\varphi \quad \text{and} \quad V_k^{n,2}(\rho, t, \varphi) = H_k^n(\rho, t)\sin n\varphi$$

are classical solutions of Problem $P1^*$ for all $k = 0, 1, \dots, \left[\frac{n}{2}\right] - 1$.

Remark 1.1. *The solutions* $V_0^{n,j}$ *are obtained by Popivanov and Schneider in* [20]. *Actually, the functions* H_k^n *appear in Khe Kan Cher* [17], *although defined there by some hypergeometric functions instead.*

Remark 1.2. *In the case of* $k = \frac{n}{2} - 1$ *the functions* $V_k^n(\rho, t, \varphi)$ *are not continuous at the origin* O.

A necessary condition for the existence of classical solution for Problem $P1$ is the orthogonality of the right-hand side function f to all functions $V_k^{n,i}(\rho, t, \varphi)$. To avoid an infinite number of necessary conditions in the frame of classical solvability, Popivanov and Schneider in [20] and [21] gave some definitions of a generalized solution of Problem $P1$ with eventually singularity on the characteristic cone S_2, or only at its vertex (0,0,0).

Definition 1.1. ([21]) *A function* $u = u(x_1, x_2, t)$ *is called a* generalized solution *of the problem P1 in* Ω, *if*:

(1) $u \in C^1(\bar{\Omega}\backslash(0,0,0))$, $u|_{S_0\backslash(0,0,0)} = 0$, $u|_{S_1} = 0$,

(2) *the identity*

$$\int_{\Omega} (u_t w_t - u_{x_1} w_{x_1} - u_{x_2} w_{x_2} - f w) dx_1\, dx_2\, dt = 0$$

holds for all $w \in C^1(\bar{\Omega})$, $w = 0$ *on* S_0 *and* $w = 0$ *in a neighborhood of* S_2.

Garabedian [10] proved the uniqueness of a classical solution of the Problem $P1$. Popivanov and Schneider [21] proved the uniqueness of a *generalized solution* in $C^1(\bar{\Omega}\backslash O)$. It is known (cf. Popivanov and Schneider [21], Aldashev [1]) that for every $n \in \mathbb{N}$, $n \geq 4$ there exists a smooth function

$$f_n(x_1, x_2, t) := f_n^1(\rho, t) \cos n\varphi \in C^{n-2}(\bar{\Omega}),$$

whose corresponding *generalized solution* of Problem $P1$ behaves like $(x_1^2 + x_2^2)^{(1-n)/2}$ around the origin O. The same phenomenon appears in the case of a more general boundary condition P_α (see [11]). These singularities of the *generalized solutions* do not propagate in the direction of the bicharacteristics on the characteristic cone. It is traditionally assumed that the wave equation whose right hand side is sufficiently smooth in $\bar{\Omega}$ cannot have a solution with an isolated singular point. For results concerning the propagation of singularities for operators of second order we refer to Hörmander [12, chap. 24.5]. For some related results in the case of the plane Darboux-Problem, see [19]. There are a lot of papers which explore the behavior of the singular solutions of Problem $P1$. However, none of them provides sufficient conditions for the nonexistence of singular solutions.

Let us fix the right-hand side as a trigonometric polynomial of order l:

$$f(x_1,x_2,t) = \sum_{n=2}^{l} \left(f_n^1(t,\rho)\cos n\varphi + f_n^2(t,\rho)\sin n\varphi \right). \tag{1.3}$$

We already know that the corresponding solution $u(x_1,x_2,t)$ may have the behavior of $(x_1^2+x_2^2)^{-(l-1)/2}$ at the point O. There are some

Open problems:

(1) To find the exact behavior of all singular solutions at the point O; or more precisely: are there *generalized solutions* for the right-hand side (1.3) with higher order of singularity, for example $(x_1^2+x_2^2)^{-k-(l-1)/2}$, $k>0$? Are there *generalized solutions* for the right-hand side (1.3) with lower order of singularity, for example $(x_1^2+x_2^2)^{k-(l-1)/2}$, $k>0$?

(2) To find some appropriate conditions for the function f under which Protter's problem has only regular solutions. We do not know any similar results even for the Problem $P1$.

(3) The a priori estimates, obtained in [11, 21], signify that the *generalized solutions* of Problems P_α and $P1$ (including the singular ones), can have at most an exponential growth as $\rho \to 0$. The natural question is: are there any singular solutions of these problems with exponential growth as $\rho \to 0$ or do all such solutions have polynomial growth?

(4) Can we prove some a priori estimates for *generalized solutions* of the Problems P_α and $P1$ with function f given by (1.3) or, more generally, by an infinite trigonometric series?

(5) Why there appears a singularity for such smooth right hand side even for the wave equation? Can we numerically explain this phenomenon?

In this paper, we give results concerning some of these questions. More precisely, for the case of Problem $P1$ we answer the Problems 1, 2, 3 and 4 above. We will prove the following results:

Theorem 1.1. *Let us suppose that $f(x_1,x_2,t) \in C^l(\bar{\Omega})$ and that*

$$\iiint_{\Omega} V_k^{n,i}(x_1,x_2,t) f(x_1,x_2,t)\,dx_1\,dx_2\,dt = 0, \tag{1.4}$$

for all $n=2,3,\dots,l$; $i=1,2$; $k=0,1,\dots,\left[\frac{n}{2}\right]-1$. Then there exists a unique generalized solution u of Problem P1. Moreover, $u \in C^{l+1}(\bar{\Omega}\backslash O)$ and it holds the a priori estimate

$$|u(x_1,x_2,t)| \le C \sum_{|\alpha|\le l-2} \max_{\bar{\Omega}} |D^\alpha f| \tag{1.5}$$

with a constant C independent of f. In addition, if $f_n^i(x_1, x_2, t) = 0$ *for all odd indexes* n *and all* $i = 1, 2$, *then* $u \in C(\bar{\Omega})$ *and*

$$|u(x_1, x_2, t)| \leq C\rho^{1/2}|\ln\rho| \sum_{|\alpha|\leq l-2} \max_{\bar{\Omega}} |D^\alpha f|. \tag{1.6}$$

Theorem 1.1 gives an a priori estimate of the *generalized solution*. The estimate (1.6) implies the continuity of the *generalized solution* at O. This feature is not necessary if $f_n^i(x_1, x_2, t) \neq 0$ for some odd number n and $i = 1$ or 2.

Theorem 1.2 provides an a priori estimate for the *generalized solution* and clarifies the significance of the above orthogonality conditions (1.4). Namely, for any couple (n, k) the corresponding condition "controls" one power-type singularity.

Theorem 1.2. *Let* $m, p \in \mathbb{N} \cup \{0\}$, $2 \leq m \leq l$ *and* $p \leq \left[\frac{m}{2}\right] - 1$. *Suppose that* $f(x_1, x_2, t) \in C^l(\bar{\Omega})$ *satisfies the orthogonality conditions* (1.4) *for any couple* (n, k), *such that* $n \in \mathbb{N}$, $n \geq 2$, $k \in \mathbb{N} \cup \{0\}$, $n - 2k \geq m - 2p + 1$ *and* $i = 1, 2$. *Then there exists a unique* generalized solution u *of Problem P1, such that:* $u \in C^{l+1}(\bar{\Omega}\backslash O)$ *and the estimate*

$$|u(x_1, x_2, t)| \leq C\rho^{-(m-2p-1)} \sum_{|\alpha|\leq l+2p-m} \max_{\bar{\Omega}} |D^\alpha f| \tag{1.7}$$

holds. If we suppose that

$$\iiint_{\Omega} V_p^{m,j}(x_1, x_2, t) f(x_1, x_2, t) dx_1\, dx_2\, dt \neq 0, \quad j = 1 \text{ or } 2, \tag{1.8}$$

then in some neighborhood of the point O one has

$$\left|u_m^j\big|_{S_2}\right| \geq c\rho^{-(m-2p-1)}, \quad c > 0, \tag{1.9}$$

where

$$u_m^1\big|_{S_2} := \int_0^{2\pi} u\big|_{t=\rho} \cos m\varphi\, d\varphi \quad \text{and} \quad u_m^2\big|_{S_2} := \int_0^{2\pi} u\big|_{t=\rho} \sin m\varphi\, d\varphi.$$

In order to explain the dependence of the singularity of the generalized solution on the orthogonality assumptions let us consider the following table:

Table 1

	l	$l-1$	$l-2$	$l-3$	...	m		...	4	3	2	1
1	...	...	...	...	...	...		...	$V_1^{4,i}$		$V_0^{2,i}$	
2	...	...	...	...	...	...		...		$V_0^{3,i}$		
3	...	...	...	...	...	...		...	$V_0^{4,i}$			
...	...	...	...	...	...	...		...				
$m-2p-1$	...	...	...	...	...	$V_p^{m,i}$	...					
$m-2p$	– –	– –	– –	– –	– –	– –						
...	...	...	...	...	...							
$l-5$	$V_2^{l,i}$		$V_1^{l-2,i}$									
$l-4$		$V_1^{l-1,i}$		$V_0^{l-3,i}$								
$l-3$	$V_1^{l,i}$		$V_0^{l-2,i}$									
$l-2$		$V_0^{l-1,i}$										
$l-1$	$V_0^{l,i}$											

Observe that both $V_k^{n,1}$ and $V_k^{n,2}$ are located in the n-th column and $(n-2k-1)$-st row. Thus, $V_0^{n,i}$ form the right most diagonal, $V_1^{n,i}$ constitute the next one, and so on. The first column designates the order of singularity of the *generalized solution*.

Theorem 1.1 asserts that the *generalized solution* of problem $P1$ is bounded if the right hand side f is orthogonal to all the functions $V_k^{n,i}$ from Table 1. Theorem 1.2 specifies that the singularity of the *generalized solution* is no worse than $(x_1^2+x_2^2)^{-(m-2p-1)/2}$ if f is orthogonal to $V_k^{n,i}$ from the triangle under the $(m-2p-1)$-st row. In other words, the functions from the k-th row of the table are "responsible" for the *generalized solutions* with behavior $(x_1^2+x_2^2)^{-k/2}$ near the origin O.

Conclusion 1. *Without any orthogonality conditions* (1.4) *on f the generalized solution u satisfies the a priori estimate*

$$|u(x_1,x_2,x_3)| \leq C\left(\max_{\bar{\Omega}} |f|\right)\rho^{-(l-1)}. \tag{1.10}$$

Similarly to the estimates from Popivanov, and Schneider [21], the aforementioned estimates (1.7) and (1.10) allow the solution to blow up at the point O. However, the estimate (1.10) is the best possible in view of (1.9) and answers Question 4 above.

The present paper consists of an introduction and five consecutive sections. In Sect. 2 are formulated the 2-D boundary problems $P2$ and $P3$, corresponding to the 3-D Problem $P1$. The main technical results are established in Sect. 3. In Sects. 4 and 5, we study the behavior of solutions of 2-D Problems $P2$ and $P3$. Finally, Sect. 6 is devoted to the proofs of Theorems 1.1 and 1.2, based on the results of the previous sections.

2. Preliminaries

Let $u(x_1,x_2,t)$ be the *generalized solution* of Problem $P1$ for the right-hand side (1.3), the existence and uniqueness of which follow from [21]. We are going to use the results from [21] in order to describe in Theorems 1.1 and 1.2 the exact behavior of u near the eventually singular point O. To this end, let us define some problems related to $P1$ and $P1^*$. Suppose that the right-hand side f of the wave equation is

$$f(\rho,t,\varphi) = f_n^1(\rho,t)\cos n\varphi + f_n^2(\rho,t)\sin n\varphi, \quad n \in \mathbb{N}.$$

We seek for solutions of the wave equation of the same form:

$$u(\rho,t,\varphi) = u_n^1(\rho,t)\cos n\varphi + u_n^2(\rho,t)\sin n\varphi.$$

Then Problem $P1$ reduces to the following:

Problem P2. To solve the equation

$$(u_n)_{\rho\rho} + \frac{1}{\rho}(u_n)_\rho - (u_n)_{tt} - \frac{n^2}{\rho^2}u_n = f_n, \tag{2.1}$$

in $\Omega_0 = \{0 < t < 1/2;\ t < \rho < 1 - t\} \subset \mathbb{R}^2$, with the boundary conditions

P2 $\quad u_n(\rho,0) = 0$ for $0 \le \rho \le 1$ and $u_n(\rho, 1-\rho) = 0$ for $1/2 \le \rho \le 1$.

Let us now introduce new coordinates $\xi = (\rho+t)/2$, $\eta = (\rho - t)/2$ and set

$$v(\xi,\eta) = \rho^{1/2}u_n(\rho,t); \quad g(\xi,\eta) = \rho^{1/2}f_n(\rho,t). \tag{2.2}$$

Denoting $\nu = n - \frac{1}{2}$ one transforms Problem $P2$ into:

Problem P3. To find a solution $v(\xi,\eta)$ of the equation

$$v_{\xi\eta} - \frac{\nu(\nu+1)}{(\xi+\eta)^2}v = g \tag{2.3}$$

in the domain $D = \{0 < \xi < 1/2; 0 < \eta < \xi\}$, with the following boundary conditions:

P3 $\quad v(\xi,\xi) = 0, \quad \xi \in (0,1/2)$ and $v(1/2,\eta) = 0, \quad \eta \in (0,1/2)$.

A basic tool for our treatment of Problem $P3$ are the Legendre's functions P_ν. Let us note that

$$R(\xi_1,\eta_1,\xi,\eta) = P_\nu\left(\frac{(\xi-\eta)(\xi_1-\eta_1) + 2\xi_1\eta_1 + 2\xi\eta}{(\xi_1+\eta_1)(\xi+\eta)}\right)$$

is a Riemann function for Eq. (2.3) (Copson [7]). Therefore, we can construct the function v as a solution of a Goursat's problem in D with boundary conditions $v(1/2,\eta) = 0$ and $v(\xi,0) = \varphi(\xi)$:

$$v(\xi,\eta) = \varphi(\xi) + \int\limits_{\xi}^{\frac{1}{2}} \varphi(\xi_1)\frac{\partial}{\partial\xi_1}R(\xi_1,0,\xi,\eta)d\xi_1 - \int\limits_{\xi}^{\frac{1}{2}}\int\limits_{0}^{\eta} R(\xi_1,\eta_1,\xi,\eta)g(\xi_1,\eta_1)d\eta_1\, d\xi_1. \tag{2.4}$$

Note that $\varphi(1/2) = 0$ because of $\varphi(\xi) := v(\xi,0)$.

Now we have to find a function $\varphi(\xi)$, such that the function $v(\xi,\eta)$, defined by (2.4), is a solution of Problem $P3$. Let us denote

$$G(\xi) = \int\limits_{\xi}^{\frac{1}{2}}\int\limits_{0}^{\xi} R(\xi_1,\eta_1,\xi,\xi)g(\xi_1,\eta_1)d\eta_1\, d\xi_1 = \int\limits_{\xi}^{\frac{1}{2}}\int\limits_{0}^{\xi} P_\nu\left(\frac{\xi_1\eta_1+\xi^2}{\xi(\xi_1+\eta_1)}\right)g(\xi_1,\eta_1)d\eta_1\, d\xi_1. \tag{2.5}$$

Then the boundary condition $v(\xi,\xi) = 0$ leads us to the equation

$$G(\xi) = \varphi(\xi) + \int\limits_{\xi}^{\frac{1}{2}} \varphi(\xi_1)\frac{\partial}{\partial\xi_1}P_\nu\left(\frac{\xi}{\xi_1}\right)d\xi_1 = -\int\limits_{\xi}^{\frac{1}{2}} \varphi'(\xi_1)P_\nu\left(\frac{\xi}{\xi_1}\right)d\xi_1, \tag{2.6}$$

for the function $\varphi(\xi)$. According to formulas (35.17), (35.28) from [23], this integral equation is invertible. In fact, we find that

$$\frac{d}{d\xi}\varphi(\xi) = -\xi\frac{d^2}{d\xi^2}\xi\int\limits_{\xi}^{\frac{1}{2}} P_\nu\left(\frac{\xi_1}{\xi}\right)\frac{G(\xi_1)}{\xi_1^2}d\xi_1 = G'(\xi) + \frac{d}{d\xi}\int\limits_{\xi}^{\frac{1}{2}} P'_\nu\left(\frac{\xi_1}{\xi}\right)\frac{G(\xi_1)}{\xi_1}d\xi_1.$$

Taking into account that $\varphi(1/2) = G(1/2) = 0$, there follows

$$\varphi(\xi) = G(\xi) + \int\limits_{\xi}^{\frac{1}{2}} P'_\nu\left(\frac{\xi_1}{\xi}\right)\frac{G(\xi_1)}{\xi_1}d\xi_1. \tag{2.7}$$

We are going to study the behavior of φ, represented by (2.7), in order to find the exact behavior of function v (respectively, u_n). Note that the argument of the Legendre's function in (2.6) belongs to the segment [0, 1], while the one from (2.7) belongs to $[1,+\infty)$.

Our treatment starts by a thorough description of the behavior of the function

$$I(\xi) := \int\limits_{\xi}^{\frac{1}{2}} P'_\nu\left(\frac{\xi_1}{\xi}\right)\frac{G(\xi_1)}{\xi_1}d\xi_1 \tag{2.8}$$

as $\xi \to +0$.

3. Estimates for the Function $I(\xi)$

Let us observe that $I(\xi)$ is smooth for $0 < \xi \leq 1/2$ when $G(\xi)$ is a smooth function and the integral $I(\xi)$ may blow up as $\xi \to +0$. Actually $P_\nu'(\xi_1/\xi)$ behave like $\xi^{1-\nu}$ for very small ξ. The eventual singularities of $I(\xi)$ are specified by the following lemma:

Lemma 3.1. *Let $\nu > 1$ be a real number with integral part $[\nu]$ and fractional part $\{\nu\} = \nu - [\nu] \neq 0$, $G \in C[0, 1/2]$, $|G(\xi)| \leq A\xi$ for some constant A and*

$$\int_0^{\frac{1}{2}} \xi^{\nu-2i-2} G(\xi) d\xi = 0 \quad \textit{for } i = 0, 1, \ldots, \left[\frac{\nu-1}{2}\right]. \tag{3.1}$$

In the case of $\nu > 2$ be impose an extra assumption

$$G \in C^{[\nu-2]}(0, 1/2], \quad |G^{(k)}(\xi)| \leq A\xi^{1-k} \quad \textit{for } k = 0, 1, \ldots, [\nu - 2]. \tag{3.2}$$

Then the function $I(\xi)$, defined by (2.8), *is $C^{[\nu]-1}(0, 1/2]$ and $I \in C[0, 1/2]$. More precisely, there is a constant C, independent of $G(\xi)$, such that:* (i) *$|I(\xi)| \leq CA\xi$, if $[\nu]$ is odd;* (ii) *$|I(\xi)| \leq CA\xi^{1-\{\nu\}}$, if $[\nu]$ is even.*

Proof: We will use the integral representation of the Legendre's functions (cf. [9])

$$P_\nu(z) = \frac{1}{\pi} \int_0^\pi (z + \sqrt{z^2 - 1} \cos t)^\nu dt, \quad z \geq 1. \tag{3.3}$$

Differentiating this expression we find

$$P_\nu'(z) = \frac{\nu(\nu+1)}{\pi} \int_0^\pi (z + \sqrt{z^2-1} \cos t)^{\nu-1} \sin^2 t \, dt.$$

Obviously, $G(\xi) \in C^k(0, 1/2]$ suffices for $I(\xi) \in C^{k+1}(0, 1/2]$. Let us define the function

$$F(\xi) := \begin{cases} \displaystyle\int_\xi^{\frac{1}{2}} \int_0^\pi \left(\xi_1 + \sqrt{\xi_1^2 - \xi^2} \cos t\right)^{\nu-1} \sin^2 t \frac{G(\xi_1)}{\xi_1} dt \, d\xi_1, & \text{for } \xi \in (0, \tfrac{1}{2}], \\ 0, & \text{for } \xi = 0. \end{cases}$$

Then our aim is to find the behavior of

$$I(\xi) = \int_\xi^{\frac{1}{2}} P_\nu'\left(\frac{\xi_1}{\xi}\right) \frac{G(\xi_1)}{\xi_1} d\xi_1 = \frac{\nu(\nu+1)}{\pi \xi^{\nu-1}} F(\xi), \tag{3.4}$$

as $\xi \to +0$. If $F(\xi) \in C^{[\nu-1]}[0, \frac{1}{2}]$, $F^{(k)}(0) = 0$ for $k = 0, 1, \ldots, [\nu - 1]$ and $F^{([\nu-1])}(\xi)$ satisfies an appropriate estimate then Maclaurin's formula

$$F(\xi) = \frac{1}{[\nu-1]!}\xi^{[\nu-1]}F^{([\nu-1])}(\xi'), \quad \xi' \in (0, \xi), \tag{3.5}$$

provides the boundness of $I(\xi)$. In order to calculate the derivatives of $F(\xi)$, let us consider the functions

$$N_k^\nu(\xi_1, \xi) := (-1)^k \frac{(\nu-2k)_{2k}}{(2k+1)!!} \int_0^\pi \left(\xi_1 + \sqrt{\xi_1^2 - \xi^2}\cos t\right)^{\nu-2k-1} \sin^{2k+2} t\, dt, \tag{3.6}$$

which are accurately defined and continuous for $(\xi_1, \xi) \in \bar{D}\backslash O$, $\nu + 1/2 > k \geq 0$. It is obvious that

$$F(\xi) = \int_\xi^{\frac{1}{2}} N_0^\nu(\xi_1, \xi)\frac{G(\xi_1)}{\xi_1}d\xi_1.$$

Note that N_k^ν are appropriately scaled in order to have concise formulae for theirs derivatives. Differentiating with respect to ξ one obtains

$$\frac{\partial}{\partial \xi}N_k^\nu(\xi_1, \xi) = \xi N_{k+1}^\nu(\xi_1, \xi). \tag{3.7}$$

Our study of $N_k^\nu(\xi_1, \xi)$ begin by the observation that for $k < \nu + 1/2$ and $\xi \leq \xi_1$ there hold:

$$N_k^\nu(\xi, \xi) = a_k\xi^{\nu-2k-1}, \quad a_k = const \neq 0, \tag{3.8}$$

$$N_k^\nu(\xi_1, 0) = b_k\xi_1^{\nu-2k-1}, \quad b_k = const \neq 0, \tag{3.9}$$

$$|N_k^\nu(\xi_1, \xi)| \leq c_k\xi_1^{\nu-2k-1}, \quad c_k = const. \tag{3.10}$$

The first and the second estimates are obvious, while (3.10) follows from (3.8) and (3.7). Now we shall calculate the derivatives of N_0^ν at $\xi = 0$. First of all, an induction provides

$$\frac{\partial^k}{\partial \xi^k}N_0^\nu(\xi_1, \xi) = \sum_{i=0}^{[\frac{k}{2}]} C_i^k \xi^{k-2i} N_{k-i}^\nu(\xi_1, \xi),$$

where the constants $C_i^k = \frac{(k-2i+1)_{2i}}{i!2^i} \neq 0$ for $i \leq k/2$.

Each of the summands can be estimated by the means of (3.10):

(i) if $\nu - 2(k-i) - 1 < 0$ then

$$|\xi^{k-2i} N^{\nu}_{k-i}(\xi_1, \xi)| \leq \xi^{k-2i} c_k \xi_1^{\nu-2(k-i)-1} \leq c_k \xi^{\nu-k-1};$$

(ii) if $\nu - 2(k-i) - 1 \geq 0$ then

$$|\xi^{k-2i} N^{\nu}_{k-i}(\xi_1, \xi)| \leq C \xi^{k-2i}.$$

Therefore, whenever $k < \nu - 1$ and $i < k/2$ there holds $\xi^{k-2i} N^{\nu}_{k-i}(\xi_1, \xi) \to 0$ as $\xi \to +0$. That amounts to the continuity of the functions $\frac{\partial^k}{\partial \xi^k} N^{\nu}_0(\xi_1, \xi)$ in $\bar{D}$ and

$$\frac{\partial^k}{\partial \xi^k} N^{\nu}_0(\xi_1, 0) = \begin{cases} 0 \quad , & \text{for an odd } k, \\ C^{\nu}_{k/2} b_{k/2} \xi_1^{\nu-k-1}, & \text{for an even } k. \end{cases} \tag{3.11}$$

On the other hand, using (3.8) one has $\xi^{k-2i} N^{\nu}_{k-i}(\xi, \xi) = a_k \xi^{\nu-k-1}$ and thus, for $k < \nu - 1$ there follows:

$$\left. \frac{\partial^k}{\partial \xi^k} N^{\nu}_0(\xi_1, \xi) \right|_{\xi_1 = \xi} = C_{\nu,k} \xi^{\nu-k-1}. \tag{3.12}$$

Now we are ready to calculate the derivatives of $F(\xi)$: An inductional argument, based on (3.8) and (3.12) for $k < \nu - 1$ and $\xi > 0$ provides

$$F^{(k)}(\xi) = \sum_{i=0}^{k-1} A^k_i G^{(i)}(\xi) \xi^{\nu-1-k+i} + \int_{\xi}^{\frac{1}{2}} \frac{\partial^k}{\partial \xi^k} N^{\nu}_0(\xi_1, \xi) \frac{G(\xi_1)}{\xi_1} d\xi_1. \tag{3.13}$$

In order to push $\xi \to +0$ in this expression, let us observe that for $k < \nu - 1$ the Assumptions (3.2) imply

$$|G^{(i)}(\xi) \xi^{\nu-1-k+i}| \leq A \xi^{\nu-k} \to 0 \quad \text{as } \xi \to 0.$$

As a result, applying (3.11) one obtains the limit of $F^{(k)}(\xi)$ as $\xi \to +0$, depending of parity of k. More precisely:

(a) if k is an odd number there holds

$$F^{(k)}(\xi) \to \int_0^{\frac{1}{2}} \frac{\partial^k}{\partial \xi^k} N^{\nu}_0(\xi_1, 0) \frac{G(\xi_1)}{\xi_1} d\xi_1 = 0; \tag{3.14}$$

(b) if k is an even number one infers that

$$F^{(k)}(\xi) \to \int_0^{\frac{1}{2}} \frac{\partial^k}{\partial \xi^k} N_0^\nu(\xi_1, 0) \frac{G(\xi_1)}{\xi_1} d\xi_1 = C \int_0^{\frac{1}{2}} \xi_1^{\nu-k-1} \frac{G(\xi_1)}{\xi_1} d\xi_1, \tag{3.15}$$

where $C = C^\nu_{k/2} b_{k/2} \neq 0$. Due to (3.14), (3.15) one concludes that $F(\xi) \in C^{[\nu-1]}[0, \frac{1}{2}]$ and $F^{(2i+1)}(0) = 0$. Moreover, $F^{(2i)}(0) = 0$ according to

$$\int_0^{\frac{1}{2}} \xi_1^{\nu-2i-2} G(\xi_1) d\xi_1 = 0 \qquad \text{for } i = 0, 1, \ldots, \left[\frac{\nu-1}{2}\right].$$

There remains to be estimated $F^{([\nu-1])}(\xi)$. Substituting $k = [\nu] - 1$ in (3.13) one gets

$$F^{([\nu]-1)}(\xi) = \sum_{i=0}^{[\nu]-2} A_i^{[\nu]-2} G^{(i)}(\xi) \xi^{i+\{\nu\}} + \int_\xi^{\frac{1}{2}} \frac{\partial^{[\nu]-1}}{\partial \xi^{[\nu]-1}} N_0^\nu(\xi_1, \xi) \frac{G(\xi_1)}{\xi_1} d\xi_1. \tag{3.16}$$

The sum is estimated straightforward by the means of the inequalities

$$|\xi^{i+\{\nu\}} G^{(i)}(\xi)| \leq A \xi^{1+\{\nu\}} \tag{3.17}$$

for $i \leq [\nu] - 2$. The integral in (3.16) is, actually,

$$\int_\xi^{\frac{1}{2}} \frac{\partial^{[\nu]-1}}{\partial \xi^{[\nu]-1}} N_0^\nu(\xi_1, \xi) \frac{G(\xi_1)}{\xi_1} d\xi_1 = \sum_{i=0}^{\left[\frac{[\nu]-1}{2}\right]} C_i^{[\nu]-1} \int_\xi^{\frac{1}{2}} \xi^{[\nu]-2i-1} N^\nu_{[\nu]-i-1}(\xi_1, \xi) \frac{G(\xi_1)}{\xi_1} d\xi_1. \tag{3.18}$$

It is easy to see that $[\nu] - 2i - 1 \geq 2$ for all $i = 0, 1, \ldots, \left[\frac{[\nu]-1}{2}\right] - 1$. Thus, using (3.10) one finds out that

$$\left|\xi^{[\nu]-2i-1} N^\nu_{[\nu]-i-1}(\xi_1, \xi)\right| \leq c_{[\nu]-1} \xi^2 \xi_1^{\{\nu\}-2}$$

for the specified values of i. Hence

$$\left|\int_\xi^{\frac{1}{2}} \xi^{[\nu]-2i-1} N^\nu_{[\nu]-i-1}(\xi_1, \xi) \frac{G(\xi_1)}{\xi_1} d\xi_1\right| \leq c_{[\nu]-1} A \xi^2 \int_\xi^{\frac{1}{2}} \xi_1^{\{\nu\}-2} d\xi_1 \leq CA\xi^{1+\{\nu\}}. \tag{3.19}$$

The last summand from (3.18) is estimated in the following way, depending on the parity of $k := [\nu] - 1$:

Case (i): k is an even number. Using the orthogonality condition (3.1) for $i = 0$, one concludes that

$$\left|\int_{\xi}^{\frac{1}{2}} N_{k/2}^{\nu}(\xi_1,\xi)\frac{G(\xi_1)}{\xi_1}d\xi_1\right| = \left|\int_{\xi}^{\frac{1}{2}} N_{k/2}^{\nu}(\xi_1,\xi)\frac{G(\xi_1)}{\xi_1}d\xi_1 - \int_{0}^{\frac{1}{2}} N_{k/2}^{\nu}(\xi_1,0)\frac{G(\xi_1)}{\xi_1}d\xi_1\right|$$

$$\leq \left|\int_{\xi}^{\frac{1}{2}} \left[N_{k/2}^{\nu}(\xi_1,\xi) - N_{k/2}^{\nu}(\xi_1,0)\right]\frac{G(\xi_1)}{\xi_1}d\xi_1\right| + \left|\int_{0}^{\xi} N_{k/2}^{\nu}(\xi_1,0)\frac{G(\xi_1)}{\xi_1}d\xi_1\right|$$

On the other hand, using (3.10) one obtain that

$$\left|N_{k/2}^{\nu}(\xi_1,\xi) - N_{k/2}^{\nu}(\xi_1,0)\right| = \left|\int_{0}^{\xi} \xi_2 N_{1+k/2}^{\nu}(\xi_1,\xi_2)d\xi_2\right| \leq \frac{c_{1+k/2}}{2}\xi^2\xi_1^{\{\nu\}-2}.$$

Finally, one concludes that

$$\left|\int_{\xi}^{\frac{1}{2}} N_{k/2}^{\nu}(\xi_1,\xi)\frac{G(\xi_1)}{\xi_1}d\xi_1\right| \leq CA\xi^{\{\nu\}+1}.$$

Combining with (3.17), (3.19), (3.5) and (3.4) one deduces that

$$|I(\xi)| \leq \frac{\nu(\nu+1)}{\pi\xi^{\nu-1}}C'\xi^{[\nu]-1}|F^{([\nu]-1)}(\xi)| \leq CA\xi,$$

if $[\nu] - 1$ is an even number.

Case (ii): $k = [\nu] - 1$ is an odd number. One needs only the estimate

$$\left|\int_{\xi}^{\frac{1}{2}} \xi N_l^{\nu}(\xi_1,\xi)\frac{G(\xi_1)}{\xi_1}d\xi_1\right| \leq \xi c_l A \int_{\xi}^{\frac{1}{2}} \xi_1^{\{\nu\}+1}d\xi_1 \leq CA\xi,$$

in order to conclude that $\left|F^{([\nu]-1)}(\xi)\right| \leq CA\xi$ and, therefore,

$$|I(\xi)| \le \frac{\nu(\nu+1)}{\pi\xi^{\nu-1}} C'\xi^{[\nu]-1}|F^{([\nu]-1)}(\xi')| \le CA\xi^{1-\{\nu\}}. \qquad \square$$

Each one of the orthogonality conditions (3.1) "controls" one power-type singularity:

Lemma 3.2. *Suppose* $\nu > 1, \{\nu\} \neq 0$, *p is a nonnegative integer*, $p \le [(\nu-1)/2]$, $G \in C[0, 1/2], |G(\xi)| \le A\xi$ *for some constant A. In the case of* $p \ge 1$ *assume also that*

$$G \in C^{2p-1}(0, 1/2], \ |G^{(k)}(\xi)| \le A\xi^{2p-[\nu]+1-k} \text{ for } k = 0, 1, \dots, 2p-1 \qquad (3.20)$$

and

$$\int_0^{\frac{1}{2}} \xi^{\nu-2i-2} G(\xi) d\xi = 0 \qquad \text{for } i = 0, 1, \dots, p-1. \qquad (3.21)$$

Then the estimate

$$|I(\xi)| \le C_1 A\xi^{-(\nu-2p-1)},$$

holds for some constant C_1 *independent of* $G(\xi)$. *Moreover, if*

$$\int_0^{\frac{1}{2}} \xi^{\nu-2p-2} G(\xi) d\xi \neq 0 \qquad (3.22)$$

then

$$|I(\xi)| \ge C_2\xi^{-(\nu-2p-1)},$$

for $C_2 > 0$ *and sufficiently small* ξ.

Remark 3.1. *In the case* $p = 0$ *the conditions* (3.20) *and* (3.21) *are vacuous and the assumptions of Lemma* 3.2 *reduce to* (3.22) *as usual.*

Proof of Lemma 3.2: The conditions (3.20) and (3.21) imply that

$$F^{(2i)}(\xi) \to 0, \qquad \text{for } \xi \to +0, \quad i = 0, 1, \dots, p-1.$$

The conditions (3.20) imposed on derivatives of $G(\xi)$ give that

$$F^{(2i+1)}(\xi) \to 0 \quad \text{for } \xi \to +0, \quad i = 0, 1, \dots, p-1.$$

Finally, (3.22) leads to

$$F^{(2p)}(\xi) \to \sum_{i=0}^{2p-1} A_i^{2p-1} G^{(i)}(\xi)\xi^{\nu-2p+i-1}\Big|_{\xi=0} + \int_0^{\frac{1}{2}} \frac{\partial^{2p}}{\partial\xi^{2p}} N_0^{\nu}(\xi_1, 0)\frac{G(\xi_1)}{\xi_1}d\xi_1$$

$$= C' \int_0^{\frac{1}{2}} \xi^{\nu-2p-2} G(\xi)d\xi = C \neq 0, \quad \text{for } \xi \to +0,$$

since the terms of the sum tend to 0 and the integral $\int_0^{\frac{1}{2}} \xi^{\nu-2p-2}G(\xi)d\xi$ converges recording to $|\xi^{\nu-2p+i-2}G^{(i)}(\xi)| \leq CA\xi^{\{\nu\}-1}$. Therefore, $F(\xi) \in C^{2p}[0,\frac{1}{2}]$ and $F^{(i)}(0) = 0$ for $i = 0, 1, \ldots, 2p-1$, as well as $F^{(2p)}(0) = C \neq 0$, if (3.22) holds. Thus, by (3.4) and Maclaurin's formula for the function $F(\xi)$ one concludes that

$$|I(\xi)| = \left|\frac{\nu(\nu+1)}{\pi\xi^{\nu-1}}F(\xi)\right| \leq C_1 A\xi^{2p-\nu+1}$$

and $|I(\xi)| \geq C_2\xi^{2p-\nu+1}, C_2 > 0$, for sufficiently small ξ, if (3.22) holds. □

4. Estimates for the Solution $v(\xi, \eta)$

The results from the previous section in the special case of $\nu = n - 1/2$ will be exploited towards the description of the behavior of $\varphi(\xi)$, defined by (2.7) as:

$$\varphi(\xi) = G(\xi) + I(\xi).$$

Corollary 4.1. *Let us assume that* $\nu > 1$, $G \in C[0,\frac{1}{2}]$, $|G(\xi)| \leq A\xi$ *for some constant* A *and there hold the orthogonality conditions* (3.1) *from Lemma* 3.1. *Moreover, for* $\nu > 2$ *suppose also that*

$$G \in C^{[\nu-2]}(0, 1/2]\ , \quad |G^{(k)}(\xi)| \leq A\xi^{1-k} \quad \text{for } k = 0, 1, \ldots, [\nu-2].$$

Then $\varphi \in C[0, 1/2]$, $\varphi \in C^{[\nu-2]}(0, 1/2]$ *in the case of* $\nu > 2$ *and:*

(i) *for an odd* $[\nu]$ *there is an estimate*

$$|\varphi(\xi)| \leq CA\xi; \tag{4.1}$$

(ii) *for an even* $[\nu]$ *there is an estimate*

$$|\varphi(\xi)| \leq CA\xi^{1/2}. \tag{4.2}$$

In both cases C is a constant, independent of $G(\xi)$.

Proof: Taking into account the representation of φ and the condition $|G(\xi)| \leq A\xi$, the estimates follow from Lemma 3.1. The smoothness of $\varphi(\xi)$

for $\xi \neq 0$ is a direct consequence of $G \in C[0, 1/2]$, or, respectively, of $G \in C^{[\nu-2]}(0, 1/2]$ for $\nu > 2$. □

Applying Lemma 3.2 instead of Lemma 3.1 one gets:

Corollary 4.2. *Suppose that* p *is a nonnegative integer,* $p \leq [(\nu-1)/2]$ *and* $G \in C(0, 1/2], |G(\xi)| \leq A\xi$ *for some constant* A. *In the case of* $p \geq 1$ *let also*

$$G \in C^{2p-1}(0, 1/2], \ |G^{(k)}(\xi)| \leq A\xi^{2p-[\nu]+1-k} \quad \text{for } k = 0, 1, \ldots, 2p-1. \tag{4.3}$$

Then under the orthogonality conditions (3.21) *from Lemma* 3.2 *the function* $\varphi \in C(0, 1/2]$, *or, respectively* $\varphi \in C^{2p-1}(0, \frac{1}{2}]$ *for* $p \geq 1$, *satisfies the estimate*

$$|\varphi(\xi)| \leq C_1 A \xi^{-(\nu-2p-1)}$$

for some constant C_1 *independent of* $G(\xi)$. *Moreover, the condition* (3.22) *implies that the lower estimate*

$$|\varphi(\xi)| \geq C_2 \xi^{-(\nu-2p-1)}$$

holds for some constant $C_2 > 0$ *and sufficiently small* ξ.

Proof: In order to prove the announced estimates, let us recall the assumption $G(\xi) \leq A\xi^{2p-[\nu]+1}$ and the fact that the singularity is of order $0 \geq 2p - [\nu] + 1 > 2p - \nu + 1$. The smoothness of $\varphi(\xi)$ for $\xi \neq 0$ follows again from the condition $G \in C(0, 1/2]$ or $G \in C^{2p-1}(0, \frac{1}{2}]$. □

Remark 4.1. *It is not hard to see that* $\varphi \in C^k(0, 1/2]$ *assuming barely* $G \in C^k(0, 1/2]$.

Now we search for some conditions on the right-hand side function $g(\xi, \eta)$ of the equation (2.3), which ensure the implementation of the conditions imposed on $G(\xi)$ in Lemma 3.1. First, let us use the formula (2.5) and calculate the integral

$$\int_0^{\frac{1}{2}} \xi^{\nu-2i-2} G(\xi) d\xi = \int_0^{\frac{1}{2}} \int_0^{\xi_1} \left(\int_{\eta_1}^{\xi_1} \xi^{\nu-2i-2} P_\nu \left(\frac{\xi_1 \eta_1 + \xi^2}{\xi(\xi_1 + \eta_1)} \right) d\xi \right) g(\xi_1, \eta_1) d\eta_1 \, d\xi_1.$$

This reduces the orthogonality conditions (3.1) to

$$\int_0^{\frac{1}{2}} \int_0^{\xi} h^\nu_{\nu-2i-2}(\xi, \eta) g(\xi, \eta) d\eta \, d\xi = 0 \quad \text{for } i = 0, 1, \ldots, \left[\frac{\nu-1}{2}\right], \tag{4.4}$$

after denoting

$$h_k^\nu(\xi,\eta)=\int\limits_\eta^\xi \tau^k P_\nu\left(\frac{\xi\eta+\tau^2}{\tau(\xi+\eta)}\right)d\tau. \tag{4.5}$$

The next result gives us some estimates for $G(\xi)$:

Lemma 4.1. (1) *Let* $g\in C(\bar D)$ *and* $|g(\xi,\eta)|\le A$ *. Then the function* $G(\xi)$, *defined by* (2.5) *satisfies:*

$$G(\xi)\in C^1[0,1/2] \quad and \quad |G^{(k)}(\xi)|\le CA\xi^{1-k} \ for \quad k=0,1.$$

(2) *Let* $q\in\mathbb{N}$, $q\le\nu$, $g\in C^{q-1}(\bar D\backslash O)$ *and*

$$|D^\alpha g(\xi,\eta)|\equiv\left|\frac{\partial^{\alpha_1+\alpha_2}}{\partial\xi^{\alpha_1}\partial\eta^{\alpha_2}}g(\xi,\eta)\right|\le A\xi^{-|\alpha|} \quad for\ |\alpha|:=\alpha_1+\alpha_2\le q-1. \tag{4.6}$$

Then $G(\xi)$ *satisfies:*

$$G(\xi)\in C^q(0,1/2] \quad and \quad |G^{(k)}(\xi)|\le CA\xi^{1-k} \quad for\ k=0,1,\dots,q. \tag{4.7}$$

In both cases the constant C *is independent on the function* $g(\xi,\eta)$.

Remark 4.2. *Due to the transform* (2.2) $g=\rho^{1/2}f_n^i$, *the condition* (4.6) *seems to be natural.*

Proof of Lemma 4.1: First, notice that $0\le\xi_1\eta_1+\xi^2\le\xi(\xi_1+\eta_1)$, since $0\le\eta_1\le\xi\le\xi_1\le 1/2$. Observe also that by induction there follows:

$$\left|\frac{\partial^k}{\partial\xi^k}P_\nu\left(\frac{\xi_1\eta_1+\xi^2}{\xi(\xi_1+\eta_1)}\right)\right|=\left|\sum_{p+q+l=k}C_{p,q,l}^k P_\nu^{(q+l)}\left(\frac{\xi_1\eta_1+\xi^2}{\xi(\xi_1+\eta_1)}\right)\frac{(\xi_1\eta_1)^q}{\xi^{2q+p}(\xi_1+\eta_1)^{q+l}}\right|\le\frac{D}{\xi^k}.$$

We are going to estimate the derivatives of $G(\xi)$. By induction

$$\begin{aligned}\frac{d^k}{d\xi^k}G(\xi)&=\sum_{i=0}^{k-2}\sum_{|\alpha|=k-i-2}A_{i,\alpha,k}\frac{1}{\xi^i}D^\alpha g(\xi,\xi)\\&+\sum_{i=0}^{k-1}\int\limits_0^\xi\left(\sum_{p+q+l+j=i}B_{p,q,l}^{j,k}\frac{\partial^j}{\partial\xi^j}\frac{\eta_1^q}{\xi^{q+p}(\xi+\eta_1)^{q+l}}\right)\frac{\partial^{k-i-1}}{\partial\xi^{k-i-1}}g(\xi,\eta_1)d\eta_1\\&+\sum_{i=0}^{k-1}\int\limits_\xi^{\frac{1}{2}}\left(\sum_{p+q+l+j=i}C_{p,q,l}^{j,k}\frac{\partial^j}{\partial\xi^j}\frac{\xi_1^q}{\xi^{q+p}(\xi_1+\xi)^{q+l}}\right)\frac{\partial^{k-i-1}}{\partial\eta^{k-i-1}}g(\xi_1,\xi)d\xi_1\end{aligned}$$

$$+\int_{\xi}^{\frac{1}{2}}\int_{0}^{\xi}\frac{\partial^k}{\partial\xi^k}P_\nu\left(\frac{\xi_1\eta_1+\xi^2}{\xi(\xi_1+\eta_1)}\right)g(\xi_1,\eta_1)d\eta_1\,d\xi_1.$$

Combining all these results with (4.6), one concludes that $|G^{(k)}(\xi)| \leq CA\xi^{1-k}$ for $k \leq q$. □

Theorem 4.1. *Let* $n \in \mathbb{N}$, $n \geq 2$ *and* $g \in C(\bar{D})$, $|g(\xi,\eta)| \leq A$. *Suppose that*

$$g \in C^{n-4}(\bar{D}\backslash O);\quad |D^\alpha g(\xi,\eta)| \leq A\xi^{-|\alpha|}\quad \textit{for } |\alpha| \leq n-4,\ \textit{if } n \geq 4 \tag{4.8}$$

and the orthogonality conditions (4.4) *are satisfied with* $\nu = n - 1/2$. *Then the solution* $v(\xi,\eta)$ *of Problem P3 satisfies the following estimates*:

(i) *if n is an even number then*

$$|v(\xi,\eta)| \leq CA\xi|\ln\xi|;$$

(ii) *if n is an odd number then*

$$|v(\xi,\eta)| \leq CA\xi^{1/2}.$$

In both cases the constant C does not depend on the function $g(\xi,\eta)$.

Proof: We have already seen that the assumptions of the present theorem imply the those ones of Corollary 4.1 and the function φ satisfies the estimates (4.1), (4.2). Therefore, the representation (2.4) of v by φ clarifies that:

(i) if n is an even number then

$$|v(\xi,\eta)| \leq |\varphi(\xi)| + \left|\int_{\xi}^{\frac{1}{2}}\varphi(\xi_1)\frac{\partial}{\partial\xi_1}P_\nu\left(\frac{(\xi-\eta)\xi_1+2\xi\eta}{\xi_1(\xi+\eta)}\right)d\xi_1\right|$$
$$+\left|\int_{\xi}^{\frac{1}{2}}\int_{0}^{\eta}P_\nu\left(\frac{(\xi-\eta)(\xi_1-\eta_1)+2\xi_1\eta_1+2\xi\eta}{(\xi_1+\eta_1)(\xi+\eta)}\right)g(\xi_1,\eta_1)d\xi_1d\eta_1\right| \leq CA\xi|\ln\xi|;$$

(ii) if n is an odd number then $|v(\xi,\eta)| \leq CA\xi^{1/2}$ □

In the case when only some of the orthogonality conditions (4.4) are satisfied Corollary 4.2 gives us:

Theorem 4.2. *Suppose that* p *is a nonnegative integer*, $p \leq [(\nu-1)/2]$ *and* $g \in C(\bar{D})$, $|g(\xi,\eta)| \leq A$. *If* $p \geq 1$ *let also*

$$g \in C^{2p-2}(\bar{D}\backslash O);\quad |D^\alpha g(\xi,\eta)| \leq A\xi^{-|\alpha|}\quad \textit{for } |\alpha| \leq 2p-2, \tag{4.9}$$

and

$$\int_0^{\frac{1}{2}} \int_0^{\xi} h^{\nu}_{\nu-2i-2}(\xi,\eta) g(\xi,\eta) d\eta\, d\xi = 0 \quad \textit{for } i = 0, 1, \dots, p-1. \tag{4.10}$$

Then the estimate

$$|v(\xi,\eta)| \le CA\xi^{-(\nu-2p-1)} \tag{4.11}$$

holds for some constant C independent of the function $g(\xi,\eta)$. *Moreover, the condition*

$$\int_0^{\frac{1}{2}} \int_0^{\xi} h^{\nu}_{\nu-2p-2}(\xi,\eta) g(\xi,\eta) d\eta\, d\xi \ne 0 \tag{4.12}$$

implies that the lower estimate

$$|v(\xi,0)| \ge c\xi^{-(\nu-2p-1)}, \tag{4.13}$$

holds for some constant $c > 0$ *and sufficiently small* ξ.

Proof: In order to prove (4.11), let us use Lemma 4.1 with $q = 2p-1$. Then the conditions (4.9), imposed on $g(\xi,\eta)$ give us

$$|G^{(k)}(\xi)| \le CA\xi^{1-k} \le CA\xi^{2p-[\nu]+2-k} \quad \text{for } k = 0, 1, \dots, 2p-1.$$

Now Corollary 4.2 can be applied to conclude that $|\varphi(\xi)| \le C_1 A\xi^{2p-\nu+1}$. Therefore $|v(\xi,\eta)| \le CA\xi^{2p-\nu+1}$. Finally, note that the last estimate (4.13) is a direct consequence of Corollary 4.2, because $v(\xi,0) = \varphi(\xi)$. □

Remark 4.3. *In view of Remark* 4.1 *and Lemma* 4.1 *one has* $v(\xi,\eta) \in C^{k+1}(\bar{D}\backslash O)$ *when* $g(\xi,\eta) \in C^k(\bar{D}\backslash O)$.

5. Estimates for the Solution $u_n(\rho, t)$

Now we return to the original coordinates (ρ, t) and the functions $f_n(\rho,t)$ and $u_n(\rho,t)$. Our first aim is to calculate the integrals $h_k(\xi,\eta)$, defined by (4.5).

Lemma 5.1. *For* $i = 0, 1, \dots, \left[\frac{\nu-1}{2}\right]$ *one has*

$$h^{\nu}_{\nu-2i-2}\left(\frac{\rho+t}{2}, \frac{\rho-t}{2}\right) = C_i \rho^{1/2} H^n_i(\rho,t), \quad C_i = const. \ne 0,$$

where $\nu = n - 1/2$ and the functions $H_i^n(\rho, t)$ are defined by (1.2) *in Lemma* 1.1.

Proof: The idea is to make use of Mellin transform. Let us introduce new variables x, z and y, defined by: $z = (\xi\eta)^{1/2}$, $2x^{1/2} = (\xi + \eta)z^{-1}$, $2y^{1/2} = (\tau z)^{-1}(z^2 + \tau^2)$. Substituting ξ, η and τ in (4.5) we apply Mellin transform over x and using formulas (1.2), (11.13(4)) and (2.10(4)) from [18] one obtains

$$h_k^\nu(x,z)z^{-k-1} \mapsto \pi \frac{\Gamma\left(\frac{1+k}{2} - s\right)\Gamma\left(-\frac{1+k}{2} - s\right)}{\Gamma\left(\frac{1-\nu}{2} - s\right)\Gamma\left(1 + \frac{\nu}{2} - s\right)},$$

where "$\mapsto$" means "transforms into". On the other hand the last one is the Mellin transform of the function

$$\frac{\pi}{\Gamma\left(\frac{\nu-k+1}{2}\right)} x^{\frac{\nu+1}{2}} I_{0+}^{\frac{k-\nu+2}{2}}\left(x^{-\frac{k+\nu+3}{2}}(x-1)_+^{\frac{\nu-k-1}{2}}\right).$$

To see this one make use of (2.2.(4)), (1.10) and (1.4) from [18] . Hence, taking $k = \nu - 2i - 2$ one has

$$h_{\nu-2i-2}^\nu(x,z)z^{-\nu+2i+1} = C_{\nu,i} x^{\frac{\nu+1}{2}} \left(\frac{d}{dx}\right)^i \left(x^{-\nu+i-1/2}(x-1)_+^{i+1/2}\right)$$

for some constants $C_{\nu,i}$. Getting back to the coordinates ρ and t, one concludes that

$$z^{\nu-2i-1} x^{\frac{\nu+1}{2}} \left(\frac{d}{dx}\right)^i \left(x^{-\nu+i-1/2}(x-1)^{i+1/2}\right) = \sum_{j=0}^{i} a_j^i \frac{t(\rho^2 - t^2)^{n-3/2-j-i}}{\rho^{n-2j-1/2}}.$$

In order to calculate the coefficients a_j^i, let us observe that each function $h_{\nu-2i-2}^\nu(\xi,\eta)$, $i = 0, 1, \ldots, \left[\frac{\nu-1}{2}\right]$, defined by (4.5) is, in fact, a solution of the homogeneous equation (2.3). Therefore the functions $\rho^{-1/2}h_{\nu-2i-2}^\nu\left(\frac{\rho+t}{2}, \frac{\rho-t}{2}\right)$ are solutions of the homogeneous equation (2.1). Substituting back in (2.1) one finds that $a_j^i = a_0^i A_j^i$ (with the constants A_j^i from Lemma 1.1). This shows

$$\rho^{-1/2}h_{\nu-2i-2}^\nu\left(\frac{\rho+t}{2}, \frac{\rho-t}{2}\right) = a_0^i H_i^n(\rho, t).$$

□

Now we are ready to state one of our main results:

Theorem 5.1. *Suppose that* $n \in \mathbb{N}$, $n \geq 2$, $f_n \in C(\bar{\Omega}_0)$, $|f_n(\rho,t)| \leq A$ *with some constant* A, *and*

$$f_n(\rho,t) \in C^{n-4}(\bar{\Omega}_0 \backslash O), |D^\alpha f_n(\rho,t)| \leq A\rho^{-|\alpha|} \quad \text{for } |\alpha| \leq n-4, \text{ if } n \geq 4. \quad (5.1)$$

Let also there hold the orthogonality conditions

$$\int_0^{\frac{1}{2}}\int_t^{1-t} H_i^n(\rho,t)f_n(\rho,t)\rho\,d\rho\,dt = 0 \quad \textit{for } i = 0,1,\ldots,\left[\frac{n}{2}\right]-1. \tag{5.2}$$

Then the solution $u_n(\rho,t)$ of Problem P2 is such that:

(i) $|u_n(\rho,t)| \leq CA\rho^{1/2}|\ln\rho|$, *if n is an even number;*

(ii) $|u_n(\rho,t)| \leq CA$, *if n is an odd number.*

In both cases the constant C does not depend on the function $f_n(\rho,t)$.

Proof: Recall (2.2), let us denote $g(\xi,\eta) = \rho^{1/2}f_n(\rho,t)$ where $\xi = (\rho+t)/2$, $\eta = (\rho-t)/2$. Then the conditions (5.1) imply that the function g satisfies the estimate $|g(\xi,\eta)| \leq A$. Moreover, if $n \geq 4$ then $|D^\alpha g(\xi,\eta)| \leq CA\xi^{-|\alpha|}$ for $|\alpha| \leq n-4$. On the other hand, by (5.2) there follows

$$\int_0^{\frac{1}{2}}\int_0^{\xi} h_{\nu-2i-2}^{\nu}(\xi,\eta)g(\xi,\eta)d\eta\,d\xi = C\int_0^{\frac{1}{2}}\int_t^{1-t} H_i^n(\rho,t)f_n(\rho,t)\rho\,d\rho\,dt = 0.$$

Using Theorem 4.1, one completes the proof. □

Theorem 5.2. *Suppose that $n,q \in \mathbb{N}\cup\{0\}$, $n \geq 2$, $q \leq \left[\frac{n}{2}\right]-1$ and $f_n \in C(\bar{\Omega}_0)$, $|f_n(\rho,t)| \leq A$. If $q \geq 1$ let also*

$$f_n \in C^{2q-2}(\bar{\Omega}_0\backslash O);\ |D^\alpha f_n(\rho,t)| \leq A\rho^{-|\alpha|} \quad \textit{for } |\alpha| \leq 2q-2, \tag{5.3}$$

and

$$\int_0^{\frac{1}{2}}\int_t^{1-t} H_i^n(\rho,t)f_n(\rho,t)\rho\,d\rho\,dt = 0 \quad \textit{for } i = 0,1,\ldots,q-1. \tag{5.4}$$

Then for the solution $u_n(\rho,t)$ of Problem P2 the upper estimate

$$|u_n(\rho,t)| \leq CA\rho^{-(n-2q-1)} \tag{5.5}$$

holds, where the constant C is independent of $f_n(\rho,t)$. If we suppose also that

$$\int_0^{\frac{1}{2}}\int_t^{1-t} H_q^n(\rho,t)f_n(\rho,t)\rho\,d\rho\,dt \neq 0, \tag{5.6}$$

then the lower estimate

$$|u_n(\rho,\rho)| \geq c\rho^{-(n-2q-1)},$$

holds for $c > 0$ *and sufficiently small* ρ.

Proof: Applying Theorem 4.2 instead of Theorem 4.1 the proof is complete. □

Remark 5.1. *As in the previous sections observe that* $u_n \in C^{k+1}(\bar{\Omega}_0 \backslash O)$ *if* $f_n \in C^k(\bar{\Omega}_0 \backslash O)$.

6. On the Singularity of the Solutions of Protter's Problem P1

Now we return to the proofs of Theorems 1.1 and 1.2. Let us suppose that the right-hand side f of the wave equation (1.1) is a trigonometric polynomial (1.3). Then we are looking for a solution of the same form:

$$u(x_1,x_2,x_3) := \sum_{n=2}^{l} \left(u_n^1(\rho,t)\cos n\varphi + u_n^2(\rho,t)\sin n\varphi\right), \tag{6.1}$$

where the functions u_n^i are solutions of the equations (2.1) with right hand side functions f_n^i and satisfy the boundary conditions $P2$. Notice that if $f \in C^l(\bar{\Omega})$, then one has $f_n^i \in C^l(\bar{\Omega}_0)$ and $|D^\alpha f_n^i| \leq CA_{|\alpha|}$ for $|\alpha| \leq l$, where $A_k = \max_{\bar{\Omega}} |f| + \sum_{|\beta|\leq k} \max_{\bar{\Omega}} |D^\beta f|$.

Proof of Theorem 1.1: First of all, let us mention that the existence and uniqueness of a *generalized solution* of Problem $P1$ have been already proved in [21]. According to Remark 5.1 the functions $u_n \in C^{l+1}(\bar{\Omega}_0 \backslash O)$, so that $u \in C^{l+1}(\bar{\Omega} \backslash O)$. Observe that (1.4) leads to the orthogonality conditions (5.2) and one can apply Theorem 5.1 to get the estimate (1.5). In other words, the solution is bounded but not necessarily continuous at point O. Theorem 5.1 provides also the estimate $|u_n^i(\rho,t)| \leq CA_l\rho^{1/2}|\ln\rho|$ for each term u_n^i from (6.1) with an even index n. Assuming $f_n^i(x_1,x_2,t) = 0$ for all odd indexes n and $i = 1,2$, then $u_n^i = 0$ for all odd n and hence (1.6) follows. This estimate implies that $u(0,0,0) = 0$ and $u \in C(\bar{\Omega})$. □

Proof of Theorem 1.2: Again Remark 5.1 forces $u \in C^{l+1}(\bar{\Omega} \backslash O)$. Let us fix some term u_n^i from (6.1). Now the conditions (1.4) leads to the orthogonality conditions (5.4) for $q - 1 = p + \left[\frac{n-m-1}{2}\right]$. Therefore, by the first conclusion of Theorem 5.2 one has

$$|u_n^i(\rho,t)| \leq CA_{2p-m+1+l}\rho^{2p-m+1},$$

whereas the condition (1.8) provides

$$|u_m^j(\rho,\rho)| \geq c\rho^{2p-m+1}.$$

□

Remark 6.1. *Obviously, the condition $f \in C^1(\bar{\Omega})$ in Theorems* 1.1 *and* 1.2 *could be reduced in view of the conditions on f_n^i in Theorems* 5.1 *and* 5.2.

Acknowledgement

This research was partially supported by the Bulgarian NSC under Grant MM-904/99.

References

[1] Aldashev, S. A.: Correctness of multi-dimensional Darboux problems for the wave equation. Ukr. Math. J. *45*, 1456–1464 (1993).
[2] Aldashev, S. A.: On Darboux problems for a class of multidimensional hyperbolic equations. Differ. Equations *34*, 65–69 (1998).
[3] Aldashev, S. A.: Some problems for a multidimensional hyperbolic integro-differential equation. Ukrainian Math. J. *52*, 590–595 (2000).
[4] Aziz, A. K., Schneider, M.: Frankl-Morawetz problems in R^3. SIAM J. Math. Anal. *10*, 913–921 (1979).
[5] Bazarbekov, Ar. B., Bazarbekov, Ak. B.: Goursat and Darboux problems for the two-dimensional wave equation, I. Differ. Equations *30*, 741–748 (1994).
[6] Bitsadze, A. V.: Some classes of partial differential equations. New York: Gordon and Breach Science 1988.
[7] Copson, E. T.: On the Riemann-Green function. J. Rath. Mech. Anal. *1*, 324–348, (1958).
[8] Edmunds, D. E., Popivanov, N. I.: A nonlocal regularization of some over-determined boundary value problems I. SIAM J. Math. Anal. *29*, 85–105 (1998).
[9] Erdélyi, A., Magnus, W., Oberhettinger, F., Tricomi, F.: Higher transcendental functions, 1. New York: McGraw-Hill 1953.
[10] Garabedian, P. R.: Partial differential equations with more than two variables in the complex domain. J. Math. Mech. *9*, 241–271 (1960).
[11] Grammatikopoulos, M. K., Hristov, T. D., Popivanov, N. I.: On the singularities of 3-D Protter's problem for the wave equation. Electron. J. Diff. Eqns.[online], Vol. 2001, 1–26 (2001). Available from: http://ejde.math.swt.edu/volumes/2001/01/
[12] Hörmander, L.: The analysis of linear partial differential operators, III. Berlin Heidelberg New York Tokyo: Springer (1985).
[13] Jong Duek Jeon, Khe Kan Cher, Ji Hyun Park, Yong Hee Jeon, Jong Bae Choi.: Protter's conjugate boundary value problems for the two-dimensional wave equation. J. Korean. Math. Soc. *33*, 857–863 (1996).
[14] Karatoprakliev, G. D.: Uniqueness of solutions of certain boundary-value problems for equations of mixed type and hyperbolic equations in space. Differ. Equations *18*, 49–53 (1982).
[15] Kharibegashvili, S.: On the solvability of a spatial problem of Darboux type for the wave equation. Georgian Math. J. *2*, 385–394 (1995).
[16] Khe Kan Cher.: Darboux-Protter problems for the multidimensional wave equation in the class of unbounded functions. Math. Notices of Jacutsk State Univ. *2*, 105–109 (1995).
[17] Khe Kan Cher.: Nontrivial solutions of some homogeneous boundary value problems for a many-dimentional hyperbolic Euler-Poisson-Darboux equation in an unbounded domain. Differ. Equations *34*, 139–142 (1998).
[18] Marichev, O. I.: Handbook of integral transforms of higher transcendental functions: Theory and algorithmic tables. Ellis Horwood Series in Mathematics and its Applications, Ellis Horwood Limited (Chichester) 1983, 336p.
[19] Nakhushev, A. M.: Criteria for continuity of the gradient of the solution to the Darboux problem for the Gellerstedt equation. Differ. Equations *28*, 1445–1457 (1992).
[20] Popivanov, N. I., Schneider, M.: The Darboux problems in $\mathbf{R}^3$ for a class of degenerated hyperbolic equations. Comptes Rend. de l'Acad. Bulg. Sci. *41*, 537–578 (1988).
[21] Popivanov, N. I., Schneider, M.: On M. H. Protter problems for the wave equation in $\mathbf{R}^3$. J. Math. Anal. Appl. *194*, 50–77 (1995).
[22] Protter, M. H.: New boundary value problems for the wave equation and equations of mixed type. J. Rat. Mech. Anal. *3*, 435–446 (1954).
[23] Samko, S. G., Kilbas, A. A., Marichev, O. I.: Fractional integrals and derivatives: Theory and applications. New York: Gordon and Breach 1993, 976 p.

[24] Tong Kwang-Chang: On a boundary-value problem for the wave equation. Science record. New Series *1*, 1–3 (1957).

Nedyu Popivanov
Todor Popov
Department of Mathematics and Informatics
University of Sofia
1164 Sofia
Bulgaria
e-mails: nedyu@fmi.uni-sofia.bg
topopover@hotmail.com

Computing [Suppl] 16, 237–244 (2002)

Computing

Construction of Shortest Line of Restricted Curvature in a Non-singly-connected Polygonal Area

S. V. Smelyakov

Abstract

In a computer aided laying-out of roads and railway lines there arises the problem of the modeling of optimal routes with restricted curvature in a given non-singly-connected polygonal (NSCP) area. However, since the triangulating network models are used now for the NSCP areas, the development of the respective AutoCADs is sufficiently hampered due to the lack of an efficient approach allowing to simulate the routes of continuous curvature with the required accuracy. Therefore, optimization of routes cannot be carried out in a proper way. For solving this nonlinear problem in compliance with the requirements of the engineering standards, a model is presented that comprises a set of basic optimization problems for the construction of the shortest route composed of segments and circular arcs. In order to provide smooth growth of acceleration for eliminating overload at high speeds, these elements are mated with the easement curves – clothoid (Euler spiral) or cubic parabola fragments. For these problems the condition of optimality and respective optimization methods are presented that show linear memory and time complexity.

AMS Subject Classifications: 53A04, 53B99.

Keywords: Non-singly-connected polygonal area, shortest curve, clothoid.

1. Introduction

We are defining a route in a given non-singly-connected polygonal (NSCP) area F on the plane as a continuous spline line of class SKC

$$p = S_1 K_1 K_1' S_2 K_2 C_2 K_2' S_3 \ldots K_n C_n K_n' S_{n+1}, \quad n \geq 1, \tag{1}$$

where S_i is a segment, K_i, K_i' – the easement curves presented by a fragment of clothoid or cubic parabola, and C_i – the circular arc, which satisfy the condition of transversality (continuity of curvature) at each point of connection. Without loss of generality, in the considerations below it is assumed that the radii of the circles, as well as the parameters of clothoid a and parabolas q remain the same for all fragments. Besides, the clothoid (parabola) fragments K_i, K_i' are congruent, their curvature varies from zero (at a point of contact with a segment) to $1/R$, and the angle of turn for the easement curve does not exceed $\pi/2$. Since in some applications (where the vehicles move at low speed) there exists no need in the easement curves, we consider also the spline lines of class SC which differs from (1) in omitting the fragments K_i, K_i'; these lines are to be smooth.

Though a direct use of conventional mathematical methods for the construction of the shortest line of type (1) in a NSCP area F is not possible, the major part of actual engineering problems of this type might be reduced to the basic problem for a fragment $\Phi = S_1CS_2$.

2. Construction of Shortest Routes without Easement Curves

Let $P_{\tau,\Lambda}[A,B]$ present the set of lines of class $\Lambda \in \{SKC, SC\}$ that belongs to the continuous family of lines $[\tau]$ connecting points A and B in the area F (Fig. 1a), and let $L(p)$ be the length of route p, and the path τ be specified by the shortest polygonal line ACB, where C is the only vertex of the polygonal boundary of area F in a large enough vicinity of the point C.

Basic problem *Z*(*SC*). Find

$$p_* = \underset{p \in P_{\tau,SC}[A,B]}{\arg\min} L(p). \tag{2}$$

Let $C(x,r)$ be a polar coordinate system with the origin in point C (Fig. 1b) where the angle x equals to zero at the ray CA. Then, any circle O with the center in point (x,R) passes through the point C. If the tangents AD, BE to such circle exist in F, they define a permissible route $p(x,R) = AD \cup \operatorname{arc} DC \cup \operatorname{arc} CE \cup EB$. In this case, denote A, B, δ the angles at the vertices A, B, C of the triangle ABC; a, b – the lengths of the sides AC, BC, and α (β) – the angle between the side AC (BC) and tangent AD (BE), respectively.

Theorem 1. *In case of $a \geq 2R$ or $b \geq 2R$, the solution to the problem* (2) *is defined by the route $p^* = p(x^*, R)$ where the angle x^* satisfies the conditions of transversality*

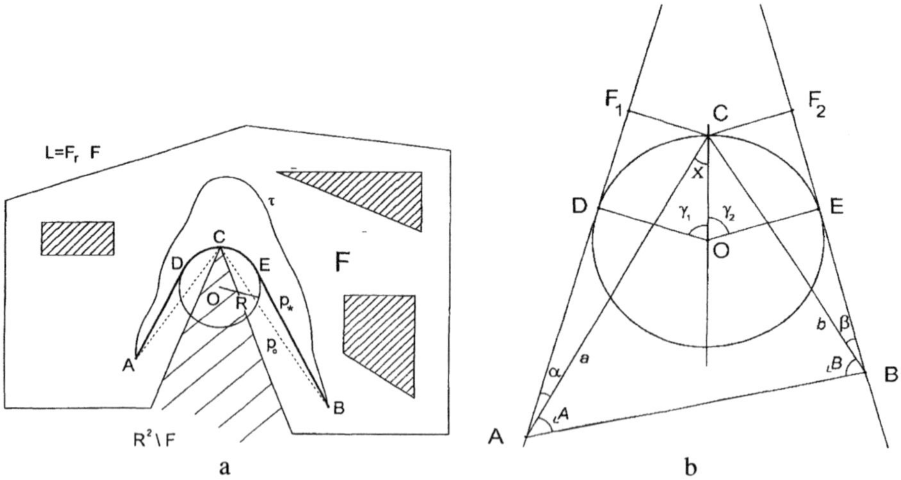

Fig. 1. Basic problem $Z(SC)$ and its solution

$$x \in X = [-\pi/2 + \delta;\ \pi/2], \tag{3}$$

$$R[1 - \sin(x - \alpha)] = a \cdot \sin(\alpha), \tag{4}$$

and the optimality condition

$$a \cdot \sin \alpha = b \cdot \sin \beta, \tag{5}$$

$$2 \cdot x = \alpha + \delta - \beta. \tag{6}$$

Here the route $p^* = p(x^*, R)$ remains optimal for any pair of points A', B' belonging to the rays DA, EB, respectively.

Note. Given $a \leq 2R$, $b \leq 2R$, the system (3)–(6) may have no solution. In this situation (which scarcely complies to the engineering standards) the solution is defined by the circle of radius R passing through the points A, B. However, if angular and/or metric boundary conditions are implied to the first (and/or last) segment of the route, they define the solution in conjunction with the transversality conditions that will take the form $x \notin X = [-\pi/2 + \delta;\ \pi/2]$, $R[1 + \sin(x - \alpha)] = a \cdot \sin(\alpha)$. So, we call the solution complying to Theorem 1 normal, and singular in the opposite case, when point C does not fit the circle in Fig. 1b. Only normal solutions are considered below.

Hence, the solution to the problem $Z(SC)$ is defined by the system

$$\begin{aligned} R[1 - \sin(x - \alpha)] - a \cdot \sin(\alpha) &= 0, \\ a \cdot \sin(\alpha) - b \cdot \sin(\delta + \alpha - 2x) &= 0. \end{aligned} \tag{7}$$

Since there exists the Jacobian of (7) and the functions in (7) are smooth, the solution to this system might easily be obtained by applying the following iterative scheme:

$$\begin{aligned} x &= \alpha + \arcsin\left[1 - \frac{a}{R}\sin \alpha\right], \\ \alpha &= \arcsin\left[\frac{b}{c}\sin(\delta + \alpha - 2x)\right]. \end{aligned} \tag{8}$$

Natural practical generalizations of the problem (2) are those ones for which a boundary condition is imposed onto angle; a solution to this problem $Z(SC_\alpha)$ is obtained directly from (4). If boundary conditions are imposed to both angles α and β, and/or to the lengths of the first and/or last segments, the system (3), (4) may become incompatible and would force us to give up the requirement for the circle O to pass through the point C, and to optimize both the position of the coordinate system $C(x, r)$ and the angle x.

By weakening the above stated assumption with respect to the absence of other boundary vertices in the vicinity of the vertex C, we come to the generalized problem $Z(SC_\ell)$ of the construction shortest route with the following presentation:

$$p_* = S_1C_1S_2C_2\dots C_nS_{n+1}, \quad n \geq 1. \tag{9}$$

Assume that the solution (9) exists and all its fragments $\Phi_i = S_iC_iS_{i+1}$, $S_i = E_{i-1}D_i$, $C_i = \operatorname{arc} D_iE_i$ of the route p_* are presenting normal solutions. Then, for each of these fragments the conditions (3)–(6) hold for all points belonging to the tangents, and solving of the problem $Z(SC_\ell)$ might be searched by iterative recalculation of the fragment parameters until the required accuracy is reached (Fig. 2).

Algorithm.

(1) Let α be an initial angle specifying the boundary condition to the problem $Z(SC_\alpha)$ for the points A and C_1, and $i = 1$. The solution of this problem uniquely defines the circle O_1 and the tangent straight line E_1, E_1' by the equations

$$x_1 = \alpha + \arcsin\left(1 - \frac{a}{R}\sin\alpha\right), \tag{10}$$

$$\operatorname{arc} C_1E_1 = \pi/2 + \alpha - x. \tag{11}$$

(2) Obtain the solution for the problem $Z(SC_\alpha)$ by taking the point E_i for A, and the angle $E_i'E_iC_{i+1}$ for α, which uniquely specifies the circle O_{i+1} that passes through C_{i+1} and the respective tangent $E_{i+1}E_{i+1}'$.

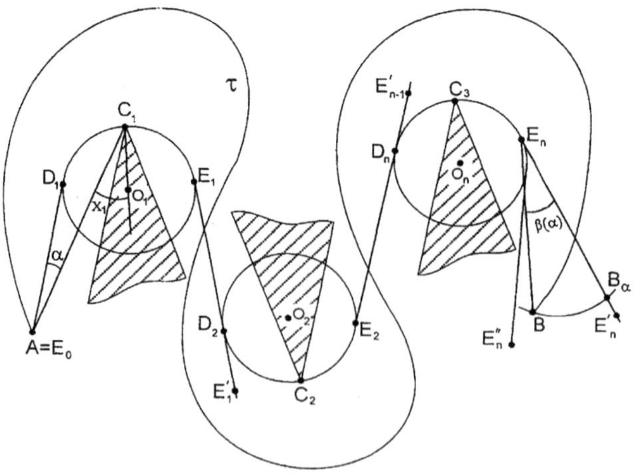

Fig. 2. Construction of the solution to the problem $Z(SC_l)$

(3) Continue step 2, until we get $E_nE'_n$. Consequently, the initial angle α defines the route $p_\alpha = AD_1 \cup \text{arc}\, D_1E_1 \cup E_1D_2 \cup \text{arc}\, D_2E_2 \cup \ldots \cup \text{arc}\, D_nE_n \cup E_nB_\alpha$, where B_α is such point on $E_nE'_n$ that $E_nB_\alpha = E_nB$.

By construction, the route p_α gives the solution to the problem $Z(SC_\ell)$ if $B_\alpha = B$. Since the angle α is taken arbitrarily, the angle $\beta(\alpha) = \angle BE_nE'_n$ does not equal to zero. However, since the angle β is a small variation and a continuous function of α, definite vicinity $\bar{\alpha} = (\alpha_1, \alpha_2)$ of the initial value α contains the value $\alpha_* \in \bar{\alpha}$, $\beta(\alpha) = 0$, specifying the solution to the problem $Z(SC_\ell)$. For obtaining the solution with the required accuracy, it is sufficient to increase the angle α from 0 with some increment Δ_α, and apply the above algorithm until the angle $\angle BE_{m+1}B_\alpha$ changes sign. If the distance between the point B and B_{α_m} or $B_{\alpha_{m+1}}$ exceeds the required limit, repeat this procedure with the smaller increment for the interval (α_m, α_{m+1}).

3. Construction of Shortest Routes with Clothoidal Easement Curves

Now, we consider the basic problem for the spline lines (1) where the same fragment of clothoid is used for the elements K_i, K'_i, and the transversality condition defines continuity of curvature. As for the class SC, we consider, firstly, the basic problem for the polygonal line $\tau = ACB$.

Basic problem *Z*(*SKC*). Find

$$p_* = \mathop{\arg\min}_{p \in P_{\tau,SKC}[A,B]} L(p). \tag{12}$$

This problem is reduced to (2) as follows. Consider a clothoid specified by the natural equation

$$\rho = \frac{a}{s}, \tag{13}$$

where a is the scale factor, s the length of the arc, and ρ is radius of curvature. Let this curve be located on the plane Zxy in a conventional way, viz. so that it crosses the axis Zx at the tangent point where its curvature equals to zero (Fig. 3). Its parametric equation in this case is as follows:

$$x = \int_0^s \cos\left(\frac{s^2}{2a}\right) ds, \qquad y = \int_0^s \sin\left(\frac{s^2}{2a}\right) ds. \tag{14}$$

Let the clothoid fragment, an arc ZM, be the tangent in point M to the circle O with radius r, and $\Delta\alpha$ be the angle of turn along the arc ZM ($\Delta\alpha \leq \pi/2$). Find the coordinate (x_M, y_M), (x_0, y_0) of point M and center of this circle.

Due to the definition of curvature k and in accordance with (13), (14) we obtain that the angle of turn along the arc ZM with the length $s = a/r$ is given by

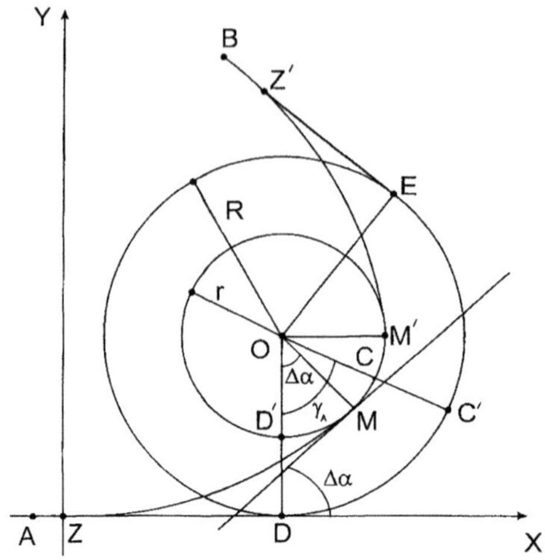

Fig. 3. Smoothing by clothoid

$$\Delta\alpha = \frac{a}{2r^2}. \tag{15}$$

Hence,

$$x_M = \int_0^{S_r} \cos\left(\frac{s^2}{2a}\right) ds, \qquad y_M = \int_0^{S_r} \sin\left(\frac{s^2}{2a}\right) ds, \quad s_r = \frac{a}{r}, \tag{16}$$

$$\begin{aligned} x_0 &= x_M - r\sin(\Delta\alpha), \\ y_0 &= y_M + r\cos(\Delta\alpha). \end{aligned} \tag{17}$$

Draw a circle around the center O with radius $R = y_0$ (D is its point of tangency with the axis Zx),

$$R = \int_a^{a/r} \sin\left(\frac{s^2}{2a}\right) ds + r\cos\left(\frac{a}{2r^2}\right). \tag{18}$$

Let A be any point on the axis Zx that lies to the left of Z, and C' – any point on the circle O with radius R, the polar angle $\gamma_A = \angle DOC$ of which satisfies condition $\Delta\alpha \le \gamma_A \le \pi/2$. Then, by construction, the line $p = AD \cup \operatorname{arc} DC' \in SC$ uniquely defines the line

$$q = AZ \cup \operatorname{arc} ZM \cup \operatorname{arc} MC \cup \operatorname{arc} CM' \cup \operatorname{arc} M'Z' \cup Z'B \in SCK. \tag{19}$$

Obtaining the line q by p we call this *smoothing* (of the line SC) *by clothoid.* At this, if the point C in the basic problem $Z(SC)$ is located within the circle of radius R at a distance $r \le R$ from its center,

$$R - r\sin(x - \alpha) = a\sin\alpha, \tag{20}$$

$$a\sin\alpha = b\sin\beta, \tag{21}$$

$$2x = \alpha + \delta - \beta. \tag{22}$$

We denote by $Z(SC_{rR})$ and p_{*rR} the basic problem $Z(SC)$ and its solution, respectively.

Theorem 2. Smoothing by clothoid of the solution p_{*rR} to the problem $Z(SC_{rR})$, that is defined by (20)–(22), gives the solution q to the problem (12). At this, R is determined by (18), whereas Z and M coordinates are defined relative to the point of tangency D and ray OD by coordinate x_0 of (17) and angle $\Delta\alpha$ of (15), respectively.

In general case (1), the problem $Z(SKC_\ell)$ arises which is similar to the problem $Z(SC_\ell)$ except that another class of lines is considered. It is clear that the proposed approach to solving problems $Z(SKC), Z(SC_\ell)$ allow us to reduce the problem $Z(SKC_\ell)$ to these ones for which we used the above described algorithm (where the problems $Z(SC_{rR})$ are considered instead of $Z(SC)$) with an iterative decreasing of the angular discrepancy.

In case the radii for all circles are the same, the calculation of the clothoid parameters is to be performed, but only once. Therefore, the application of the algorithm may require about $\kappa \approx (30 + 2\kappa_\varphi) \cdot n = 70 \cdot n$ (operations), whereas the iterative elimination of the angular discrepancy may require about $\kappa^* \approx 200 \cdot n/\Delta$ (operations).

4. Construction of Shortest Routes with Parabolic Easement Curves

Consider the spline (1) where the transversality condition defines continuity of curvature, and the same fragment of cubic parabola

$$y = \frac{x^3}{6q}, \quad x \leq x_{\max} = \sqrt[4]{4q^2/5} \tag{23}$$

is used for the elements K_i, K_i'; here q is the scale factor, and $x_{\max}$ is the limit of the monotonic growth of curvature. As for the basic problem $Z(SKC)$, we consider the polygonal line $\tau .= ACB$.

Basic problem $Z(SPC)$ Find

$$p_* = \underset{p \in P_{\tau,SPC}[A,B]}{\arg\min}\; L(p). \tag{24}$$

In compliance with the suggested approach to solving problems $Z(SKC)$, $Z(SKC_\ell)$, it is clear from Fig. 3, that it may also be used for solving the analogous problems with parabolic easement curves and with the same time and memory complexity, except that radius R, as well as the Z and M coordinates are to be defined with respect to (23).

S. V. Smelyakov
Astronomicheskaya, Str. 35 – V, Apt. 52
310035 Kharkov
Ukraine
e-mail: pluto@velton.kharkov.ua

Springer-Verlag
and the Environment